Morphology and Internal Mixing of Atmospheric Particles

Morphology and Internal Mixing of Atmospheric Particles

Special Issue Editors

Swarup China
Claudio Mazzoleni

MDPI • Basel • Beijing • Wuhan • Barcelona • Belgrade

MDPI

Special Issue Editors

Swarup China
Pacific Northwest National Laboratory
USA

Claudio Mazzoleni
Michigan Technological University
USA

Editorial Office
MDPI
St. Alban-Anlage 66
Basel, Switzerland

This is a reprint of articles from the Special Issue published online in the open access journal *Atmosphere* (ISSN 2073-4433) from 2017 to 2018 (available at: http://www.mdpi.com/journal/atmosphere/special_issues/atmospheric_particles)

For citation purposes, cite each article independently as indicated on the article page online and as indicated below:

LastName, A.A.; LastName, B.B.; LastName, C.C. Article Title. *Journal Name* **Year**, *Article Number, Page Range.*

ISBN 978-3-03897-133-7 (Pbk)
ISBN 978-3-03897-134-4 (PDF)

Contents

About the Special Issue Editors

Swarup China is a postdoctoral researcher at the Environmental Molecular Sciences Laboratory at the Pacific Northwest National Laboratory. He completed his MS degree in Civil and Environmental Engineering from University of Nevada Las Vegas in 2008. He moved to Michigan Technological University and completed his Ph.D. in Atmospheric Sciences in 2014. He started his postdoctoral research in 2015 at the Pacific Northwest National Laboratory. His major areas of research interest are the morphological and optical properties of atmospheric aerosols, atmospheric aerosol chemistry and heterogeneous ice nucleation. He aims to better understand the physical chemistry of atmospheric particles controlling aerosol-cloud interactions.

Claudio Mazzoleni is a Professor in the department of Physics and a member of the Atmospheric Sciences Program of the Michigan Technological University, U.S.A. He earned a Laurea in Physics from the University of Trento, Italy, in 1995. In 1999, he transferred to the U.S.A. to pursue a Ph.D. in Atmospheric Sciences at the Desert Research Institute, part of the University of Reno, Nevada. He earned his Ph.D. in 2003. From 2005 until 2008 he was a Director's postdoctoral fellow at the Los Alamos National Laboratory. In 2008, he started his academic career at the Michigan Technological University where he is currently managing the Environmental Optics Laboratory, while teaching classes in physics and atmospheric sciences. His research interest is the study of atmospheric particles, their physical and optical properties and their impacts on human health and climate.

atmosphere

MDPI

Editorial

Preface: Morphology and Internal Mixing of Atmospheric Particles

Swarup China [1,*] and Claudio Mazzoleni [2,*]

[1]   Environmental Molecular Sciences Laboratory, Pacific Northwest National Laboratory,
     Richland, WA 99354, USA
[2]   Physics Department and Atmospheric Sciences Program, Michigan Technological University,
     Houghton, MI 49931, USA
*   Correspondence: swarup.china@pnnl.gov (S.C.); cmazzoleni@mtu.edu (C.M.)

Received: 26 June 2018; Accepted: 2 July 2018; Published: 4 July 2018

Keywords: aerosol; mixing state; morphology; black carbon; soot; lifecycle; optical properties; cloud condensation nuclei; ice-nucleating particle; radiative forcing

1. Introduction

The properties of atmospheric particles (often also termed aerosol) have been the subject of scientific studies for several decades because of their effects on air quality, health, visibility, propagation of electromagnetic radiation in the atmosphere, and climate. However, despite intense research efforts, several knowledge gaps remain to be filled, the reason being related to the complexity of the physical and chemical characteristics of these particles and of their dynamic interactions with the surrounding environment [1]. One of the frontier topics in this scientific endeavor is the subject of this special issue: the morphology and mixing of atmospheric aerosol at the single-particle level. With aerosol "morphology" and "mixing", here, we refer to three broad properties of a single particle suspended in the atmosphere: (1) the shape and size of a particle; (2) the geometrical distribution of components with different physico-chemical characteristics within a particle; and (3) the topology of a particle (e.g., the existence of convex or concave regions on the particle's surface, or fractal-like structures). Figure 1 shows some examples of the complex morphology and mixing state of atmospheric particles. These three properties affect important atmospheric processes; for example: (a) The shape and size of a particle affects its optical and aerodynamic properties, determining their effectiveness when interacting with electromagnetic radiation, their settling velocity, their ability to penetrate deeply into the lungs, and the sampling efficiency of instrumentation, inlets, or sampling lines. Even the simple concept of size becomes ill-defined (or ambiguous) for particles that have nontrivial shapes (e.g., different from spheres, spheroids, cylinders, or cubes); (b) The geometrical distribution of different components affects the particles' ability to interact with other atmospheric components, including water, and their effectiveness to scatter or absorb electromagnetic radiation. Therefore, particles with similar size and similar components' mass fractions, but different geometric distributions of these components, can have different cloud condensation and ice nucleation properties, can promote different heterogeneous reactions, can experience different aging processes, and can exert different radiative forcing; (c) The topology of the particle can affect its ability to nucleate water droplets and ice crystals, or favor the condensation of other material from the gas phase, determining, for example, the mass growth rate of secondary organic aerosol, and it can also affect heterogeneous chemical reactions and the particle toxicity. An important example where the topology has a key role is for fractal-like, combustion-generated soot particles (often also referred to as black carbon). These particles have a key role in determining the overall radiative forcing of anthropogenic aerosol [2–4]. Their initial structure is typically well described by a fractal formalism, and their mass scales with a measure of their length through a power law—with an exponent different from that of convex objects [5].

This structure changes over time in the atmosphere and, therefore, the properties of these particles are very dynamic [3,6–11].

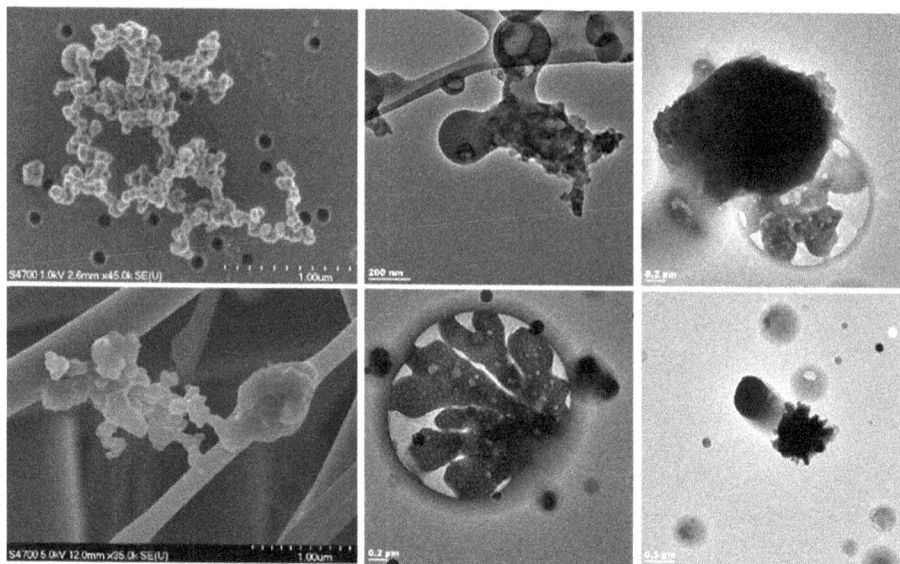

Figure 1. Complex morphology and internal mixing of atmospheric particles.

In this special issue, the authors discuss several of these aspects and present recent advances in this field, including ambient and laboratory characterizations, as well as theoretical and numerical treatments. In the next section, we will briefly summarize the individual contributions to this issue, categorized into three broad topics: (1) sampling and laboratory techniques; (2) analyses of atmospheric particles; and (3) theoretical and numerical studies.

2. Summary of This Special Issue

2.1. Sampling and Laboratory Techniques

Chen et al. [12] report a detailed study of the modification of the particle morphology deposited on different substrates and subjected to various conditions during storage, handling, and analysis. For their study, they generated, collected, and analyzed three types of particles: sodium chloride, sulfuric acid, and soot coated with sulfuric acid. Depending on substrates and conditions, they observe that morphological changes of the deposited particles could vary from negligible to severe. They, therefore, recommend caution during each step of the specimen lifetime from the collection to the analysis. They also provide specific conditions and sampling media that can work better for the specific problem and particle type investigated.

Bhandari et al. [13] analyze fresh soot particles generated in the laboratory using acetylene and methane flames, before and after a thermodenuder. Thermodenuders are often used to estimate the effect that coating material has on the soot particles' optical properties through the "lensing" effect. Freshly generated soot has typically a fractal-like (lacy) structure, but such a structure can collapse to a more compact one upon aging. Compaction, as coating, affects the soot optical properties as well. If the goal of using a thermodenuder is to assess the effect of coating only, then the thermodenuder itself should cause minimal or no compaction. The study verifies that indeed that is the case.

Kulkarni et al. [14] demonstrate a laboratory-based experimental method to investigate the immersion freezing of ice residuals using two continuous-flow diffusion chambers and a pumped counterflow virtual impactor. Ice is nucleated in immersion freezing mode in the first diffusion chamber. The larger ice crystals are separated and sublimated using the virtual impactor and a heat exchanger. The ice residuals are transferred to the second ice chamber to investigate the immersion freezing properties of ice residuals. The results from this study show that not all the ice residuals nucleate ice in the second chamber. The transformation of morphology and chemical composition of ice residuals during the freezing–sublimation process can influence the freezing properties of the ice residuals.

Brus et al. [15] present laminar flow tube measurements of sulfuric acid diffusion coefficients as a function of relative humidity, temperature, and concentration. They use a chemical ionization mass spectrometer to monitor sulfuric acid concentrations at different positions along the flow tube and calculate the effective sulfuric acid diffusion coefficient from the wall loss rate. They apply a computational fluid dynamics model to investigate the laminar flow in the tube. The authors find that the effective sulfuric acid diffusion coefficients linearly decrease with increasing relative humidity, while they show a power dependence with respect to temperature. They further use clustering kinetics simulations to investigate the effective diffusion coefficients. They suggest that the attachment of sulfuric acid molecules with base molecules may be responsible for a higher temperature dependence.

2.2. Analyses of Atmospheric Particles

Kiriya et al. [16] report a study of aerosol surface area during a field study in Japan, in 2015–2016. Often size distributions are reported in terms of number concentrations per size bin; the authors here, instead, study the aerosol surface area distributions. The motivation is that the surface area might be particularly important for the toxicity of the particles and their ability to react with humans' and animals' cells and other pollutants. They find that the surface area correlated with the black carbon concentrations. Freshly emitted black carbon particles are typically fractal-like aggregates of small monomers and have a relatively large surface area. This surface area can decrease over time due to aging and coating processes. Therefore, the authors interpret their result as evidence that the black carbon transported to the site was mostly uncoated.

Mahish et al. [17] report an analysis of different methods used to calculate cloud condensation nuclei spectra and assess the effect of various assumptions and simplifications, including those regarding the mixing state of atmospheric aerosol. For their analysis, they use a large dataset collected at the Southern Great Planes site in Oklahoma from the U.S. Department of Energy, Atmospheric Radiation Measurement program. They first make a baseline estimate of the cloud condensation nuclei spectra with κ-Köhler theory and using all the data available without averaging. Then, they compare several estimates using different assumptions and they find the best agreement with their baseline when they include size-dependent internal mixture hygroscopicity information.

Xu et al. [18] discuss a single particle study using electron microscopy and spectroscopy on samples collected in the Northeast of China. They characterize the particles based on their elemental composition, mixing, and morphology. From their analysis, they suggest that coal combustion in low-efficiency stoves, used for household heating, is a dominant source in the area. The authors also suggest that biomass burning is of secondary importance, although not negligible. They conclude that the anthropogenic emissions from rural regions can be transported to urban areas and substantially add to the local pollution, contributing to regional haze episodes.

Wang et al. [19] investigate the morphology and mixing state of individual atmospheric particles in the megacity of Beijing during the 2015 China Victory Day parade, using transmission electron microscopy coupled with energy dispersive X-ray spectrometry. They classify particles into two broader groups and within each group, they define different subcategories (primary: mineral dust, soot, and organic; and secondary: homogeneous mixed S-rich, and organic coated S-rich particles) based on their morphology and mixing state. They find that secondary particles dominate (~79%) the total particle population. They also observe that the average diameter of secondary particles increases with increasing relative humidity. They suggest that organic coated S-rich particles may be formed by condensation of secondary organic aerosol on seed S-rich particles.

Fraund et al. [20] present a microspectroscopy analysis of single particles collected in the Amazon basin from three sites with different proximity from the city of Manaus. The authors quantitatively combine two complementary microspectroscopy techniques, Scanning Transmission X-ray Microscopy/Near-Edge Fine Structure Spectroscopy and Scanning Electron microscopy/Energy Dispersive X-ray Spectroscopy. They estimate the particle-specific mass fraction and calculate the bulk and individual particle diversity parameters. They utilize the mass fraction data for k-means clustering analysis to identify several particle classes. They use the diversity parameter to quantify the mixing state (i.e., the mixing state index) of the particle population. The mixing state index varies from 0 (completely externally mixed) to 1 (completely internally mixed). The results of this study suggest that the background site contains less cluster variety and fewer anthropogenic clusters than samples collected at the sites nearer the city.

2.3. Theoretical and Numerical Studies

Sorensen et al. [21] provide a detailed review of the power of Q-space analysis that offers general insights on the optical properties of particles and allows the discovery of patterns useful for the interpretation of these properties. After a general introduction of the basic concepts using Mie and Rayleigh scattering theories, the authors expand the analysis to the more complex topic of the optical properties of irregular particles. Particle types discussed include dust, abrasive powders, fractal aggregates, and ice crystals, covering a wide range of sizes, shapes, and index of refractions, and making the work relevant to several branches of science, even beyond purely atmospheric applications. For their discussion, the authors use experimental and theoretical data (e.g., numerical methods such as T-matrix and the discrete dipole approximation). Among several interesting findings, what stands out is the clear distinction between fractal and nonfractal particles, especially at some regimes.

Chin et al. [22] present a numerical modeling analysis of how the mixing state of black carbon affects its atmospheric removal through nucleation scavenging. Nucleation scavenging, where a water droplet grows by condensation on a particle, is possibly the most important mechanism for the removal of black carbon from the atmosphere, determining the lifetime of this type of particle. Because black carbon strongly absorbs solar radiation, an understanding of its lifecycle is key to accurately modelling the effect of aerosol on climate. For their work, the authors use a detailed particle-resolved model that can follow the aerosol mixing (internal and external) at the single-particle level. Based on their analysis, they suggest that typically employed models that ignore (or simplify) the intricacy of the single-particle mixing, can significantly overestimate the scavenged black carbon mass fraction, especially for lower supersaturation conditions.

Hughes et al. [23] apply the machine learning approach to predict the global distribution of the aerosol mixing state and its implications on hygroscopicity, as quantified by a mixing state metric. Machine learning is an emerging tool in atmospheric science; it utilizes a set of algorithms to recognize patterns in large datasets and make predictions. Global climate models use a simplified representation of aerosol mixing state, which can introduce large uncertainties. In this study, the authors utilize a large ensemble of particle-resolved box model simulations and the machine learning approach to determine mixing state matrices and identify the regions where the external and internal mixing state assumption can be applied in global climate models. They find that the mixing state metric varies between 20% and 100%. They also report how the mixing state metric varies with particle diameter, geographical location, and season. This study shows how machine learning can be applied to link detailed particle-resolved models to large-scale global climate models.

3. Conclusions

This special issue assembles a dozen contributions discussing a wide range of aspects on the morphology and mixing state of individual particles by internationally recognized experts in the field. The results and techniques discussed and proposed in this special issue will be of interest to experimentalists as well as global climate modelers, and guide the improvement of numerical simulations of past and future climate changes. However, we also hope that these results will be useful to other research communities, including those that study air quality, visibility, particle toxicity, combustion processes, atmospheric optics, and chemistry, and possibly those that study particles for technological applications. Finally, we hope that the research presented here will spark new ideas and indicate future research directions.

Acknowledgments: We would like to thank all the contributors to this issue that made it "special", as well as the Editorial team of Atmosphere. We acknowledge the support from the Environmental Molecular Sciences Laboratory (EMSL), a national scientific user facility located at the Pacific Northwest National Laboratory (PNNL) and sponsored by the Office of Biological and Environmental Research of the U.S. Department of Energy (U.S. DOE). PNNL is operated by the U.S. DOE by the Battelle Memorial Institute under contract DEAC05-76RL0 1830.

Conflicts of Interest: The authors declare no conflict of interest.

References

1. Boucher, O.; Randall, D.; Artaxo, P.; Bretherton, C.; Feingold, G.; Forster, P.; Kerminen, V.-M.; Kondo, Y.; Liao, H.; Lohmann, U.; et al. Clouds and aerosols. In *Climate Change 2013: The Physical Science Basis. Contribution of Working Group I to the Fifth Assessment Report of the Intergovernmental Panel on Climate Change*; Cambridge University Press: Cambridge, UK; New York, NY, USA, 2013.

2. Ramanathan, V.; Carmichael, G. Global and regional climate changes due to black carbon. *Nat. Geosci.* **2008**, *1*, 221–227. [CrossRef]

3. Jacobson, M.Z. Strong radiative heating due to the mixing state of black carbon in atmospheric aerosols. *Nature* **2001**, *409*, 695–697. [CrossRef] [PubMed]

4. Bond, C.T.; Bergstrom, W.R. Light absorption by carbonaceous particles: An investigative review. *Aerosol. Sci. Technol.* **2006**, *40*, 27–67. [CrossRef]

5. Sorensen, C.M. Light scattering by fractal aggregates: A review. *Aerosol. Sci. Technol.* **2001**, *35*, 648–687. [CrossRef]

6. China, S.; Scarnato, B.; Owen, R.C.; Zhang, B.; Ampadu, M.T.; Kumar, S.; Dzepina, K.; Dziobak, M.P.; Fialho, P.; Perlinger, J.A.; et al. Morphology and mixing state of aged soot particles at a remote marine free troposphere site: Implications for optical properties. *Geophys. Res. Lett.* **2015**, *42*, 1243–1250. [CrossRef]

7. Cappa, C.D.; Onasch, T.B.; Massoli, P.; Worsnop, D.R.; Bates, T.S.; Cross, E.S.; Davidovits, P.; Hakala, J.; Hayden, K.L.; Jobson, B.T.; et al. Radiative absorption enhancements due to the mixing state of atmospheric black carbon. *Science* **2012**, *337*, 1078–1081. [CrossRef] [PubMed]

8. Liu, S.; Aiken, A.C.; Gorkowski, K.; Dubey, M.K.; Cappa, C.D.; Williams, L.R.; Herndon, S.C.; Massoli, P.; Fortner, E.C.; Chhabra, P.S.; et al. Enhanced light absorption by mixed source black and brown carbon particles in UK winter. *Nat. Commun.* **2015**, *6*, 8435. [CrossRef] [PubMed]

9. Mahrt, F.; Marcolli, C.; David, R.O.; Grönquist, P.; Barthazy Meier, E.J.; Lohmann, U.; Kanji, Z.A. Ice nucleation abilities of soot particles determined with the horizontal ice nucleation chamber. *Atmos. Chem. Phys. Discuss.* **2018**, *41*. [CrossRef]

10. Adachi, K.; Buseck, P.R. Changes of ns-soot mixing states and shapes in an urban area during calnex. *J. Geophys. Res. Atmos.* **2013**, *118*, 3723–3730. [CrossRef]

11. Adachi, K.; Chung, S.H.; Buseck, P.R. Shapes of soot aerosol particles and implications for their effects on climate. *J. Geophys. Res. Atmos.* **2010**, *115*. [CrossRef]

12. Chen, C.; Enekwizu, O.; Ma, Y.; Zakharov, D.; Khalizov, A. The impact of sampling medium and environment on particle morphology. *Atmosphere* **2017**, *8*, 162. [CrossRef]

13. Bhandari, J.; China, S.; Onasch, T.; Wolff, L.; Lambe, A.; Davidovits, P.; Cross, E.; Ahern, A.; Olfert, J.; Dubey, M.; et al. Effect of thermodenuding on the structure of nascent flame soot aggregates. *Atmosphere* **2017**, *8*, 166. [CrossRef]

14. Kulkarni, G. Immersion freezing of total ambient aerosols and ice residuals. *Atmosphere* **2018**, *9*, 55. [CrossRef]

15. Brus, D.; Škrabalová, L.; Herrmann, E.; Olenius, T.; Trávničková, T.; Makkonen, U.; Merikanto, J. Temperature-Dependent Diffusion of H_2SO_4 in Air at Atmospherically Relevant Conditions: Laboratory Measurements Using Laminar Flow Technique. *Atmosphere* **2017**, *8*, 7. [CrossRef]

16. Kiriya, M.; Okuda, T.; Yamazaki, H.; Hatoya, K.; Kaneyasu, N.; Uno, I.; Nishita, C.; Hara, K.; Hayashi, M.; Funato, K.; et al. Monthly and diurnal variation of the concentrations of aerosol surface area in Fukuoka, Japan, measured by diffusion charging method. *Atmosphere* **2017**, *8*, 114. [CrossRef]

17. Mahish, M.; Jefferson, A.; Collins, D. Influence of common assumptions regarding aerosol composition and mixing state on predicted CCN concentration. *Atmosphere* **2018**, *9*, 54. [CrossRef]

18. Xu, L.; Liu, L.; Zhang, J.; Zhang, Y.; Ren, Y.; Wang, X.; Li, W. Morphology, composition, and mixing state of individual aerosol particles in northeast china during wintertime. *Atmosphere* **2017**, *8*, 47. [CrossRef]

19. Wang, W.; Shao, L.; Xing, J.; Li, J.; Chang, L.; Li, W. Physicochemical characteristics of individual aerosol particles during the 2015 China victory day parade in Beijing. *Atmosphere* **2018**, *9*, 40. [CrossRef]

20. Fraund, M.; Pham, D.; Bonanno, D.; Harder, T.; Wang, B.; Brito, J.; de Sá, S.; Carbone, S.; China, S.; Artaxo, P.; et al. Elemental mixing state of aerosol particles collected in central amazonia during goamazon2014/15. *Atmosphere* **2017**, *8*, 173. [CrossRef]

21. Sorensen, C.; Heinson, Y.; Heinson, W.; Maughan, J.; Chakrabarti, A. Q-space analysis of the light scattering phase function of particles with any shape. *Atmosphere* **2017**, *8*, 68. [CrossRef]

22. Ching, J.; West, M.; Riemer, N. Quantifying impacts of aerosol mixing state on nucleation-scavenging of black carbon aerosol particles. *Atmosphere* **2018**, *9*, 17. [CrossRef]

23. Hughes, M.; Kodros, J.; Pierce, J.; West, M.; Riemer, N. Machine learning to predict the global distribution of aerosol mixing state metrics. *Atmosphere* **2018**, *9*, 15. [CrossRef]

atmosphere

MDPI

Article

The Impact of Sampling Medium and Environment on Particle Morphology

Chao Chen [1,2], Ogochukwu Y. Enekwizu [2,3], Yan Ma [1], Dmitry Zakharov [4] and
Alexei F. Khalizov [2,3,*]

[1]   Collaborative Innovation Center of Atmospheric Environment and Equipment Technology,
    School of Environmental Science and Engineering, Nanjing University of Information Science & Technology,
    Nanjing 210044, China; achao_corn@163.com (C.C.); my_nj@163.com (Y.M.)
[2]   Department of Chemistry and Environmental Science, New Jersey Institute of Technology, Newark,
    NJ 07102, USA; oye2@njit.edu
[3]   Department of Chemical Biological and Pharmaceutical Engineering, New Jersey Institute of Technology,
    Newark, NJ 07102, USA
[4]   Center for Functional Nanomaterials, Brookhaven National Laboratory, Upton, NY 11973, USA;
    dzakharov@bnl.gov
*   Correspondence: khalizov@njit.edu; Tel.: +1-973-596-3853

Received: 23 July 2017; Accepted: 26 August 2017; Published: 29 August 2017

Abstract: Sampling on different substrates is commonly used in laboratory and field studies
to investigate the morphology and mixing state of aerosol particles. Our focus was on the
transformations that can occur to the collected particles during storage, handling, and analysis.
Particle samples were prepared by electrostatic deposition of size-classified sodium chloride,
sulfuric acid, and coated soot aerosols on different substrates. The samples were inspected by
electron microscopy before and after exposure to various environments. For coated soot, the imaging
results were compared against mass-mobility measurements of airborne particles that underwent
similar treatments. The extent of sample alteration ranged from negligible to major, depending on
the environment, substrate, and particle composition. We discussed the implications of our findings
for cases where morphology and the mixing state of particles must be preserved, and cases where
particle transformations are desirable.

Keywords: substrate; morphology; electron microscopy; aerosols; soot; sodium chloride; sulfuric
acid; sampling

1. Introduction

Atmospheric aerosols play a major role in regional air quality and global climate [1–4]. The effects
of aerosols are highly dependent on the particle number concentration, and also on the particle size,
composition, mixing state, phase state, and morphology. The knowledge of these properties is crucial
for an accurate prediction of the environmental impacts of aerosols.

Although some current instrumentation is capable of online analysis of aerosol particles [5–10],
the aerosol community is heavily reliant on off-line sampling, due to lower costs and a more
comprehensive range of particle characterization methods available. Typically, a series of samples
are collected during laboratory experiments or field measurements, with several replicates to ensure
sufficient statistics. The samples are stored for hours, days, or even months, before being transported
to the analysis facility, either by the researcher or over a commercial carrier. Finally, the samples
are analyzed by the researcher or by the facility staff. Electron microscopy (EM) is a widely used
off-line technique for particle analysis because it can provide a direct evaluation of the particle size and

morphology [11]; also, particle composition can be assessed when EM is augmented by such methods as Energy-Dispersive X-ray Spectroscopy (EDX) or Electron Energy Loss Spectroscopy (EELS).

At every step in the above sequence, from collection through analysis, sample alteration is possible to a degree that the collected particles may fail to represent the original airborne particles, both individually and statistically. For instance, the collection of samples is often conducted by aerodynamic deposition of airborne particles on various substrates in a low-pressure cascade impactor [12]. This method may discriminate against solid particles, which tend to bounce off the substrate surface [13,14]. Greasing the substrate reduces particle bouncing, but is not always applicable when the sample is intended for EM imaging and chemical analysis. Since the impaction of particles occurs at a high velocity, the fragmentation of loosely connected agglomerates is possible, such as those produced by the Brownian coagulation of soot aggregates originally emitted by diesel engines; the original aggregates, however, are strongly bound and do not fragment upon impact [15]. Although the presence of a small amount of organic adsorbates stabilizes coagulation-produced agglomerates [15], thick liquid coatings could themselves be lost ("shaken off") upon impaction of particles against substrate [16,17]. Collection approaches using electrostatic deposition [18,19], diffusion [15], or thermophoresis [20] are more gentle, but there are a number of factors which limit their use. Collection by filtration on porous membranes is by itself non-damaging [21], but typically requires a sputtered metal overcoat to make the sample electrically conductive, if it is intended for EM analysis. Also, such membrane samples by design are only suitable for examination by Scanning Electron Microscopy (SEM), but not for Transmission Electron Microscopy (TEM) because the membrane is not transparent to the electron beam. Thus, the choice of the particle collection and analysis methods imposes significant constraints on the choice of sampling substrates. While any sufficiently flat substrate that can be made electrically conductive by metal sputtering is suitable for SEM analysis, the application of TEM is limited to thin-film or web-like substrates that are sufficiently transparent for the electron beam.

A common limitation of standard TEM and SEM approaches comes from the need to place the particle sample under high vacuum, where evaporation of volatile and semi-volatile particle components is possible [22,23]. This problem is exacerbated by exposure of particles to the electron beam, leading to thermal evaporation and surface charging, reducing the image resolution and contrast. A lower accelerating voltage can be used to minimize this problem, but at the expense of a reduced resolution. One way to minimize sample damage is by using Environmental EM methods (i.e., ETEM and ESEM), where the sample is held at several millibar or even atmospheric pressure. Also, sample encapsulation under a thin membrane can be used to protect particles from high vacuum conditions [24].

Another frequently overlooked aspect of sample analysis involves substrate-particle interactions, which may depend strongly on the particle and substrate composition. For instance, interaction with the surface of beryllium substrate has been proposed to explain the failure to observe sulfuric acid particles after deposition [17]. The deliquescence point of water soluble particles [25] and mixing state of organic/inorganic aerosol particles [26] have been shown to depend on the hydrophilic properties of substrates. Hydrophilicity, or generally, the surface energy of the substrate, controls the ability of the particle material to wet the substrate surface, and hence the shape of liquid particles. At the limit of perfect wetting, a particle may lose its spherical shape entirely by spreading in a thin layer. Similarly, a core-and-shell particle may lose its liquid coat to the surface of substrate. The rate of transport from the particle to the surface is slow for thin coats [27]. However, thickly coated particles may potentially experience a significant coating loss within hours or even minutes, depending on the viscosity of the liquid layer [27], which depend on the degree of particle aging. To reduce photo-induced changes, particle samples are commonly kept refrigerated in the dark [21], but chemical changes are still possible through dark oxidation by molecular oxygen and neutralization of particle-phase acids by ubiquitous ammonia. Additionally, temperature and humidity swings during storage and transport present yet another possible cause of sample modification. For instance, for a sample sealed at 25 °C and 50%

relative humidity (RH), a 10 °C drop in temperature would increase the RH to 93%, resulting in the deliquescence of most inorganic particle constituents and associated morphological changes in the particles. Indeed, freezing and thawing cycles of collected particles have been shown to cause severe particle agglomeration due to water uptake [24]. Finally, even for perfect samples, the human factor may contribute to a bias during image analysis through particle discrimination based on size and shape, as discussed previously [28–30].

The goal of this study was to investigate the role of particle-substrate interactions and changing environmental conditions on the outcome of morphological analysis for several types of submicron aerosols. We examined aerosol particles composed of sodium chloride, sulfuric acid, untreated soot, and sulfuric acid-coated soot. Sodium chloride is a crystalline material with deliquescence and efflorescence RH of 75 and 45%, respectively. Sulfuric acid is a low volatile, highly hygroscopic liquid whose water content depends on the ambient RH. Coated soot particles are aggregates of hydrophobic graphitic spheres with a thin layer of hygroscopic sulfuric acid. The particles were collected on several types of substrates (lacey grids, untreated silicon, hydrophobic silicon, and silicon nitride), exposed to varying temperature/humidity conditions, and analyzed to elucidate factors leading to significant morphological changes.

2. Experiments

2.1. Particle Generation, Processing, and Mass-Mobility Analysis

Soot aerosol was produced by the combustion of natural gas in an inverted diffusion burner [31]. A global flame equivalence ratio of 0.5 was used to form fractal particles with a negligible fraction of organic carbon [19,32]. Sodium chloride and sulfuric acid aerosols were produced by nebulization of the corresponding aqueous solutions in a constant output atomizer (Aerosol Generator 3076, TSI Inc., Shoreview, MN, USA). In all cases, the generated aerosol was diluted with particle-free, purified air, and then passed through a diffusion drier filled with silica gel, a Nafion drier (PD-07018T-24MSS, Perma Pure LLC., Lakewood, NJ, USA), and a bipolar diffusion charger (Po-210, 400 µCi, NRD Staticmaster, New York, NY, USA). An integrated system (Figure S1 in Supplementary Materials) consisting of two differential mobility analyzers (DMA 3081, TSI Inc., Shoreview, MN, USA), an aerosol particle mass analyzer (APM 3601, Kanomax Inc., Andover, NJ, USA), and a condensation particle counter (CPC 3772, TSI Inc., Shoreview, MN, USA) was used to size-classify and characterize aerosols [33]. A more detailed description of the aerosol system can be found in [19].

Sodium chloride and sulfuric acid particles were directly collected on substrates after size-classification. In the case of soot particles, processing could be applied before particle collection; this involved the coating of particles with sulfuric acid and/or exposure to elevated relative humidity. For coating, the soot aerosol was passed through a pick-up chamber half-filled with sulfuric acid (80 wt %) that was maintained at a controlled temperature (52 ± 2 °C). Following the coating application, the aerosol was either directly collected in the precipitator or passed through a drier/humidifier (PD-07018T-24MSS, Perma Pure LLC., Lakewood, NJ, USA), and then collected. The amount of sulfuric acid on soot particles and the associated change in the particle mobility diameter induced by coating and humidification were monitored through Tandem DMA and DMA-APM measurements, and reported as growth factors by diameter (Gfd) and mass (Gfm), respectively. The Gfd and Gfm were calculated from the ratios of the processed particle mobility diameter and particle mass (D_p, m_p) over the initial diameter and mass (D_o, m_o), respectively. In all cases, following size-classification, soot aerosol was passed through a thermal denuder maintained at 300 °C to remove residual organics.

2.2. Substrates, Collection Procedures, and Sample Processing

Particles were collected in a custom-made electrostatic precipitator [19] similar to the one described earlier [18]. Lacey grids (01883, Ted Pella, Inc., Redding, CA, USA), silicon wafer chips (16008,

Ted Pella, Inc., Redding, CA, USA), and silicon nitride window chips (301.3892, Hummingbird Scientific, Lacey, WA, USA) were used as collection substrates. Lacey grids have a Formvar/Carbon layer with web-like structure on a 300 mesh copper grid. Upon collection, some particles attached themselves to the fibers where the particle-fiber contact area was relatively small, while other particles landed on flat areas. Lacey grids were used as is, without any treatment. Silicon chips were sonicated in methanol to remove possible contamination, and dried with high purity nitrogen before use. Cleaned untreated silicon chips were hydrophilic. To obtain hydrophobic silicon chips, cleaned chips were immersed in a dichlorodimethylsilane solution (5% DCDMS in toluene, Sigma-Aldrich, St. Louis, MO, USA), rinsed sequentially with toluene and methanol, and dried with high-purity nitrogen. To verify the effect of the DCDMS treatment, a 3-µL water droplet was placed on the surface of both DCDMS -treated and untreated silicon chips. Images of the droplets (Figure 1) taken by a digital camera (eheV1-USBplus, Oasis Scientific, Taylors, SC, USA) indicated that the average contact angle increased from 45° to 92° after the DCDMS treatment (Table 1), confirming that the wettability of the silicon surface was reduced significantly (for reference, the wetting angle for water on Teflon was 104°). The wetting angle was determined from digital images using image-processing software (Adobe Photoshop, Adobe Systems Inc., Mountain View, CA, USA). Silicon nitride chips in ETEM experiments were used as received. These chips have a narrow-slit window covered with a 50 nm thick silicon nitride membrane. A regular chip and a spacer chip were sandwiched together to create a controlled environment for ETEM imaging of particle samples.

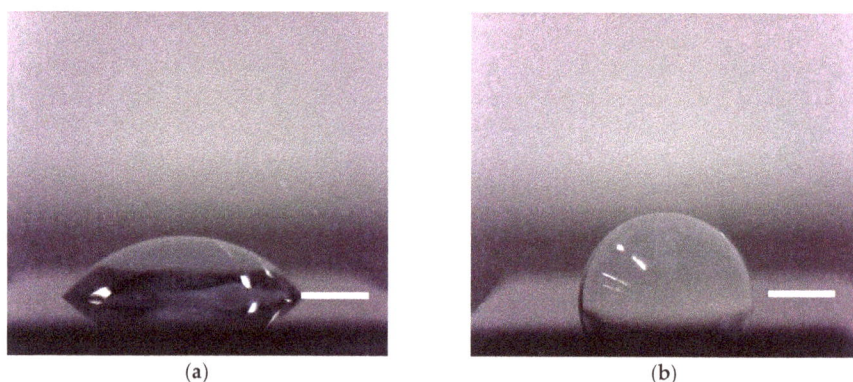

| (a) | (b) |

Figure 1. A droplet of water on the surface of (**a**) untreated hydrophilic and (**b**) hydrophobically treated silicon chip. Scale bar for both images is 1 mm.

Table 1. Wetting angle of liquids on different surfaces.

Chemical	Wetting Angle, Degrees		Silicon Nitride	Graphite
	Silicon			
	Hydrophilic	Hydrophobic		
Water	45	92		86
Sulfuric acid (20%)			69	73
Sulfuric acid (50%)			61	
Sulfuric acid (80%)	20 [1]	92		16
Sulfuric acid (98%)			20 [1]	

[1] Because of effective surface wetting the angle cannot be measured precisely.

Particle samples collected on lacey grids and silicon chips were processed off-line and then imaged by SEM. The conditions corresponding to each of the three types of processing environments are described in Table 2. During Type I processing, substrates with collected particles were exposed to a 92–96% RH inside a small stainless steel cylindrical chamber for 20 min. Such exposure might correspond to a moderate change in environmental conditions during sample transport or contact with the operator's breath. The relative humidity inside the chamber was maintained using a humidified flow obtained by passing particle-free, purified air through a single 30 cm long Nafion tube (TT-050, 0.042" ID, 0.053" OD, Perma Pure LLC., Lakewood, NJ, USA) immersed in warm distilled water. The relative humidity was measured by a RH sensor (SRH77A, Cooper Atkins, Middlefield, CT, USA). Type II processing involved exposing the samples to a 55% RH in an open holder box (PELCO®X-TREME, Ted Pella, Inc., Redding, CA, USA) placed inside a plastic bag. Next, the box was hermetically sealed and transferred to a freezer maintained at −20 °C. After 12 h of storage, the box was removed from the freezer, brought to room temperature, and then opened to expose the samples to ambient air (RH 16–55%, as noted in each case). Type III processing was similar to Type II, but after removal from the freezer, the box was opened immediately, allowing ambient water vapor to condense on particle samples. Condensation and surface flooding could be observed visually, with the amount of condensed water strongly dependent on the ambient humidity (RH 16–20% for Type IIIa and RH 50% for Type IIIb). The samples collected on silicon nitride chips were humidified in situ during ETEM imaging, using a 0.5 mL/min flow of helium that passed through the single-tube Nafion humidifier described above.

Table 2. Different types of environments used to process collected particles.

Environment	Description
Type I (Humidified)	Exposed to a 92–96% relative humidity (RH)
Type II (Cold storage)	Exposed to a 55% RH at room temperature, sealed in a container, chilled in a freezer to −20 °C, removed from the freezer, brought to room temperature, and then exposed to ambient air (RH 16 to 50%)
Type IIIa (Moderately flooded)	Exposed to a 55% RH at room temperature, sealed in a container, chilled in a freezer to −20 °C, removed from the freezer and exposed to ambient air while still cold; ambient air RH is 16–20%.
Type IIIb (Severely flooded)	Exposed to a 55% RH at room temperature, sealed in a container, chilled in a freezer to −20 °C, removed from the freezer and exposed to ambient air while still cold; ambient air RH is 50%.

2.3. Electron Microscopy and Image Processing

Most of the particle samples were imaged with a LEO 1530VP Field Emission Scanning Electron Microscope (FE-SEM), using a 5 kV accelerating voltage. A limited number of samples were studied by FEI Titan 80–300 ETEM located in the Center for Functional Nanomaterials at Brookhaven National Lab, using a 300 kV accelerating voltage. No metal or carbon film coating was applied to the particle samples before imaging. The SEM and ETEM images were manually pre-processed using Adobe Photoshop to adjust contrast and/or gray level in order to separate particles from the substrate background. For soot aggregates on silicon and silicon nitride, adjusting the levels was sufficient, but for soot aggregates on lacey grid samples, a manually drawn outline was required because the aggregates and lacey fibers had a comparable level of gray and/or contrast. For sulfuric acid and sodium chloride particles on all substrates, adjustment of gray level was sufficient to separate the particles from the image background.

Processed images were used to measure particle dimensions and morphology. In the case of soot, the aggregate convexity (Figure S2) was determined [21,34]. Convexity is the ratio of the aggregate projected area A_a over the area of the convex hull polygon $A_{polygon}$, as defined by Equation (1):

$$\text{Convexity} = \frac{A_a}{A_{polygon}} \tag{1}$$

where A_a and $A_{polygon}$ were determined from the images using MATLAB-based code written in-house (Text S2). The convexity characterizes the compactness of soot aggregates, varying between 0 and 1, with the larger value corresponding to more compact aggregates.

3. Results and Discussion

3.1. Sodium Chloride

Figure 2 shows sodium chloride particles with an initial mobility diameter of 200 nm that were collected on several different substrates and treated following procedures described in Table 2. As expected, the untreated particles appeared as cubes with well-defined edges (Figure 2a) because the highest RH experienced by these particles during sample manipulation never exceeded 50%, which is well below the 75% deliquescence RH of sodium chloride. For treated particles, the appearance depended on several factors, including the procedure, the maximum level of RH, and the type of substrate. The particles were stable under the electron beam at magnifications below ×100 k. At magnifications exceeding ×700 k, the resistive heating from the electric current delivered by the electron beam caused particle melting and then re-crystallization into a set of smaller particles.

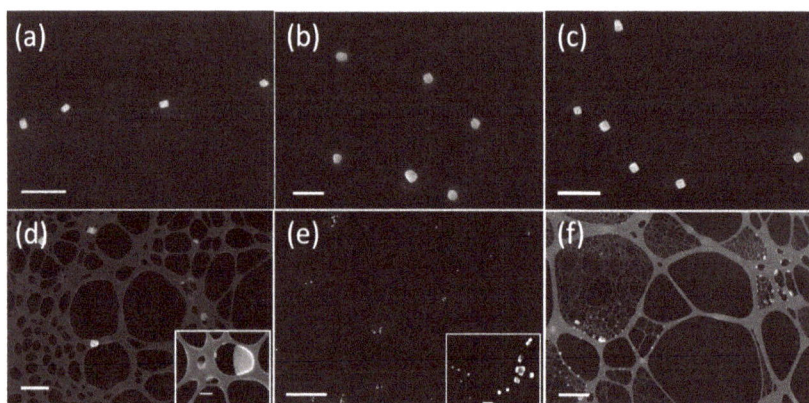

Figure 2. Scanning electron microscopy (SEM) images of 200 nm initial mobility diameter sodium chloride particles deposited on different substrates: (**a**) untreated silicon chip, dry conditions; (**b**) untreated silicon chip, subjected to Type I treatment; (**c**) hydrophobic silicon chip, subjected to Type I treatment; (**d**) lacey support film, subjected to Type I treatment; (**e**) untreated silicon chip, subjected to Type IIIa treatment (16–20% ambient RH); and (**f**) lacey support film, subjected to Type IIIb treatment (45% ambient RH). Inset in (**d**) shows recrystallized NaCl within holes of the lacey grid. Inset in (**e**) shows polydisperse recrystallized particles on untreated silicon chip. Scale bar for images (**a–d,f**), and inset (**e**) is 1 μm. Scale bar for inset (**d**) is 200 nm. Scale bar for image (**e**) is 20 μm.

For sodium chloride particles subjected to Type I treatment (94% RH), morphological changes varied from negligible to major, depending on the substrate. Particles supported by untreated silicon chips lost their sharp edges, becoming roundish; some particles even acquired non-cubic shapes

(Figure 2b). Also, light circular spots developed around some of the particles. Presumably, these spots were composed of a thin deposit of sodium chloride, which led to surface charging around the particles due to the poor electrical conductivity of crystalline sodium chloride. From this behavior, we inferred that the sodium chloride crystals were converted to aqueous droplets at high humidity, and the droplets partially spread over the surface. After water evaporation, most droplets re-crystallized to form individual particles, but the shape of those particles was significantly affected by the interactions between the aqueous droplets and the hydrophilic substrate. On the other hand, most particles that were collected on hydrophobic silicon restored their original cubic structure (Figure 2c) because of poor surface wetting. Overall, particles subjected to Type I treatment on hydrophobic silicon chips appeared nearly indistinguishable from the original dry sodium chloride particles (Figure 2a).

The response of sodium chloride collected on lacey grids depended strongly on the particle location within the grid. Particles adhering to the fibers, far away from lacey holes, maintained their cubic morphology or became somewhat roundish after Type I treatment (Figure 2d). However, particles located near small lacey holes changed their morphology entirely. It appears that after deliquescence, the aqueous solution was drawn into the holes, and when the water evaporated, sodium chloride re-crystallized within the holes, forming solid plugs (Figure 2d inset).

Subjecting samples of sodium chloride to Type II treatment (Cold storage) caused only negligible changes in the particle morphology. The Type II treatment was designed to reproduce a typical sampling routine, whereby a particle sample is sealed at a 55% indoor RH, transferred to the freezer, and upon removal from the freezer is allowed to equilibrate thermally before exposure to ambient air. The lack of discernible changes in the particle morphology on different substrates indicates that the particles were never exposed to an RH above the deliquescence point of sodium chloride. We believe this was caused by the freeze-drying effect due to the container walls being cooled faster than the particle sample, which was mounted in the middle of the container on a thermally non-conductive support. Water vapor diffused from the warmer sample region to the colder wall region, where it condensed, keeping the RH near the sample relatively low. If particle samples were mounted directly on the container wall, the impact would have been significant, as shown in [24], where the sample cooling and thawing at the same rate as cooling and thawing of the container walls resulted in water condensation, deliquescence, and coalescence of the particles.

Under the Type III treatment (Flooding), the sample box was opened immediately after removal from the freezer, allowing no temperature equilibration. Evidently, ambient RH was a key factor in determining the extent of sample change in this case. When exposed to a 50% RH (Type IIIb), the surface of the cold substrates was heavily flooded by the water condensate and no particles could be observed afterwards on either untreated silicon or hydrophobic silicon. Apparently, the particles were dissolved and washed away completely by the thick water layer. When ambient RH was lower than 20% (Type IIIa), some particles could be observed on the substrates, but their appearance was different from that of the original particles. As shown in Figure 2e, these new particles were of non-uniform sizes and clustered in groups. In this case, the level of water saturation was sufficient to activate sodium chloride particles into supermicron-sized droplets, but insufficient to flood the surface. The droplets grew in size and coalesced to form continuous areas while the sample was still cold. When the sample temperature increased and water evaporated, the dissolved sodium chloride re-crystallized randomly on the surface defects, forming particles of a broad range of sizes (Figure 2e inset), with some particles as large as several micrometers. A similar behavior was seen in [24] where freezing or thawing particles without humidity regulation caused morphology changes, while controlling the humidity decreased the severity of change observed in particle size and shape.

When Type III treatment was applied to lacey grids, particles were recovered in all cases and no difference in the particle appearance was noted between the experiments conducted at a 16 or 45% RH. The particles were not washed off because the extent of surface flooding was prevented by the Kelvin effect, due to the small radius of curvature of the fibers. Figure 2f shows that Type IIIb treatment of lacey grids triggered a sequence of droplet activation, coalescence, and re-crystallization, producing random groups of sodium chloride particles of a broad size range. As seen previously with Type I treatment, some of the sodium chloride particles re-crystallized within the lacey holes.

3.2. Sulfuric Acid

Experiments with sulfuric acid aerosol were conducted using DMA1 set to 200 nm. Sulfuric acid droplets collected on hydrophilic and hydrophobic silicon chips appeared as nearly perfect spheres with similar average diameters of 229 ± 21 nm (Figure 3a) and 237 ± 15 nm, respectively. A small number of droplets with multiple charges and larger sizes were also observed, but those were excluded from the average diameter calculation. For samples prepared on a lacey grid, a significant difference was observed between the droplets on the fibers and the flat area, 204 ± 21 nm and 292 ± 20 nm, respectively. The agreement between the SEM and DMA diameters for droplets on fibers indicates that they remained nearly spherical after deposition. In other cases, some flattening occurred from the droplet-surface interaction. The behavior of submicron droplets of concentrated sulfuric acid was drastically different from that of millimeter-scale droplets, which wet the silicon chip surface almost perfectly and spread to form a thin film, due to a stronger gravitational force (Table 1).

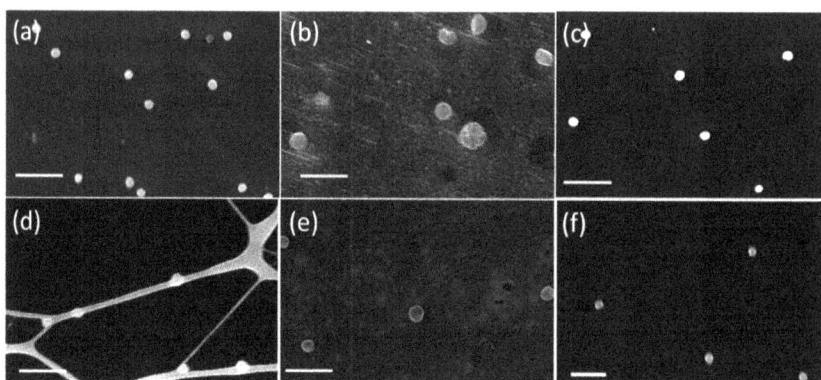

Figure 3. SEM images of 200 nm initial mobility diameter sulfuric acid particles on different substrates. The scale bar is 1 μm. The particles were (**a**) deposited onto a untreated hydrophilic silicon chip and kept dry, (**b**) deposited onto a hydrophilic silicon chip and subjected to Type I treatment, (**c**) deposited onto a hydrophobic silicon chip and subjected to Type I treatment, (**d**) deposited on lacey grid fibers and subjected to Type I treatment, (**e**) deposited onto an untreated silicon chip and subjected to Type II treatment, (**f**) deposited onto a hydrophobic silicon chip and subjected to Type II treatment.

The 200 nm droplets of sulfuric acid were stable under the electron beam at lower magnifications, but evaporated within seconds at magnifications exceeding ×200 k. Exposure of the droplets deposited on hydrophilic silicon to Type I treatment (Humidified) resulted in roundish particles of non-uniform shape and a significantly larger final diameter, 387 ± 36 nm (Figure 3b). The increase in the average diameter relative to the unprocessed sample was caused by droplet flattening after undergoing the humidification/drying cycle. First, at high RH the droplet footprint may have doubled from water absorption, because the hygroscopic growth factor of sulfuric acid droplets between 5 and 94% RH is about 2. The coalescence of adjacent droplets may have caused an additional increase in size. Next, when water was lost through evaporation upon drying, the droplets did not retreat to their original sizes, but flattened instead because of the adhesion to the substrate surface. The presence of darker spots around the droplets, probably the residue of sulfuric acid after the droplet retreat, confirms this hypothesis. The uneven shape of processed droplets could be caused by the submicron-scale surface roughness. Also, it is possible that sulfuric acid in the droplets was partially or completely neutralized by ammonia, which is ubiquitous in ambient air. We estimated that in the presence of mere 10 parts per billion, a common indoors ammonia level [35,36], it takes only few minutes to neutralize sulfuric acid inside the 200 nm droplets. The resulting salts, ammonium bisulfate (efflorescence RH 10%) and ammonium sulfate (efflorescence RH 39%), would crystallize upon drying under the SEM vacuum to form non-spherical particles. It has been reported that neutralization of sulfuric acid particles by ammonia makes them more stable under the electron beam [17].

For droplets on hydrophobic silicon chips and lacey grid fibers, the diameter increased to 259 ± 26 nm and 265 ± 23 nm, respectively after Type I treatment (Figure 3c,d). Less significant changes in the droplet appearance and size on hydrophobic silicon may have resulted from a larger wetting angle of dilute sulfuric acid (Table 1), leading to less droplet spreading at high RH. In the case of fibers, the lesser impact may be due to a lower contact area with the droplets.

On all substrates, Type II treatment (Cold storage) produced less significant changes to sulfuric acid droplets than Type I treatment (Figure 3e,f). For example, the droplet diameter increased to 321 ± 48 nm and 248 ± 8 nm on hydrophilic and hydrophobic silicon, respectively. However, after Type III treatment (Flooding), no droplets could be observed on either hydrophilic or hydrophobic silicon substrates at any ambient RH. It appears that flooding re-distributed sulfuric acid over the entire surface of the silicon substrate, and the acid remained on the surface in the form of a thin layer, even after the evaporation of water. However, for lacey grids samples, groups of non-uniform spheres could be still observed after flooding and drying, because of the limited dispersion of sulfuric acid along the fibers.

3.3. Soot Aggregates

Soot particles with an initial mobility diameter of 350 nm deposited on various substrates are shown in Figure 4. Uncoated soot particles collected on both types of silicon chips appeared as fractal aggregates of graphitic spherical monomers (Figure 4a). The monomer diameter was 28 ± 7 nm, showing no clear dependence on the overall aggregate size. A small fraction of larger aggregates with multiple charges was present, but no smaller fragments were evident, indicating that no fracturing occurred upon deposition of the aggregates on the collection substrate. For unprocessed soot of 350 nm mobility diameter, chosen in our study for most experiments, the aggregate convexity was 0.56 ± 0.05.

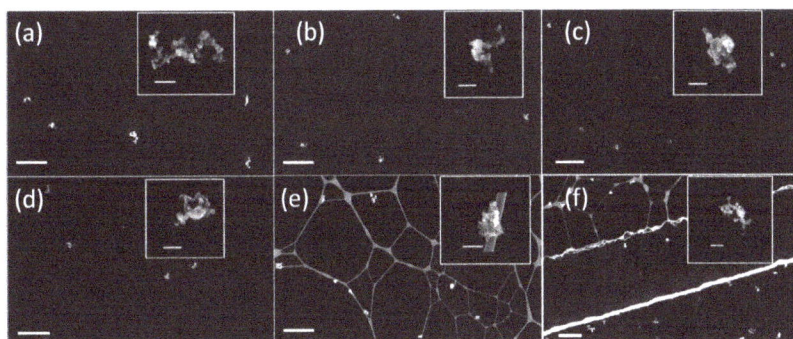

Figure 4. SEM images of uncoated, coated, and coated/humidified soot particles deposited on different substrates ((**a–d**) correspond to untreated silicon chips and (**e,f**) to lacey TEM grids). Initial particle mobility diameter is 350 nm and the scale bar is 2 μm. Insets show magnified individual particles with a scale bar of 200 nm. The particles were (**a**) uncoated, (**b**) coated by H_2SO_4, (**c**) coated by H_2SO_4 and humidified to 88% prior to deposition, (**d**) coated by H_2SO_4, deposited onto a silicon chip, and subjected to Type I treatment, (**e**) coated by H_2SO_4, deposited onto a lacey grid (fibers area), and subjected to Type I treatment, (**f**) coated by H_2SO_4, deposited onto a lacey grid (flat area), and subjected to Type I treatment.

In the atmosphere, fractal soot aggregates age rapidly upon exposure to condensable vapors of sulfuric acid and oxidized organics [37–42]. Sufficiently aged soot may undergo significant structural changes, making the aggregates less fractal [16]. The restructuring is promoted in the presence of hygroscopic materials, which absorb water and increase in volume at elevated RH [33]. It is possible that airborne and surface-bound aggregates respond differently to humidification. For instance, the interaction with the surface may anchor the aggregate, limiting the extent of structural changes. To address this knowledge gap, we investigated morphologies of soot aggregates coated by sulfuric acid and exposed to elevated RH, both in an airborne state and after deposition onto different substrates.

Structural changes in the airborne soot were investigated using the Tandem Differential Mobility Analyzer (TDMA) system, similarly as described earlier [33]. Size-classified soot aerosol (350 nm) emerging from DMA1 was coated by sulfuric acid, humidified, thermally denuded, and then scanned by DMA2 to determine the growth factor, Gfd. Since the coating was removed from the particles before they entered DMA2, the reported Gfd reflects the structural change in the soot backbone induced by the coating material and water, but excludes the contribution from the coating material itself. In all the experiments, the mass fraction of sulfuric acid coating was maintained at 0.26, corresponding to a Gfm of 1.35, as measured by the APM. Figure 5 shows the evolution in Gfd with the increase in RH. The first point with a Gfd of 0.92 corresponds to soot aerosol coated by sulfuric acid and stabilized at 5% RH to minimize coating evaporation. Only a small decrease in Gfd occurred when RH was increased from 5 to 50%, in agreement with previous observations [33]. The most significant decrease in Gfd occurred at RH above 50%, reaching a value of 0.74 at 88% RH. The latter Gfd corresponded to the most compact morphology attainable by the 350 nm soot aggregates generated in our burner [19].

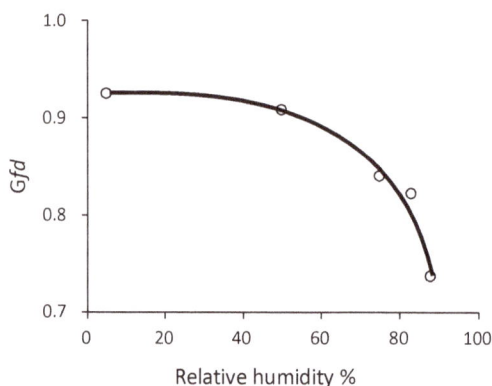

Figure 5. The decrease in the diameter growth factor (*Gfd*) of H_2SO_4-coated soot aggregates with an increase in the relative humidity (RH). The initial mobility diameter of soot is 350 nm and the coating mass fraction is 0.26. Following humidification, the particles were thermally denuded to remove all of the coating material.

To investigate the role of substrate, two sets of experiments were performed, using soot aggregates deposited on hydrophilic silicon chips, hydrophobic silicon chips, and lacey grids. In the first set of experiments, airborne soot aggregates were coated by sulfuric acid, humidified, and only then collected on the substrates. Figure 4b shows an image of the coated soot deposited on a hydrophilic silicon chip at 5% RH. While the presence of sulfuric acid on the aggregates was not apparent, their morphology was clearly more compact than that of unprocessed soot. As shown in Figure 6 (5% RH), convexity was 0.64 ± 0.08, 0.65 ± 0.08, and 0.67 ± 0.07 for coated aggregates on hydrophilic silicon, hydrophobic silicon, and lacey grid fibers, respectively. A somewhat higher convexity observed on hydrophobic silicon chips and grid fibers may reflect a lower interaction of the aggregates with those supports, resulting in a partial restructuring when the samples were exposed to 40–45% ambient RH during transfer to the SEM chamber. Convexity was significantly higher (0.74 ± 0.11) when the coated airborne soot aggregates were humidified to 88% RH before deposition (Figure 4c). An even higher convexity (0.81 ± 0.05) was observed when these coated and humidified aggregates were denuded before deposition, suggesting that the evaporation of water and sulfuric acid off airborne particles could promote their further compaction, as suggested earlier [43,44].

In the second set of experiments, the coated soot aggregates were first deposited on three different substrates and only then humidified, using the protocols described in Table 2. Figure 4d–f shows examples of particles subjected to Type I treatment (94% RH) on hydrophilic silicon and lacey grids (fibers and flat areas separately). The images of particles on hydrophobic silicon are not included because they were essentially the same as on hydrophilic silicon. There was a marked difference in the extent of restructuring experienced by the aggregates located on flat substrates and on fibers (Figure 6). The largest convexity (0.78 ± 0.05) was observed for the aggregates deposited on fibers, close to the convexity of coated aggregates that were exposed to comparable humidity in the airborne state. The convexities of particles located on the flat area of lacey grids, hydrophilic silicon, and hydrophobic silicon were notably lower and of comparable magnitudes, 0.62 ± 0.08, 0.68 ± 0.07, 0.69 ± 0.09, respectively.

As with sodium chloride, Type II treatment (Cold storage) of the coated soot particles made no discernible change in their morphology on any tested substrate. In the case of Type IIIa treatment (Moderately flooded), significant restructuring occurred to soot aggregates attached to the fibers of the lacey grid (0.79 ± 0.04), but the samples prepared on both types of silicon remained practically unchanged (0.62 ± 0.06), as shown in Figure 6.

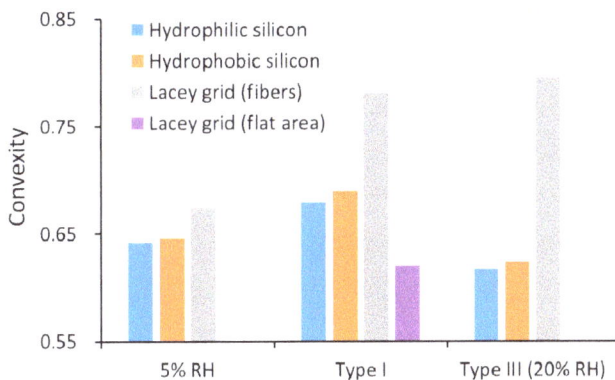

Figure 6. Morphology of coated soot aggregates deposited on different substrates and then exposed to different environmental conditions. The initial mobility diameter was 350 nm and the coating mass fraction 0.26 ± 0.1. Convexities of uncoated soot and coated airborne soot exposed to 88% RH were 0.56 and 0.74, respectively.

Experiments with coated soot clearly show that particle-substrate interactions hinder aggregate restructuring. The hindering effect depends on the contact area, being the lowest for aggregates attached to thin fibers and the largest for aggregates sitting on flat surfaces. Two factors may be responsible for this effect, the reduced mobility of the aggregate branches anchored to the surface and the escape of liquid coating material (sulfuric acid) from the coated aggregate to the substrate, due to capillary action. To obtain some insight on the origin of the hindering effect, we investigated the restructuring of coated soot aggregates in situ, using ETEM.

3.4. In Situ Processing of Soot Aggregates

Figure 7a shows an ETEM image of the coated soot aggregate (0.26 mass fraction of sulfuric acid) under a flow of dry helium. The aggregate structure remained unchanged over several minutes of continuous exposure to the electron beam. Next, the electron beam was diverted, and dry helium was replaced with humidified helium. The first image of the humidified sample was taken when the RH at the exit from the sample cell reached 83 ± 2%. Figure 7b shows that several distinct changes occurred in the appearance of the aggregate, and also on the surface of silicon nitride. First, under dry conditions, no coating material could be observed, whereas upon humidification the aggregate, not only did the surface become visibly embedded in an aqueous coating, but it also developed a dark outline on the substrate surface. Second, some branches of the aggregate experienced a minor re-arrangement. Third, a large spot originating at the aggregate and reaching across the entire frame became visible on the substrate. Over the next 55 seconds, the spot contracted in size (Figure 7c–e) and then remained nearly stable for over four minutes (Figure 7f). The outline around the aggregate became lighter and the aggregate appeared as if it lost some of the aqueous coating.

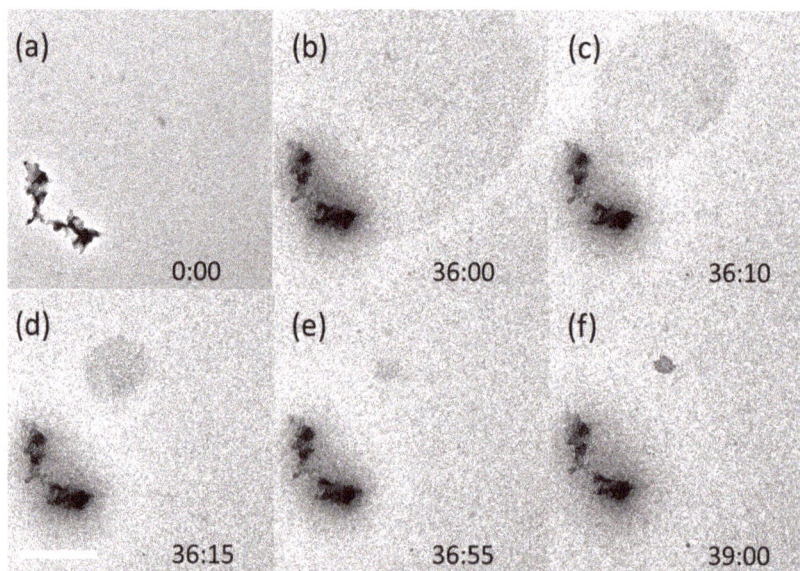

Figure 7. The evolution in the appearance of a soot aggregate coated by sulfuric acid and of the surface of silicon nitride substrate upon humidification and exposure to the electron beam in environmental transmission electron microscopy (ETEM): (**a**) dry initial conditions, (**b**) sample humidification begins, (**c–f**) the sample evolves from the joint impact of humidity and exposure to the electron beam. The sample was not exposed to the electron beam between 0 and 36 min; the exposure was continuous after 36 min. The time stamp is minutes and seconds; the scale bar is 400 nm.

The evolution in the appearance of the aggregate and substrate can be interpreted by considering the joint effect of humidification and exposure to the electron beam. At high humidity and in the absence of imaging, the sulfuric acid coating absorbed water, forming a larger shell around the aggregate. Some of the aqueous sulfuric acid, driven by capillary action, migrated to the silicon nitride surface, forming an outline in close vicinity of the aggregate, and also spreading to a larger surface area. As shown in Table 1, both concentrated and aqueous sulfuric acid could wet the surface of silicon nitride. This behavior closely resembled the behavior of the sodium chloride and sulfuric acid particles upon humidification. The migration of aqueous sulfuric acid to the substrate surface stripped the aggregate of its coating shell, causing it to retain its backbone morphology nearly unchanged even after humidification. When the electron beam was focused on the aggregate, the heat produced by the beam raised the local temperature, reducing RH and leading to water evaporation. We also noted that after switching back to dry helium flow, it was impossible to reset the aggregates precisely to their original state (not shown). The appearance of processed soot pointed to a minor damage in the graphitic monomers, probably due to oxidation by OH radicals generated from water radiolysis.

4. Conclusions

We investigated morphological changes in three types of particles subjected to several types of environments, which may be commonly experienced by particle samples during handling, transport, and imaging. The particles were composed of sodium chloride, a material with a sharp deliquescence transition at a defined RH; sulfuric acid, which adjusts its water content continuously with variation in RH; and soot aggregates coated by sulfuric acid, in which the hygroscopic coating was expected to absorb water and induce structural changes in the soot backbone at elevated RH. The samples were prepared by electrostatic deposition of particles on wafer chips and lacey grids. Overall, four types of particle-substrate combinations were investigated, including two for silicon (original hydrophilic and processed hydrophobic) and two for lacey grids (fibers and flat areas). The environmental conditions included those to which samples are often subjected during normal handling, such as refrigeration or freezing in tightly sealed containers, and also extreme cases, such as the exposure of a cold sample to ambient air.

Based on our findings summarized in Table 3, several major points can be drawn. The most damaging impact was caused by the Type IIIb treatment, when frozen samples were exposed to ambient air with RH in excess of 50%. The substrate became heavily flooded with liquid water, which washed off the particles. However, at an ambient RH below 20% (Type IIIa), differentiation in the extent of sample alteration was observed, depending on the particle composition and the substrate. Whereas sulfuric acid particle samples were all defaced, sodium chloride and coated soot survived with minor structural alterations on selected substrates. Type III treatment is equivalent to a major mishap during sample handling, but even then, certain particle-substrate combinations appear to survive nearly unchanged.

Table 3. Magnitude of structural change in aerosol particles on different substrates.

Aerosol	Environment [1]	Untreated Silicon	Hydrophobic Silicon	Lacey Grid	
				Fiber	Flat
Coated Soot	Type I (Humidified)	Minimal	Minimal	Major	Minimal
	Type II (Cold storage)	Minimal	Minimal	Minimal	Minimal
	Type IIIa (Moderately flooded)	Minimal	Minimal	Major	-
	Type IIIb (Severely flooded)	Major	Major	Major	-
Sodium Chloride	Type I (Humidified)	Moderate	Minimal	Moderate	-
	Type II (Frozen)	Minimal	Minimal	Minimal	-
	Type IIIa (Moderately flooded)	Major	Minimal	Moderate	-
	Type IIIb (Severely flooded)	Major	Major	Major	-
Sulfuric Acid	Untreated	Minimal	Minimal	Minimal	Moderate
	Type I (Humidified)	Major	Moderate	Moderate	Major
	Type II (Frozen)	Major	Minimal	Minimal	-
	Type IIIa (Moderately flooded)	Major	Major	Major	-
	Type IIIb (Severely flooded)	Major	Major	Major	-

[1] See Table 2 for the description of different types of environments.

The Type II treatment is equivalent to routine sample handling, when substrates are placed in sealed cases, cooled to low temperature, and then brought up to ambient temperature while still sealed [21]. Comparable temperature swings may arise during refrigerated/frozen storage, or during transportation. In all the cases, with the exception of sulfuric acid on hydrophilic silicon chips, this treatment introduced minor or negligible changes to the particles.

The Type I treatment involved the exposure of samples to a high RH (94%), but below the water saturation level. Such conditions may be experienced by the samples during collection (e.g., during rain, in early morning or late evening) or handling (an inadvertent exposure to the operator's breath or transfer to a room with a significantly different temperature/humidity). The impact of this treatment varied broadly, depending on the nature of particles and substrate. Coated soot on any type of silicon chips and sodium chloride on hydrophobic silicon were not affected. On the other hand, coated soot on lacey fibers and sulfuric acid on hydrophilic silicon experienced significant changes.

The extent of the particle modification depends on an interplay between several factors, including the actual level of RH experienced by the particles, the physical and chemical properties of the particle material, and the interaction between the particle and substrate (both surface energy and contact area). For any given particle-substrate combination, additional factors such as the rate of sample cooling and relative diffusivity of water vapor versus gas also play a role. Samples must therefore be handled with the utmost care to prevent conditions when morphology changes are induced by increased humidity (from operator's breath) or humidity swings (flooding). Humidity control is important for ensuring collected particles maintain most of their morphology from source to analysis [24].

Deciding on the best type of substrate for use in aerosol sampling is not only dependent on the factors described previously, but also on the objective of the investigation. In some studies, the environment is changed intentionally in a controlled way to examine the response of the particles. Our results had direct implications for cases where morphological changes in the deposited particles are desired, such as upon exposure to an elevated RH, using ETEM, ESEM, or Atomic Force Microscopy (AFM) [45–49]. Clearly, interaction with substrate may severely hinder or alter the extent of change in the particle morphology. Hence, one must strive to reduce the interaction by minimizing the substrate surface energy and the surface area in direct contact with the particle. Finally, it must be noted that appropriate handling of the sample may still not prevent additional measurement bias introduced via exposure to vacuum, electron beam, or AFM sample tip.

Supplementary Materials: The following are available online at www.mdpi.com/2073-4433/8/9/162/s1, Figure S1: An integrated system for aerosol generation, processing and analysis, Figure S2: Calculation of convexity.

Acknowledgments: This work was supported by the National Science Foundation (AGS 1463702) and the Center for Functional Nanomaterials at Brookhaven National Laboratory (34050). C.C. acknowledges scholarship from the China Scholarship Council. The authors thank Abraham Kupperman from the Bergen County Technical High School for help with wetting angle measurements.

Author Contributions: C.C., A.F.K., and Y.M. conceived and designed the experiments; C.C., O.Y.E., A.F.K., and D.Z. performed the experiments; C.C., A.F.K., and O.Y.E. analyzed the data; C.C., A.F.K., and O.Y.E. wrote the paper.

Conflicts of Interest: The authors declare no conflict of interest. The founding sponsors had no role in the design of the study; in the collection, analyses, or interpretation of data; in the writing of the manuscript, and in the decision to publish the results.

References

1. Lohmann, U.; Feichter, J. Global indirect aerosol effects: A review. *Atmos. Chem. Phys.* **2005**, *5*, 715–737. [CrossRef]
2. Pöschl, U. Atmospheric aerosols: Composition, transformation, climate and health effects. *Cheminform* **2005**, *44*, 7520. [CrossRef] [PubMed]
3. Haywood, J.; Boucher, O. Estimates of the direct and indirect radiative forcing due to tropospheric aerosols: A review. *Rev. Geophys.* **2000**, *38*, 513–543. [CrossRef]
4. Tao, W.K.; Chen, J.P.; Li, Z.; Wang, C.; Zhang, C. Impact of aerosols on convective clouds and precipitation. *Rev. Geophys.* **2012**, *50*, RG2001. [CrossRef]

5. Jayne, J.T.; Leard, D.C.; Zhang, X.F.; Davidovits, P.; Smith, K.A.; Kolb, C.E.; Worsnop, D.R. Development of an aerosol mass spectrometer for size and composition analysis of submicron particles. *Aerosol Sci. Technol.* **2000**, *33*, 49–70. [CrossRef]

6. DeCarlo, P.F.; Kimmel, J.R.; Trimborn, A.; Northway, M.J.; Jayne, J.T.; Aiken, A.C.; Gonin, M.; Fuhrer, K.; Horvath, T.; Docherty, K.S.; et al. Field-deployable, high-resolution, time-of-flight aerosol mass spectrometer. *Anal. Chem.* **2006**, *78*, 8281–8289. [CrossRef] [PubMed]

7. China, S.; Kulkarni, G.; Scarnato, B.; Sharma, N.; Pekour, M.; Shilling, J.; Wilson, J.; Zelenyuk, A.; Chand, D.; Liu, S.; et al. Morphology of diesel soot residuals from supercooled water droplets and ice crystals: Implications for optical properties. *Environ. Res. Lett.* **2015**, *10*, 114010. [CrossRef]

8. Beranek, J.; Imre, D.; Zelenyuk, A. Real-time shape-based particle separation and detailed in situ particle shape characterization. *Anal. Chem.* **2012**, *84*, 1459–1465. [CrossRef] [PubMed]

9. McMurry, P.H.; Wang, X.; Park, K.; Ehara, K. The relationship between mass and mobility for atmospheric particles: A new technique for measuring particle density. *Aerosol Sci. Technol.* **2002**, *36*, 227–238. [CrossRef]

10. Levy, M.E.; Zhang, R.; Khalizov, A.F.; Zheng, J.; Collins, D.R.; Glen, C.R.; Wang, Y.; Yu, X.Y.; Luke, W.; Jayne, J.T.; et al. Measurements of submicron aerosols in houston, texas during the 2009 sharp field campaign. *J. Geophys. Res. Atmos.* **2013**, *118*, 10518–10534. [CrossRef]

11. Posfai, M.; Buseck, P.R. Nature and climate effects of individual tropospheric aerosol particles. In *Annual Review of Earth and Planetary Sciences*; Jeanloz, R., Freeman, K.H., Eds.; Annual Reviews: Palo Alto, CA, USA, 2010; Volume 38, pp. 17–43.

12. Lee, R.E. The size of suspended particulate matter in air. *Science* **1972**, *178*, 567–575. [CrossRef] [PubMed]

13. Rao, A.K.; Whitby, K.T. Non-ideal collection characteristics of inertial impactors—II. Cascade impactors. *J. Aerosol Sci.* **1978**, *9*, 87–100. [CrossRef]

14. Cheng, Y.S.; Yeh, H.C. Particle bounce in cascade impactors. *Environ. Sci. Technol.* **1979**, *13*, 1392–1396. [CrossRef]

15. Rothenbacher, S.; Messerer, A.; Kasper, G. Fragmentation and bond strength of airborne diesel soot agglomerates. *Part. Fibre Toxicol.* **2008**, *5*, 9. [CrossRef] [PubMed]

16. Zhang, R.; Khalizov, A.F.; Pagels, J.; Zhang, D.; Xue, H.; McMurry, P.H. Variability in morphology, hygroscopicity, and optical properties of soot aerosols during atmospheric processing. *Proc. Natl. Acad. Sci. USA* **2008**, *105*, 10291–10296. [CrossRef] [PubMed]

17. Huang, P.-F.; Turpin, B. Reduction of sampling and analytical errors for electron microscopic analysis of atmospheric aerosols. *Atmos. Environ.* **1996**, *30*, 4137–4148. [CrossRef]

18. Dixkens, J.; Fissan, H. Development of an electrostatic precipitator for off-line particle analysis. *Aerosol Sci. Technol.* **1999**, *30*, 438–453. [CrossRef]

19. Chen, C.; Fan, X.; Shaltout, T.; Qiu, C.; Ma, Y.; Goldman, A.; Khalizov, A.F. An unexpected restructuring of combustion soot aggregates by subnanometer coatings of polycyclic aromatic hydrocarbons. *Geophys. Res. Lett.* **2016**, *43*, 11080–11088. [CrossRef]

20. Soewono, A.; Rogak, S. Morphology and raman spectra of engine-emitted particulates. *Aerosol Sci. Technol.* **2011**, *45*, 1206–1216. [CrossRef]

21. China, S.; Salvadori, N.; Mazzoleni, C. Effect of traffic and driving characteristics on morphology of atmospheric soot particles at freeway on-ramps. *Environ. Sci. Technol.* **2014**, *48*, 3128–3135. [CrossRef] [PubMed]

22. Chakrabarty, R.K.; Moosmüller, H.; Garro, M.A.; Arnott, W.P.; Walker, J.; Susott, R.A.; Babbitt, R.E.; Wold, C.E.; Lincoln, E.N.; Hao, W.M. Emissions from the laboratory combustion of wildland fuels: Particle morphology and size. *J. Geophys. Res. Atmos.* **2006**, *111*, D07204. [CrossRef]

23. Bambha, R.P.; Dansson, M.A.; Schrader, P.E.; Michelsen, H.A. Effects of volatile coatings and coating removal mechanisms on the morphology of graphitic soot. *Carbon* **2013**, *61*, 80–96. [CrossRef]

24. Laskina, O.; Morris, H.S.; Grandquist, J.R.; Estillore, A.D.; Stone, E.A.; Grassian, V.H.; Tivanski, A.V. Substrate-deposited sea spray aerosol particles: Influence of analytical method, substrate, and storage conditions on particle size, phase, and morphology. *Environ. Sci. Technol.* **2015**, *49*, 13447–13453. [CrossRef] [PubMed]

25. Eom, H.-J.; Gupta, D.; Li, X.; Jung, H.-J.; Kim, H.; Ro, C.-U. Influence of collecting substrates on the characterization of hygroscopic properties of inorganic aerosol particles. *Anal. Chem.* **2014**, *86*, 2648–2656. [CrossRef] [PubMed]

26. Zhou, Q.; Pang, S.-F.; Wang, Y.; Ma, J.-B.; Zhang, Y.-H. Confocal raman studies of the evolution of the physical state of mixed phthalic acid/ammonium sulfate aerosol droplets and the effect of substrates. *J. Phys. Chem. B* **2014**, *118*, 6198–6205. [CrossRef] [PubMed]

27. Gao, C.; Bhushan, B. Tribological performance of magnetic thin-film glass disks: Its relation to surface roughness and lubricant structure and its thickness. *Wear* **1995**, *190*, 60–75. [CrossRef]

28. Ghazi, R.; Tjong, H.; Soewono, A.; Rogak, S.N.; Olfert, J.S. Mass, mobility, volatility, and morphology of soot particles generated by a mckenna and inverted burner. *Aerosol Sci. Technol.* **2013**, *47*, 395–405. [CrossRef]

29. Pyrz, W.D.; Buttrey, D.J. Particle size determination using tem: A discussion of image acquisition and analysis for the novice microscopist. *Langmuir* **2008**, *24*, 11350–11360. [CrossRef] [PubMed]

30. Rice, S.B.; Chan, C.; Brown, S.C.; Eschbach, P.; Han, L.; Ensor, D.S.; Stefaniak, A.B.; Bonevich, J.; Vladár, A.E.; Hight Walker, A.R.; et al. Particle size distributions by transmission electron microscopy: An interlaboratory comparison case study. *Metrologia* **2013**, *50*, 663–678. [CrossRef] [PubMed]

31. Stipe, C.B.; Higgins, B.S.; Lucas, D.; Koshland, C.P.; Sawyer, R.F. Inverted co-flow diffusion flame for producing soot. *Rev. Sci. Instrum.* **2005**, *76*, 023908. [CrossRef]

32. Ghazi, R.; Olfert, J. Coating mass dependence of soot aggregate restructuring due to coatings of oleic acid and dioctyl sebacate. *Aerosol Sci. Technol.* **2013**, *47*, 192–200. [CrossRef]

33. Khalizov, A.F.; Zhang, R.; Zhang, D.; Xue, H.; Pagels, J.; McMurry, P.H. Formation of highly hygroscopic soot aerosols upon internal mixing with sulfuric acid vapor. *J. Geophys. Res. Atmos.* **2009**, *114*, D05208. [CrossRef]

34. Chakrabarty, R.K.; Moosmüller, H.; Arnott, W.P.; Garro, M.A.; Walker, J. Structural and fractal properties of particles emitted from spark ignition engines. *Environ. Sci. Technol.* **2006**, *40*, 6647–6654. [CrossRef] [PubMed]

35. Fischer, M.L.; Littlejohn, D.; Lunden, M.M.; Brown, N.J. Automated measurements of ammonia and nitric acid in indoor and outdoor air. *Environ. Sci. Technol.* **2003**, *37*, 2114–2119. [CrossRef] [PubMed]

36. Tidy, G.; Neil Cape, J. Ammonia concentrations in houses and public buildings. *Atmos. Environ. A Gen. Top.* **1993**, *27*, 2235–2237. [CrossRef]

37. Peng, J.; Hu, M.; Guo, S.; Du, Z.; Zheng, J.; Shang, D.; Levy Zamora, M.; Zeng, L.; Shao, M.; Wu, Y.-S.; et al. Markedly enhanced absorption and direct radiative forcing of black carbon under polluted urban environments. *Proc. Natl. Acad. Sci. USA* **2016**, *113*, 4266–4271. [CrossRef] [PubMed]

38. Adachi, K.; Chung, S.H.; Buseck, P.R. Shapes of soot aerosol particles and implications for their effects on climate. *J. Geophys. Res.* **2010**, *115*, D15. [CrossRef]

39. Cappa, C.D.; Onasch, T.B.; Massoli, P.; Worsnop, D.R.; Bates, T.S.; Cross, E.S.; Davidovits, P.; Hakala, J.; Hayden, K.L.; Jobson, B.T.; et al. Radiative absorption enhancements due to the mixing state of atmospheric black carbon. *Science* **2012**, *337*, 1078–1081. [CrossRef] [PubMed]

40. Liu, D.; Whitehead, J.; Alfarra, M.R.; Reyes-Villegas, E.; Spracklen, D.V.; Reddington, C.L.; Kong, S.; Williams, P.I.; Ting, Y.-C.; Haslett, S.; et al. Black-carbon absorption enhancement in the atmosphere determined by particle mixing state. *Nat. Geosci.* **2017**, *10*, 184–188. [CrossRef]

41. Liu, S.; Aiken, A.C.; Gorkowski, K.; Dubey, M.K.; Cappa, C.D.; Williams, L.R.; Herndon, S.C.; Massoli, P.; Fortner, E.C.; Chhabra, P.S.; et al. Enhanced light absorption by mixed source black and brown carbon particles in UK winter. *Nat. Commun.* **2015**, *6*, 8435. [CrossRef] [PubMed]

42. China, S.; Scarnato, B.; Owen, R.C.; Zhang, B.; Ampadu, M.T.; Kumar, S.; Dzepina, K.; Dziobak, M.P.; Fialho, P.; Perlinger, J.A.; et al. Morphology and mixing state of aged soot particles at a remote marine free troposphere site: Implications for optical properties. *Geophys. Res. Letts.* **2015**, *42*, 1243–1250. [CrossRef]

43. Mikhailov, E.F.; Vlasenko, S.S.; Kiselev, A.A.; Ryshkevich, T.I. Restructuring factors of soot particles. *Izv. Atmos. Ocean Phys.* **1998**, *34*, 307–317.

44. Ma, X.; Zangmeister, C.D.; Gigault, J.; Mulholland, G.W.; Zachariah, M.R. Soot aggregate restructuring during water processing. *J. Aerosol Sci.* **2013**, *66*, 209–219. [CrossRef]

45. Huang, P.F.; Turpin, B.J.; Pipho, M.J.; Kittelson, D.B.; McMurry, P.H. Effects of water condensation and evaporation on diesel chain-agglomerate morphology. *J. Aerosol Sci.* **1994**, *25*, 447–459. [CrossRef]

46. Semeniuk, T.A.; Wise, M.E.; Martin, S.T.; Russell, L.M.; Buseck, P.R. Hygroscopic behavior of aerosol particles from biomass fires using environmental transmission electron microscopy. *J. Atmos. Chem.* **2007**, *56*, 259–273. [CrossRef]

47. Hiranuma, N.; Brooks, S.D.; Auvermann, B.W.; Littleton, R. Using environmental scanning electron microscopy to determine the hygroscopic properties of agricultural aerosols. *Atmos. Environ.* **2008**, *42*, 1983–1994. [CrossRef]

48. Köllensperger, G.; Friedbacher, G.; Kotzick, R.; Niessner, R.; Grasserbauer, M. In-situ atomic force microscopy investigation of aerosols exposed to different humidities. *Anal. Bioanal. Chem.* **1999**, *364*, 296–304. [CrossRef]

49. Ebert, M.; Inerle-Hof, M.; Weinbruch, S. Environmental scanning electron microscopy as a new technique to determine the hygroscopic behaviour of individual aerosol particles. *Atmos. Environ.* **2002**, *36*, 5909–5916. [CrossRef]

atmosphere

MDPI

Article

Effect of Thermodenuding on the Structure of Nascent Flame Soot Aggregates

Janarjan Bhandari [1,*], Swarup China [1,†], Timothy Onasch [2,3], Lindsay Wolff [3], Andrew Lambe [2,3], Paul Davidovits [3], Eben Cross [2], Adam Ahern [4], Jason Olfert [5], Manvendra Dubey [6] and Claudio Mazzoleni [1,*]

[1] Department of Physics and Atmospheric Sciences Program, Michigan Technological University, Houghton, MI 49931, USA; schina@mtu.edu

[2] Aerodyne Research Inc., Billerica, MA 01821, USA; onasch@aerodyne.com (T.O.); lambe@aerodyne.com (A.L.); escross@aerodyne.com (E.C.)

[3] Chemistry Department, Boston College, Chestnut Hill, MA 02467, USA; lrwolff@aerodyne.com (L.W.); davidovi@bc.edu (P.D.)

[4] Centre for Atmospheric Particle Studies, Carnegie Mellon University, Pittsburgh, PA 15232, USA; aahern@andrew.cmu.edu

[5] Department of Mechanical Engineering, University of Alberta, Edmonton, AB T6G 2G8, Canada; jolfert@ualberta.ca

[6] Earth and Environmental Sciences Division, Los Alamos National Laboratory, Los Alamos, NM 87545, USA; dubey@lanl.gov

* Correspondence: jbhandar@mtu.edu (J.B.); cmazzoleni@mtu.edu (C.M.)

† Current address: Pacific Northwest National Laboratory, Richland, WA 99354, USA.

Received: 27 July 2017; Accepted: 31 August 2017; Published: 6 September 2017

Abstract: The optical properties (absorption and scattering) of soot particles depend on soot size and index of refraction, but also on the soot complex morphology and the internal mixing with materials that can condense on a freshly emitted (nascent) soot particle and coat it. This coating can affect the soot optical properties by refracting light, or by changing the soot aggregate structure. A common approach to studying the effect of coating on soot optical properties is to measure the absorption and scattering coefficients in ambient air, and then measure them again after removing the coating using a thermodenuder. In this approach, it is assumed that: (1) most of the coating material is removed; (2) charred organic coating does not add to the refractory carbon; (3) oxidation of soot is negligible; and, (4) the structure of the pre-existing soot core is left unaltered, despite the potential oxidation of the core at elevated temperatures. In this study, we investigated the validity of the last assumption, by studying the effect of thermodenuding on the morphology of nascent soot. To this end, we analyzed the morphological properties of laboratory generated nascent soot, before and after thermodenuding. Our investigation shows that there is only minor restructuring of nascent soot by thermodenuding.

Keywords: thermodenuding; soot morphology; aggregates; compaction; restructuring

1. Introduction

Soot particles are mostly composed of refractory carbonaceous material that forms from incomplete combustion during burning activities [1]. A nascent soot particle appears as a fractal-like (sometimes referred to as a lacy) aggregate of small spherules (called nanospheres or monomers) [2] and its structure is scale invariant [3]. Here, the term "nascent" is used to refer to freshly emitted soot particles that have a negligible coating on the monomers. The diameter of these monomers varies in a range from 10 nm to more than 50 nm, depending on the fuel source and combustion conditions [4–7]. During the atmospheric processing, soot particles interact with organic and inorganic materials (in the form of aerosol or condensable vapors). During these interactions, soot undergoes morphological changes including compaction, coagulation, and coating [8]. Combustion generated nascent soot aggregates often have different kinds of polycyclic aromatic hydrocarbons (PAHs) that thinly coat the monomers, depending upon the flaming conditions and fuel types, even in very controlled combustions [9]. In this case, coating is acquired at the source, and not added later through atmospheric processing. The degree of coating in atmospheric particles is very variable [6,8,10–14]. In some studies, thinly coated soot particles have been found in large fractions in the atmosphere. For example, in the study by China et al. [6], a large fraction (by number) of freshly emitted soot particles collected on freeway on ramps were thinly coated (72%). In another field study, carried out at Pico Mountain Observatory in the Azores, China et al. [8] found that 37% of the soot particles in one sample, were thinly coated, even after days of atmospheric processing during the long range transport in the free troposphere from the source. These two studies were carried out at very different locations (very near the source in the first study, and very far from the source in the second study), showing that thinly coated soot particles can exist in the atmosphere in different environments and geographic locations. In another study, in two plumes one sampled from Mexico City and one from outside of Mexico City, Adachi and Buseck [4] found that 7% of particles were soot without coating. Coating or internal mixing in general, changes the optical properties of soot, even when the structure of the refractory components remains the same. These changes consequently affect the radiative forcing of soot [15–22]. Several studies have also shown that coating of soot by partially-absorbing or non-absorbing materials increases the absorption and scattering cross sections [13,18,23–25]. These increases are termed "absorption and scattering enhancements" (E_{abs} and E_{sca}). The enhancement is typically calculated as the ratio of the light absorption or scattering coefficient of the coated soot to the light absorption or scattering coefficient of the soot core, after the coating material has been removed [18].

Thermodenuders (TDs) that remove the coating by evaporation, are often used in the field and in the laboratory to study and quantify these optical enhancements [16,26–28]. During the thermodenuding process, coated soot particles are passed through a heated column, typically at ~200–300 °C to evaporate the volatile coating material while leaving behind the refractory soot [29,30]. The temperature gradient within the TD can result in particle losses due to thermophoretic forces, though these losses can be measured and accounted for [29]. To correctly assess the E_{abs} and E_{sca} using a TD, one needs to make the following assumptions: (1) most of the coating material is removed from the soot by the TD; (2) organic coating material does not transform into refractory carbon due to charring; (3) refractory carbon is not oxidized to a substantial extent under elevated temperatures; and, (4) the structure of the refractory soot particle is unaffected, meaning that the thermodenuding process alone does not induce restructuring of the core lacy soot aggregates.

Contrary to assumption (1), thermodenuding may not remove refractory particulate material, such as some inorganic salts, and may not remove all of the non-refractory material from soot particles [23]. For example, Liu et al. [13] observed that denuded soot still contained heavily coated soot particles, although, in a smaller fraction with respect to ambient particles, suggesting that the TD may not completely remove the coating. Healy et al. [12] found that, on average, only 74% of the mass of coating material was removed from soot samples after thermodenuding. The mass removal efficiency by the TD was even less (approximately 60%) for wildfire emission samples. Swanson and Kittelson [31] have also cautioned about semi-volatile particle artifacts due to incomplete removal of evaporated compounds in the TD. Similarly, Knox et al. [32] found that there was no significant difference in the mass absorption cross section between themodenuded and non-thermodenuded aged-soot particles as compared to fresh soot, due to the incomplete removal of coating materials from aged soot particles. On the other hand, Khalizov et al. [33] hypothesized that the thermodenuder may remove all of the coating material from ambient soot, including the coating acquired during atmospheric processing, as well as the nascent coating present on soot at the source, and therefore, they suggested that the denuded particles cannot represent the nascent soot morphology.

Next, we briefly review assumptions (2) and (3). The elevated temperatures during the thermodenuding process may cause charring of some organic matter into refractory, elemental carbon, and/or some oxidation of the carbonaceous matter. The charring of organic particulate material into elemental carbon is a known phenomenon under the elevated temperatures employed in organic carbon and elemental carbon (OCEC) analyses [34,35]. Issues that influence the charring include temperatures and residence times, as well as chemical composition. Charring is likely to be more of an issue for oxidized organics, such as biomass burning or secondary organic aerosol (SOA), than reduced organics, such as efficient combustion products (i.e., diesel and laboratory flame soot) [36,37]. Two significant differences between OCEC and thermodenuding include: (1) OCEC techniques typically operate at higher temperatures than TDs, and (2) OCEC charring occurs in a helium atmosphere, whereas thermodenuding occurs in air (i.e., oxidizing environment). Thus, at a low temperature (<300 °C), thermodenuding will be less likely to char, and the particles may be more susceptible to oxidation due to an oxidizing environment. Oxidation of refractory carbon soot typically occurs at significantly elevated temperatures, but can occur at lower temperatures as well, especially if catalyzed by impurities in the soot [38]. Soot oxidation is likely limited in thermodenuding due to the low temperatures and the relatively short residence time, but this issue will require more study in the future.

We finally discuss the assumption (4), which is the focus of our study. Previous studies have shown that nascent soot particles can restructure during the condensation or evaporation of the coating material, depending on their surface tension [39–41]. Xue et al. [28] found significant restructuring of soot particles when the particles were first coated with glutaric acid and then denuded. Ghazi and Olfert [16] reported the dependence of soot restructuring on the mass amount of different coating material types. This restructure alone can affect the optical properties of soot particles. For example, a laboratory study was performed on soot compacted upon humidification; the study measured modest changes in the absorption cross-section (5% to 14%), but the extinction cross-section was much more sensitive to compaction, with changes of more than 30% [42]. Similarly, China et al. [8,43], using numerical simulations, predicted small changes in the absorption cross-section (a few percent), but a much more substantial change in the total scattering cross section (up to ~300%) upon soot compaction. In addition to affecting the optical properties, changes in the soot structure can also affect the results of laser induced incandescence measurements [5]. Finally, the condensation of secondary organic matter preferentially takes place in empty pores on soot particles, and therefore, it is possible that compaction will affect secondary organic condensation on soot [44]. Two potential explanations for the coated soot restructuring detected during these studies can be: (1) Soot might be compacted during condensation of the coating materials due to surface tension effects [14,41,45,46]. (2) The removal of the coating material during the subsequent thermodenuding may cause compaction when the coating evaporates, still due to surface tension effects [39,40]. However, we hypothesize a third potential pathway, in which soot restructuring might take place solely due to the thermodenuding process, through the added thermal energy. Coating alone might not cause full compaction (i.e., completely collapsed structure). For example, in an experiment, Leung et al. [47] found that the coating did not restructure the soot aggregate even when the aggregate was completely covered by the coating material. Some coating material indeed results in substantial or even maximum compaction, but other coating materials actually result in negligible compaction. A clear example is shown in a laboratory study where particles coated with sulfuric acid did undergo severe restructuring, while the soot particles coated by dioctyl sebacate showed only minimal or no compaction [9]. Also, from Mexico City samples, Adachi and Buseck [4] found coated soot but with lacy structure. For the case of coated but yet only partially compacted soot, thermodenuding may facilitate further restructuring. The coating material can become less viscous at the elevated temperatures during the thermodenuding and restructure the soot core [47].

Next, we will discuss some lines of evidence that the thermodenuding process alone can in some cases, favor the compaction of lacy aggregates of various materials, even in the absence of coating material that condenses onto the primary aggregates. If a similar process happens for ambient soot, such a process would potentially bias the measured properties (e.g., absorption or scattering enhancements) of soot when a thermodenuder is used. The main objective of our study is to test this hypothesis, to assure that the themodenuding process alone does not introduce this bias. A couple of potential restructuring processes induced during thermodenuding, are discussed next:

(a) When heated, fractal-like aggregates of metal nanoparticles, such as silver, copper, and metallic oxides (e.g., titania), have been found to restructure to more compact morphologies at temperatures well below the bulk material melting points. For example, thermal restructuring has been found in silver aggregates, even at temperatures as low as 100 °C, with full compaction at just 350 °C (much below the melting temperature of silver), while the primary particle size remained unchanged [48,49]. Another study found that aggregates of titania started to collapse when temperatures reached 700 °C [50]. These authors speculated that the heating causes the weakest branches in an aggregate to rotate around their contact points, resulting in the aggregate restructuring. Alternatively, Schmidt-Ott [51] hypothesized that the monomers in silver nanoparticles aggregates might slide on each other when heated, also causing compaction. Both processes would restructure the aggregates without a complete breakage of the bonds between the monomers due to Van der Waals forces.

(b) As mentioned earlier, nascent soot aggregates typically have polycyclic aromatic hydrocarbons thinly coating them. This nascent coating could play a role in determining the soot structure if the coating properties (i.e., viscosity and surface tension) change at the higher temperature of the thermodenuder. Chen et al. [52] found that some polycyclic aromatic hydrocarbons, like phenanthrene and flouranthene, when present as a subnanometer layer on soot, behaved as subcooled liquid that weakened the bonds between the monomers, allowing them to slide and roll over each other and resulting in soot restructuring. Rothenbacher et al. [53] provided evidence that thermodenuding might make a difference in the strength of the adhesive bonds between the monomers. For aged soot, they found a higher degree of fragmentation for thermodenuded particles (75% at 280 °C) than for untreated (not thermodenuded) particles (60%) when impacted at ~200 m/s. The degree of fragmentation was defined as the fraction of broken bonds in an aggregate. Although the process involved both the effect of coating and impaction, the higher degree of fragmentation for thermodenuded particles suggests that the thermal energy has a role on the increased degree of fragmentation.

These lines of evidence motivated us to study the potential effects of thermodenuding on the specific case of nascent soot. With this goal in mind, we analyzed the structure of laboratory generated nascent soot particles produced from two different fuel sources (ethylene flame and methane flame) and size selected at different mobility diameters before and after thermodenuding. This assessment is important for evaluating the potential biases that might be introduced by thermodenuding while, for example, estimating the absorption or scattering enhancements of laboratory or ambient soot particles.

2. Experiments

2.1. Experimental Setup and Sample Collection

We analyzed five pairs of mobility-selected soot samples collected during two different experiments: the Boston College Experiment 2 (BC2) and the Boston College Experiment 4 (BC4). The sampling schematics are shown in the Figure 1a,b. None of the soot particles were coated with additional external coating material, and the minimal coating present on the nascent soot was solely due to the fuel residuals accumulated during the combustion and dilution processes.

Three soot sample sets were collected during BC2 from the combustion of ethylene and oxygen using a premixed flat flame burner [9]. The fuel equivalence ratio (∅) for all the three sample sets was 2.1. A TD [29] was used to remove volatile components from the nascent soot particles. The heating section of the TD was set at 250 °C to vaporize the non-refractory soot components, which were absorbed by a charcoal section maintained at room temperature. Particles for a range of mobility diameters (d_m) were selected to investigate the effect of thermodenuding on particle size. For our investigation, we selected three sets of nascent vs. nascent-denuded soot particles with d_m = 153 nm, 181 nm and 250 nm for nascent and d_m = 151 nm, 175 nm, and 241 nm for the corresponding denuded soot particles. Soot particles were collected on 13 mm diameter Nuclepore polycarbonate filters having a pore size of 0.3 µm (Whatman Inc., Chicago, IL, USA). Additional details regarding the BC2 experimental set-up are provided elsewhere [9].

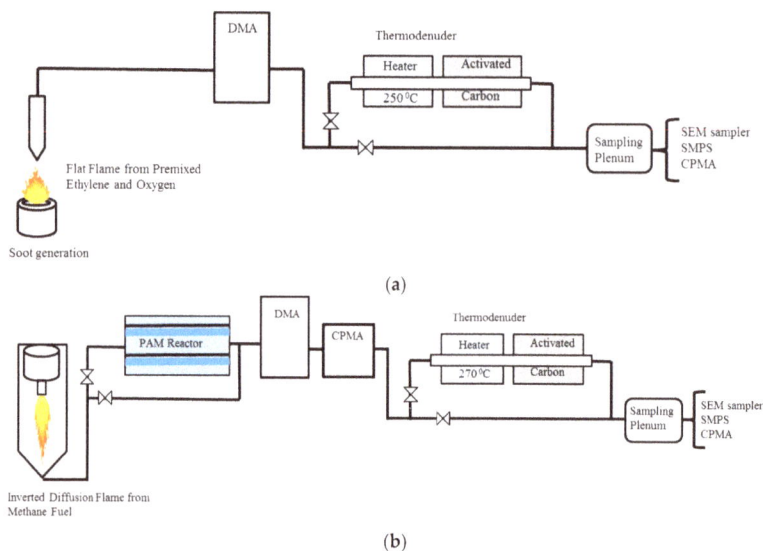

Figure 1. Soot generation and sampling set-ups in (**a**) Boston College Experiment 2 (BC2) and (**b**) Boston College Experiment 4 (BC4).

In addition, we selected two sets of soot samples generated during BC4 from the combustion of methane in an inverted diffusion flame burner (methane and O_2 mixture) at a d_m = 253 nm and 252 nm for nascent and d_m = 253 nm, and 251 nm for the corresponding denuded soot particles. The global ∅ for both sample sets was about 0.7, but the actual value of ∅ is unknown. In the diffusion flame, the fuel burns in excess of air making the value of ∅ less than 1. Effluent from the flame burner was passed through separate annular denuders loaded with molecular sieves and activated charcoal to remove water vapor and volatile organic compounds from the sample flow. As in BC2, a Huffman TD (heating section set at 270 °C) was used to remove the volatile components. For both experiments, the sample flow rate through the TD was 2 LPM, resulting in a residence time of 5 s in the heating section and 4 s in the denuder section. During BC4, unlike during BC2, particles were first mobility size selected by a Differential Mobility Analyzer (DMA) (TSI Inc., Saint Paul, MN, USA) and the mass was selected by a Centrifugal Particle Mass Analyzer (CPMA) (Cambustion Ltd., Cambridge, UK) before thermodenuding. The first set of samples consisted of nascent and nascent-denuded soot, while the second set consisted of nascent-oxidized and nascent-oxidized-denuded soot. Soot was oxidized by exposure to ozone (O_3) and hydroxyl (OH) radicals in a Potential Aerosol Mass (PAM) oxidation flow reactor [54], at input O_3 and H_2O mixing ratios of ~15 ppm and ~1%, and UV actinic flux ~2 × 10^{12} ph cm^{-2} s^{-1} (λ = 254 nm). These operating conditions correspond to an integrated OH exposure of approximately 2 × 10^{12} molec cm^{-3} s [55], and likely generate highly oxygenated organic molecules, such as carboxylic acids on the surface of the nascent-oxidized soot particles [56]. The nascent-oxidized soot was thermo-denuded at a temperature of 270 °C. The set of nascent-oxidized soot samples was included here to investigate if the thermodenuding effect is different for nascent versus nascent-oxidized soot. During BC4, soot particles were collected on 13 mm diameter Nuclepore filters having a pore size of 0.1 μm diameter (Whatman Inc., Chicago, IL, USA).

All the filters were coated with 1.8 (±10%) nm thick layer of Pt/Pd alloy in a sputter coater (Hummer® 6.2, Anatech USA, Union city, CA, USA) and imaged with a field emission scanning electron microscope (FE-SEM) (Hitachi S-4700, Tokyo, Japan). From the FE-SEM images, several morphological parameters were evaluated [6] using the image processing software ImageJ (National Institutes of Health, Bethesda, MD, USA) [57].

2.2. Soot Morphological Parameters

As mentioned in the introduction, soot particles are aggregates of monomers that exhibit scale-invariant fractal structures [58,59]. Soot aggregates can therefore be characterized by a fractal dimension (D_f), in which the mass of the aggregate M (proportional to the number of monomers N in the aggregate) scales with the ratio of the radius of gyration (R_g) to the radius of the monomers (R_p), as in M (or N) $\propto (R_g/R_p)^{D_f}$, [60]. D_f is a commonly used parameter to quantify the soot morphology. Lacy soot particles have low D_f values, while compact soot particles have higher D_f values. The D_f of an ensemble of soot particles can be calculated by plotting N vs. R_g (or a surrogate for it). N scales with R_g as a power law with exponent D_f [61]:

$$N = k_g \left(\frac{R_g}{R_p} \right)^{D_f} \tag{1}$$

where k_g is a pre-factor whose value depends on the overlap between the monomers in the aggregate. The relation formulated by Köylü et al. [61] was used to estimate D_f with the geometric mean diameter of the aggregate, $\sqrt{LW}$, as a surrogate for $2R_g$:

$$N = k_{LW} \left(\frac{\sqrt{LW}}{2R_p} \right)^{D_f} \tag{2}$$

where L is the maximum length and W is the maximum width (orthogonal to L), K_{LW} is a prefactor and R_p is calculated from the mean of the projected area of the monomer. In general, it is difficult to measure N using an SEM image alone, because only two-dimensional (2-D) projections of the soot particles are typically available. Therefore, N is often estimated from the projected area of the soot aggregate A_p and the mean projected area of the monomers A_m using the relation provided by Oh and Sorensen [62]:

$$N = k_a \left(\frac{A_p}{A_m} \right)^{\alpha} \tag{3}$$

where α and k_a are constants that depend on the overlap between monomers in the 2-D projected image of the particle. In our case, we used $K_a = 1.15$ and $\alpha = 1.09$ for all of our nascent and nascent-denuded soot aggregates [61]. This selection of K_a and α values is reasonable since we only studied nascent soot particles that are minimally coated.

In addition to D_f, several other 2-D morphological parameters were calculated from the FE-SEM images to investigate potential changes due to thermodenuding. The calculated parameters included roundness, convexity, aspect ratio (AR), and area equivalent diameter (D_{Aeq}). Figure 2a shows the definition of some of these parameters. D_{Aeq} is the diameter of a spherical particle with a projected area equivalent to the projected area of the aggregate. Roundness is calculated from the ratio of the projected area of the aggregate to the area of the circle having a diameter equal to the maximum projected length L, and fully inscribing the projected image of the aggregate (Figure 2b). Convexity (sometimes termed solidity) is the ratio of the projected area of the particle to the area of the smallest convex hull polygon, in which the 2-D projection of the aggregate is inscribed (Figure 2c). AR is calculated as the ratio of L to W. Higher values of roundness and convexity or lower AR often corresponds to more compact soot particles. However, it has to be noted that D_f, roundness and convexity are parameters with very different meanings and definitions. The first is a scaling factor, the second is a geometric property, and the third is a measure of the particle topology. That is why we investigated all of these three parameters to characterize the morphology of soot rather than looking at a single one. We analyzed a total of 1223 images of individual soot particles.

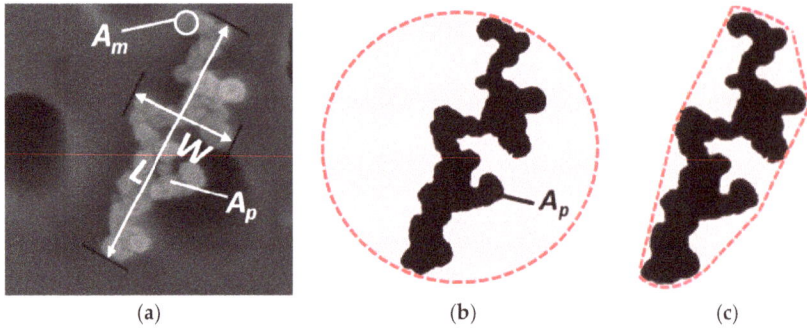

Figure 2. (**a**) Example of SEM image of a soot particle showing the definition of several parameters measured from the projected image: maximum projected length L, maximum projected width W, projected area of monomer A_m and projected area of particle A_p. (**b**) Schematic representation of the roundness calculation for the same soot particle shown in (**a**). (**c**) Schematic representation of the convexity calculation for the same soot particle shown in (**a**). The pink shades in (**b**,**c**) represent the equivalent area for a circle and the convex hull, respectively, for the binary image of the soot particle shown in (**a**).

3. Results and Discussion

As mentioned earlier, we analyzed images from four sets of nascent and nascent-denuded soot sample pairs of different sizes and a fifth set of nascent-oxidized denuded soot. Examples of images of soot particles before and after thermodenuding are shown in Figure 3.

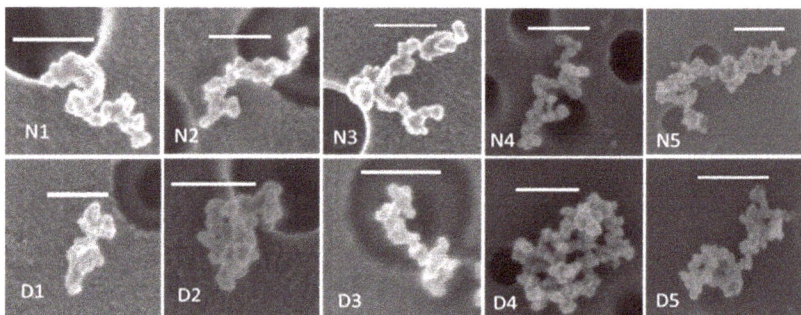

Figure 3. SEM micrographs of nascent (*N*) and thermodenuded (*D*) soot particles. The white horizontal bar in each micrograph represents a length scale of 200 nm. Dark circles are the holes in the filter.

N1, *N2*, *N3*, and *N4* are four differently sized nascent soot samples and *D1*, *D2*, *D3*, and *D4* are the corresponding nascent-denuded sets. *N5-D5* is a pair of nascent-oxidized soot before and after thermodenuding. Table 1 summarizes the features of the analyzed soot particles. Sets *N1-D1*, *N2-D2*, and *N3-D3* are the three sets from BC2, while sets *N4-D4*, and *N5-D5* are from BC4.

Table 1. Summary of physical and morphological parameters for the soot particles analyzed.

Experiment	Statistics	BC2						BC4			
Sample		N1	D1	N2	D2	N3	D3	N4	D4	N5	D5
Fuel type		E	E	E	E	E	E	M	M	M	M
#Particles analyzed		108	151	113	163	114	109	113	105	122	125
	Mean	41	55	121	104	110	153	158	188	155	166
N	S.D	16	26	65	53	44	90	96	87	75	106
	S.E	2	2	6	4	4	9	9	8	7	9
d_m (nm)	Mean	153	151	181	175	250	241	253	253	252	251
M_{CPMA} (fg)	Mean	1.02	0.78	1.52	1.08	2.85	2.20	2.37	2.34	2.41	2.18
	S.D.	(0.03)	(0.03)	(0.05)	(0.04)	(0.14)	(0.13)	(0.11)	(0.13)	(0.11)	(0.11)
D_f	Fit slope	1.86	1.84	1.73	1.72	1.78	1.79	1.80	1.76	1.65	1.80
	S.E.	(0.05)	(0.04)	(0.05)	(0.06)	(0.08)	(0.05)	(0.05)	(0.06)	(0.05)	(0.05)
K_g	Fit intercept	1.78	1.98	2.50	2.50	2.22	2.00	2.10	2.56	2.87	2.16
	S.E.	(0.04)	(0.03)	(0.05)	(0.05)	(0.08)	(0.06)	(0.06)	(0.07)	(0.06)	(0.06)
	Mean	33.5	31.8	26.8	25.7	32.1	30.3	23.5	22.8	23.9	23.1
d_p (nm)	Median	33.5	32.4	26.5	25.9	32.1	28.9	23.2	22.5	23.7	23.0
	S.D.	(2.1)	(3.3)	(2.7)	(2.6)	(2.1)	(6.9)	(3.1)	(2.2)	(2.5)	(3.4)
	S.E.	(0.21)	(0.27)	(0.26)	(0.21)	(0.20)	(0.66)	(0.30)	(0.22)	(0.23)	(0.31)
	Mean	0.41	0.43	0.36	0.34	0.38	0.31	0.31	0.35	0.33	0.33
Roundness	Median	0.42	0.42	0.35	0.35	0.35	0.30	0.30	0.34	0.32	0.31
	S.D.	(0.12)	(0.12)	(0.11)	(0.10)	(0.12)	(0.09)	(0.11)	(0.12)	(0.11)	(0.11)
	S.E.	(0.01)	(0.01)	(0.01)	(0.01)	(0.01)	(0.01)	(0.01)	(0.01)	(0.01)	(0.01)
	Mean	0.72	0.75	0.66	0.66	0.62	0.59	0.61	0.66	0.61	0.63
Convexity	Median	0.73	0.74	0.66	0.65	0.62	0.58	0.61	0.66	0.61	0.62
	S.D.	(0.09)	(0.08)	(0.09)	(0.10)	(0.09)	(0.10)	(0.10)	(0.11)	(0.12)	(0.11)
	S.E.	(0.01)	(0.01)	(0.01)	(0.01)	(0.01)	(0.01)	(0.01)	(0.01)	(0.01)	(0.01)
	Mean	169	181	220	196	255	262	215	230	219	214
D_{Aeq} (nm)	Median	171	175	208	189	262	260	199	220	213	202
	S.D.	(33)	(35)	(55)	(41)	(46)	(49)	(54)	(56)	(50)	(59)
	S.E.	(3)	(3)	(5)	(3)	(4)	(5)	(5)	(5)	(5)	(5)
	Mean	1.79	1.73	1.84	1.92	1.78	1.85	1.99	1.95	1.85	1.88
AR	Median	1.66	1.62	1.70	1.78	1.68	1.72	1.85	1.82	1.80	1.83
	S.D.	(0.51)	(0.42)	(0.49)	(0.51)	(0.57)	(0.50)	(0.60)	(0.60)	(0.50)	(0.50)
	S.E.	(0.05)	(0.03)	(0.05)	(0.04)	(0.05)	(0.05)	(0.06)	(0.06)	(0.05)	(0.04)

In Table 1, E = ethylene and M = methane represent the fuel type. N is the average number of monomers per aggregate, estimated in each sample using Equation (3). K_g values have been estimated using the relation $K_g = K_{LW} \cdot (1.17)^{D_f}$ where $\sqrt{LW}/2R_g = 1.17$ has been taken from Köylü et al. [61] and the values of K_{LW} and D_f have been calculated from a log-log plot using Equation (2). dp is the mean diameter of the monomers in an aggregate, d_m is the mean mobility diameter (in nm) and M_{CPMA} represents the mean mass of the particle (in fg) as measured by the CPMA. For D_f the term in parenthesis is the standard error (S.E.) calculated from the power fit using Equation (2), for the other quantities, it is the S.E. (standard deviation of the mean) and the standard deviation (S.D.).

The largest decrease in the mean value of d_p (by 5.6%) after thermodenuding is found for the N3-D3 set. The decrease in d_p could be due to the partial removal of material volatile at the TD temperature and present on the nascent soot. A decrease in the monomer size after thermal treatment was previously observed when soot samples were heated at higher temperatures (400–900 °C) due to the removal of a part of the nascent PAH layers from the monomers surface [63]. Also, the mean d_p size, as well as the differences in the mean values of d_p after thermodenuding are smaller for the inverted diffusion flame with respect to those of the McKenna flame. These changes suggest that there was less volatile material present in the nascent soot generated from the inverted flame. This effect could be due to the different type of fuel, as well as different $\varnothing$. In a study of ethylene flame generated soot from a McKenna burner, the size of d_p in thermodenuded soot particles was found larger for higher $\varnothing$ [64]. This is consistent with the study by Ghazi and Olfert [16] that generated soot by an inverted diffusion flame and found no measurable amount of volatile material when the mass was measured after thermodenuding. While, Slowik et al. [64], using a McKenna flame, found that thermodenuding removed only about 0.05 mass fraction of volatile material for the nascent soot containing 0.1 mass fraction of non-refractory material (at $\varnothing = 2.1$) from an ethylene flat flame.

To investigate whether the soot aggregates restructured after thermodenuding, we first analyze the changes in D_f as summarized in Figure 4.

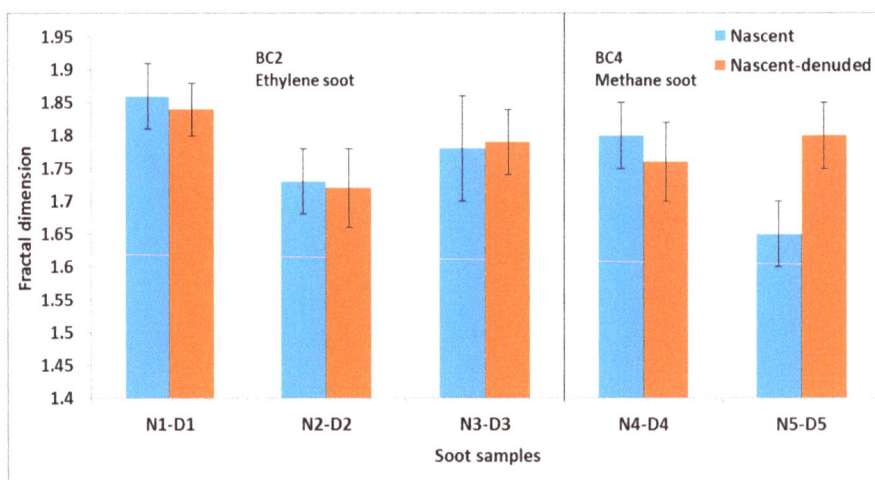

Figure 4. Fractal dimension of nascent (in blue) and nascent-denuded (in orange) soot pairs of different mobility sizes. The error bars represent the standard errors.

For all five sample sets, D_f lies between 1.65 and 1.86 (Table 1). The derivation of D_f and plots for all of the samples are shown in the supplementary material (Figure S1). These values of D_f are in agreement with the observations made in previous studies on nascent soot particles produced from different fuel sources [3,65]. Also for all nascent vs. denuded pairs (except for the nascent-oxidized pair: N5-D5), there is no significant change (within 1σ) in D_f after thermodenuding (Figure 4). For the N5-D5 pair, the D_f changes by about 9% (from 1.65 to 1.80), whereas for all other cases, the change is less than 2.3%. The CPMA data for the BC4 sample shows that the mass decreased from 2.37 to 2.34 fg for nascent soot, while for the nascent-oxidized soot of the same mobility size, the mass decreased from 2.41 to 2.18 fg after thermodenuding. The larger decrease in mass for the nascent oxidized soot suggests that the coating material on the oxidized soot was removed during thermodenuding. A possible explanation for the increase of D_f after thermodenuding the oxidized soot might be that the soot structure was slightly modified during the evaporation of the coating material. Interestingly, for the BC2 soot samples, there is no significant change in D_f despite the significant change in mass (up to ~29%) of soot after thermodenuding (see CPMA data in Table 1). This result suggests that for the BC2 sample sets, the removal of the coating present on nascent soot did not affect the structure of soot. This is most probably due to the chemical composition of the organics that were removed by TD. This result is consistent with the thermodenuding experiment of uncoated soot (fractal soot generated at lower $\varnothing = 2.1$) by Slowik et al. [64] that found no change in D_f (derived from mass-mobility relation in their case) after denuding. They suggested that the removal of organics from the uncoated soot during denuding cannot change the skeletal framework of soot. Cross et al. [9] observed only minor restructuring of soot when dioctyl sebacate coating was removed by thermodenuding, suggesting that the removal of organic coating may have little impact on the restructuring of soot. For soot from a flat flame burner, Slowik et al. [66] found that the organic carbon (OC) content (mass fraction of 0.1) was composed of a comparable amount of aliphatic and aromatic compounds at a lower $\varnothing$ ($\varnothing = 1.85$), but at a higher $\varnothing$ ($\varnothing > 4$), the OC content (mass fraction of 0.55) had only a minor fraction of aliphatic compounds. We thus hypothesize that the nascent organics on the soot from the BC2 experiments considered here consisted in a large fraction of aliphatic compounds.

To account for the mass change after thermodenuding on the coating of soot particle, we calculated coating thickness (ΔR_{ve}) in terms of volume equivalent radius (R_{ve}). The difference between the volume equivalent radius of nascent soot and the thermodenuded soot particle was used to estimate the thickness of the coating material. For the case of maximum mass loss (~29%), coating thickness was estimated to be 8.4 nm. (See supplementary material for the calculation).

To further investigate possible morphological changes after thermodenuding, we studied the convexity and roundness of soot particles for all five sample sets. The maximum change in the mean value of roundness occurs for set *N3-D3* (about 18%), followed by the set *N4-D4* (about 13%). For the other sets, the mean value of roundness changes by less than 10%. For the case of convexity, the maximum change in the mean value occurs for set *N4-D4* (about 8%). For all other sets, the mean value of convexity changes by less than 5%. The larger changes in roundness and convexity for these sample sets are statistically significant (at 1σ) although still minor.

We should point out, however, that image acquisition conditions (e.g., due to different magnifications, scan rates or over/under focusing) and image processing biases (e.g., image thresholding) can introduce additional errors in roundness, convexity, and D_f. In some cases, these errors are larger than the statistical errors provided in Table 1. To quantify these uncertainties, we acquired multiple images of six individual particles (from sample *N5*) and processed them under different conditions, as mentioned above. We estimated the uncertainties due to image acquisition and image processing biases in roundness and convexity to be 0.01 and 0.02, respectively. Similarly, uncertainties in N and d_p were estimated to be 16% and 13% (3.4 nm), respectively, which resulted in an error of 0.08 in D_f.

In Figure 5a,b we show box and whisker plots for the convexity and the roundness, respectively of the soot particles before and after thermodenuding. The convexity ranges from 0.37 to 0.91, while the roundness ranges from 0.09 to 0.75 (see Table 1 for details). No substantial changes in roundness or convexity are evident after thermodenuding.

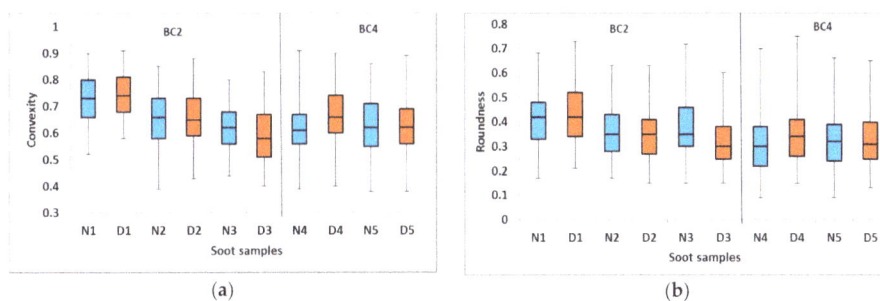

(a)　　　　　　　　(b)

Figure 5. Box and whisker plots of (**a**) convexity and (**b**) roundness. Blue boxes represent the nascent soot and orange boxes represent the nascent-denuded soot. The horizontal bar inside the box represents the median value while the lower part and upper part of the box separated by the horizontal bar represent the first and third quartiles, respectively. The lower and upper extremities of the whiskers represent the minimum and maximum values, respectively.

In Figure 6, we show the probability distributions of convexity and roundness for all nascent and denuded soot pairs. The solid and the dashed lines represent the mean values for nascent and denuded soot, respectively, while the shaded color bands in blue and orange represent one standard deviation. These means and uncertainty bands were calculated with a bootstrap approach, resampling with replacement from the raw data and constructing 100,000 frequency distributions [67].

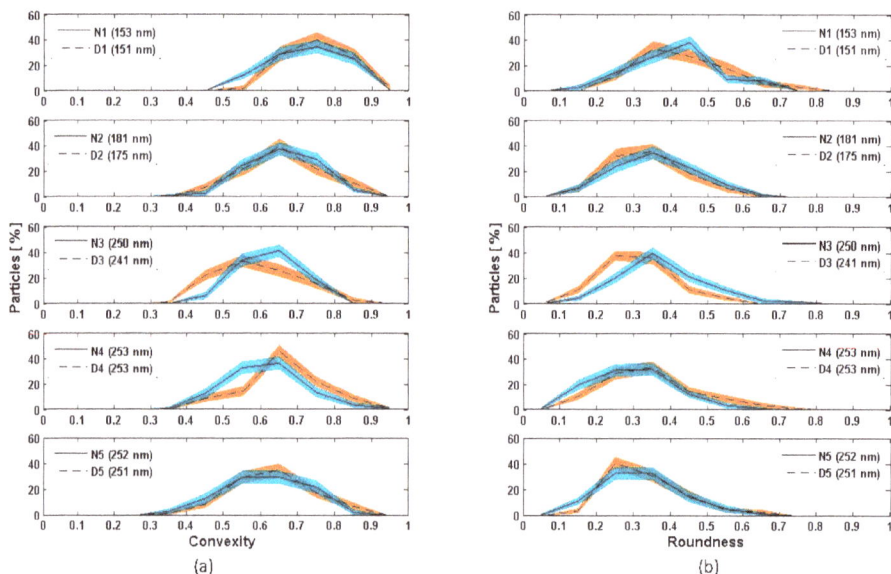

Figure 6. Distributions of (**a**) convexity and (**b**) roundness for nascent and nascent-denuded soot particles of different sizes (the mobility diameter is reported in parenthesis in the legends).

For the *N3-D3* pair, the distribution of convexity and roundness peaks at slightly lower values after thermodenuding. The convexity of particles decreases slightly with the increasing value of the mobility diameter for both nascent and denuded particles. This suggests that the smaller soot particles are more compact when compared to larger particles, in agreement with previous studies [68,69]. Figure 6a also suggests that for smaller mobility diameters, the convexity of soot from the ethylene diffusion flame might be less affected by thermodenuding as compared to the larger sized particles. With the methane diffusion flame (*N4-D4* and *N5-D5* sets) particles showed negligible changes in roundness and convexity after thermodenuding, for both nascent and nascent-oxidized soot (Figure 6a *N4-D4* and *N5-D5*, respectively).

For completeness, we also investigated the changes in *AR* and D_{Aeq}. Both show only small changes after thermodenuding (Table 1). Our observations on the five sets of soot pairs show only minor changes in the morphology of nascent soot after thermodenuding.

To study the potential effect of PAM on nascent soot prior to thermodenuding, we compared the parameters between *N4* (nascent soot without oxidation in PAM) and *N5* (nascent soot with oxidation in PAM) samples. *N4* and *N5* have comparable masses of 2.37 fg and 2.41 fg, respectively, and a similar mobility diameter ~250 nm. A total of 113 and 122 individual soot particles were analyzed for *N4* and *N5*, respectively. Both samples show nearly the same number of monomers in the soot particles imaged. *N4* has 158 and *N5* has 155 monomers on average. Also, the mean diameter of monomers is similar in the two samples, 23.5 nm for *N4* and 23.9 nm for *N5*. The similar key properties of the soot particles in the two experiments suggest that *N4* and *N5* are suitable samples to make a comparison of nascent soot experiments with and without PAM reactor without thermodenuding.

The roundness for *N4* (0.31) and for *N5* (0.33) and the convexity (0.61 for both *N4* and *N5*) are within the error bars. However, D_f for *N5* (1.65) is smaller than for *N4* (1.80). The value of D_f for *N5* is somehow smaller than the values typically found for nascent soot (1.7–1.9), but lie within the limit when the imaging and thresholding uncertainties discussed above are added in quadrature to the statistical errors. However, in the downstream of the thermodenuder, *D5* and *D4* (samples with and without PAM treatment, respectively) show comparable values of roundness (0.35 for *D4* and 0.33

for *D5*), convexity (0.66 for *D4*, and 0.63 for *D5*), and D_f (1.76 for *D4* and 1.80 for *D5*). Since we have only one set for the nascent-oxidized soot, we are unable to draw a firm conclusion on the effect of thermodenuding on such particles. Although at the time, we have no clear explanation for the minor difference in D_f, it is possible that the different nanophysical properties of the nascent-oxidized soot might indeed result in a higher sensitivity to thermodenuding.

From a study of young and mature soot particles under high-resolution transmission electron microscopy (HRTEM), Alfè et al. [70] found no significant difference in the nanostructure of soot monomers. In addition, they found that the change in the H/C ratio is smaller for methane soot when compared to that of other fuels. In another study, Vander Wal and Tomasek [71], also using HRTEM, reported that the oxidation rate of nascent soot depends upon the nanostructure, for example, the length of graphene segments, curvature, and its orientation. Ishiguro et al. [72], Song et al. [73], Müller et al. [74], also showed a relation between the monomers nanostructure and the soot oxidation from different fuel sources. Other studies showed negligible influence of ozone on soot oxidation [75,76], as compared to the OH radical. In another study of soot oxidation [77], both the ozone and OH at atmospherically relevant levels were found to have no effect on the oxidation of the elemental carbon (EC) fraction in soot. In our case, the CPMA data showed that the main fraction (>90%) of methane-generated soot consisted of EC, suggesting that the oxidation in the PAM chamber might have a negligible effect on the overall morphology of soot.

In a study on the fragmentation and bond strength of diesel soot, Rothenbacher et al. [53] made a comparison between nascent soot treated with and without a TD as a function of impact velocity, and found no substantial change in the degree of fragmentation of nascent soot aggregates due to the thermodenuding. A low-pressure impactor was used to impart velocities of up to 300 m/s to the soot particles. The TD used in their study had a residence time of 0.43 s, and the sample was heated to 280 °C. In another study by Raj et al. [63], soot fragmentation was observed after thermodenuding in the temperature range of 400–900 °C on diesel soot and commercial soot (Printex-U). However, in the lower temperature range, below 500 °C, they found a minor effect on soot fragmentation. Bambha et al. [26] noticed only a small effect of thermodenuding at 410 °C (transit time of ~34 s) on the morphology of soot during the removal of oleic acid coating. In another study, Slowik et al. [64] did not observe any measurable change in the structure of soot when fresh soot (generated at ∅ = 2.1 and 3.5) was thermodenuded at 200 °C. Our results of the negligible or minor restructuring of thermodenuded soot particles are in agreement with these previous studies suggesting that these results are robust and reproducible.

4. Conclusions

In this study, we used scanning electron microscopy to investigate the morphology of nascent soot aggregates prior to, and after, thermodenuding in a low-temperature regime (<270 °C). Despite mass losses of up to ~29% in the nascent soot (removal of ~8 nm coating layer from the soot surface), we detected only minor effects on the soot structure after thermodenuding, irrespective of the fuel type and particle size. We observed no significant change in the fractal dimension, although roundness and convexity showed some minor changes in our case. Future work should focus on the effect on the structure of nascent soot of higher thermodenuding temperatures.

Supplementary Materials: The following are available online at www.mdpi.com/2073-4433/8/9/166/s1, Figure S1: Plots of fractal dimension of nascent-denuded soot pairs. The solid line and dashed line in each plot represent the slope for nascent and denuded soot respectively.

Acknowledgments: This work was supported in part by the Office of Science (BER), Department of Energy (Atmospheric System Research) Grant no. DE-SC0011935 and no. DE-SC0010019, and the Atmospheric Chemistry program of the National Science Foundation Grant no. AGS-1536939 to Boston College, 1537446 to Aerodyne Research Inc. S. China was partially supported by a NASA Earth and Space Science Graduate Fellowships no. NNX12AN97H.

Author Contributions: This manuscript describes the analysis of soot samples obtained in two experimental projects performed in the laboratories of Paul Davidovits at Boston College. Paul Davidovits and Tim Onasch participated in the planning, setting up and supervising of BC2 and BC4. Adam Ahern and Jason Olfert assisted in the experimental design and the experiments. Tim Onasch and Eben Cross led the BC2 experiments including operation of all experimental aspects of the project and analysis of the SMPS and CPMA data used here. Tim Onasch, Lindsay Wolff and Andrew Lambe led the BC4 experiment including operation of all experimental aspects of the project. Swarup China and Claudio Mazzoleni collected the samples during BC4 and BC2, respectively. Manvendra Dubey participated in the experiments and provided the instrumentation used for sampling during BC2. Janarjan Bhandari and Swarup China performed the SEM analysis. Janarjan Bhandari, Claudio Mazzoleni and Swarup China wrote most of the paper with significant contributions and edits from all the coauthors.

Conflicts of Interest: The authors declare no conflict of interest.

References

1. Haynes, B.S.; Wagner, H.G. Soot formation. *Prog. Energy Combust. Sci.* **1981**, *7*, 229–273. [CrossRef]
2. Buseck, P.R.; Adachi, K.; Gelencsér, A.; Tompa, É.; Pósfai, M. Ns-soot: A material-based term for strongly light-absorbing carbonaceous particles. *Aerosol Sci. Technol.* **2014**, *48*, 777–788. [CrossRef]
3. Sorensen, C. Light scattering by fractal aggregates: A review. *Aerosol Sci. Technol.* **2001**, *35*, 648–687. [CrossRef]
4. Adachi, K.; Buseck, P. Internally mixed soot, sulfates, and organic matter in aerosol particles from Mexico city. *Atmos. Chem. Phys.* **2008**, *8*, 6469–6481. [CrossRef]
5. Bambha, R.; Dansson, M.A.; Schrader, P.E.; Michelsen, H.A. *Effects of Volatile Coatings on the Morphology and Optical Detection of Combustion-Generated Black Carbon Particles*; Sandia National Laboratories (SNL-CA): Livermore, CA, USA, 2013. Available online: http://prod.sandia.gov/techlib/access-control.cgi/2013/137660.pdf (accessed on 27 July 2017).
6. China, S.; Salvadori, N.; Mazzoleni, C. Effect of traffic and driving characteristics on morphology of atmospheric soot particles at freeway on-ramps. *Environ. Sci. Technol.* **2014**, *48*, 3128–3135. [CrossRef] [PubMed]
7. Park, K.; Kittelson, D.B.; McMurry, P.H. Structural properties of diesel exhaust particles measured by transmission electron microscopy (TEM): Relationships to particle mass and mobility. *Aerosol Sci. Technol.* **2004**, *38*, 881–889. [CrossRef]
8. China, S.; Scarnato, B.; Owen, R.C.; Zhang, B.; Ampadu, M.T.; Kumar, S.; Dzepina, K.; Dziobak, M.P.; Fialho, P.; Perlinger, J.A.; et al. Morphology and mixing state of aged soot particles at a remote marine free troposphere site: Implications for optical properties. *Geophys. Res. Lett.* **2015**, *42*, 1243–1250. [CrossRef]
9. Cross, E.S.; Onasch, T.B.; Ahern, A.; Wrobel, W.; Slowik, J.G.; Olfert, J.; Lack, D.A.; Massoli, P.; Cappa, C.D.; Schwarz, J.P.; et al. Soot particle studies—Instrument inter-comparison—project overview. *Aerosol Sci. Technol.* **2010**, *44*, 592–611. [CrossRef]
10. Adachi, K.; Chung, S.H.; Buseck, P.R. Shapes of soot aerosol particles and implications for their effects on climate. *J. Geophys. Res. Atmos.* **2010**, *115*, D15. [CrossRef]
11. China, S.; Mazzoleni, C.; Gorkowski, K.; Aiken, A.C.; Dubey, M.K. Morphology and mixing state of individual freshly emitted wildfire carbonaceous particles. *Nat. Commun.* **2013**, *4*, 2122. [CrossRef] [PubMed]
12. Healy, R.M.; Wang, J.M.; Jeong, C.H.; Lee, A.K.; Willis, M.D.; Jaroudi, E.; Zimmerman, N.; Hilker, N.; Murphy, M.; Eckhardt, S.; et al. Light-absorbing properties of ambient black carbon and brown carbon from fossil fuel and biomass burning sources. *J. Geophys. Res. Atmos.* **2015**, *120*, 6619–6633. [CrossRef]
13. Liu, S.; Aiken, A.C.; Gorkowski, K.; Dubey, M.K.; Cappa, C.D.; Williams, L.R.; Herndon, S.C.; Massoli, P.; Fortner, E.C.; Chhabra, P.S.; et al. Enhanced light absorption by mixed source black and brown carbon particles in UK winter. *Nat. Commun.* **2015**, *6*, 8435. [CrossRef] [PubMed]
14. Zhang, R.; Khalizov, A.F.; Pagels, J.; Zhang, D.; Xue, H.; McMurry, P.H. Variability in morphology, hygroscopicity, and optical properties of soot aerosols during atmospheric processing. *Proc. Natl. Acad. Sci. USA* **2008**, *105*, 10291–10296. [CrossRef] [PubMed]
15. Adachi, K.; Buseck, P.R. Changes of ns-soot mixing states and shapes in an urban area during CalNex. *J. Geophys. Res. Atmos.* **2013**, *118*, 3723–3730. [CrossRef]
16. Ghazi, R.; Olfert, J. Coating mass dependence of soot aggregate restructuring due to coatings of oleic acid and dioctyl sebacate. *Aerosol Sci. Technol.* **2013**, *47*, 192–200. [CrossRef]

17. Jacobson, M.Z. Strong radiative heating due to the mixing state of black carbon in atmospheric aerosols. *Nature* **2001**, *409*, 695–697. [CrossRef] [PubMed]
18. Lack, D.; Cappa, C. Impact of brown and clear carbon on light absorption enhancement, single scatter albedo and absorption wavelength dependence of black carbon. *Atmos. Chem. Phys.* **2010**, *10*, 4207–4220. [CrossRef]
19. Liu, D.; Taylor, J.W.; Young, D.E.; Flynn, M.J.; Coe, H.; Allan, J.D. The effect of complex black carbon microphysics on the determination of the optical properties of brown carbon. *Geophys. Res. Lett.* **2015**, *42*, 613–619. [CrossRef]
20. Schnitzler, E.G.; Dutt, A.; Charbonneau, A.M.; Olfert, J.S.; Jäger, W. Soot aggregate restructuring due to coatings of secondary organic aerosol derived from aromatic precursors. *Environ. Sci. Technol.* **2014**, *48*, 14309–14316. [CrossRef] [PubMed]
21. Van Poppel, L.H.; Friedrich, H.; Spinsby, J.; Chung, S.H.; Seinfeld, J.H.; Buseck, P.R. Electron tomography of nanoparticle clusters: Implications for atmospheric lifetimes and radiative forcing of soot. *Geophys. Res. Lett.* **2005**, *32*, L24811. [CrossRef]
22. Westcott, S.L.; Zhang, J.; Shelton, R.K.; Bruce, N.M.; Gupta, S.; Keen, S.L.; Tillman, J.W.; Wald, L.B.; Strecker, B.N.; Rosenberger, A.T.; et al. Broadband optical absorbance spectroscopy using a whispering gallery mode microsphere resonator. *Rev. Sci. Instrum.* **2008**, *79*, 033106. [CrossRef] [PubMed]
23. Cappa, C.D.; Onasch, T.B.; Massoli, P.; Worsnop, D.R.; Bates, T.S.; Cross, E.S.; Davidovits, P.; Hakala, J.; Hayden, K.L.; Jobson, B.T.; et al. Radiative absorption enhancements due to the mixing state of atmospheric black carbon. *Science* **2012**, *337*, 1078–1081. [CrossRef] [PubMed]
24. Fuller, K.A.; Malm, W.C.; Kreidenweis, S.M. Effects of mixing on extinction by carbonaceous particles. *J. Geophys. Res. Atmos.* **1999**, *104*, 15941–15954. [CrossRef]
25. Khalizov, A.F.; Xue, H.; Wang, L.; Zheng, J.; Zhang, R. Enhanced light absorption and scattering by carbon soot aerosol internally mixed with sulfuric acid. *J. Phys. Chem. A* **2009**, *113*, 1066–1074. [CrossRef] [PubMed]
26. Bambha, R.P.; Dansson, M.A.; Schrader, P.E.; Michelsen, H.A. Effects of volatile coatings and coating removal mechanisms on the morphology of graphitic soot. *Carbon* **2013**, *61*, 80–96. [CrossRef]
27. Lack, D.A.; Langridge, J.M.; Bahreini, R.; Cappa, C.D.; Middlebrook, A.M.; Schwarz, J.P. Brown carbon and internal mixing in biomass burning particles. *Proc. Natl. Acad. Sci. USA* **2012**, *109*, 14802–14807. [CrossRef] [PubMed]
28. Xue, H.; Khalizov, A.F.; Wang, L.; Zheng, J.; Zhang, R. Effects of coating of dicarboxylic acids on the mass—Mobility relationship of soot particles. *Environ. Sci. Technol.* **2009**, *43*, 2787–2792. [CrossRef] [PubMed]
29. Huffman, J.A.; Ziemann, P.J.; Jayne, J.T.; Worsnop, D.R.; Jimenez, J.L. Development and characterization of a fast-stepping/scanning thermodenuder for chemically-resolved aerosol volatility measurements. *Aerosol Sci. Technol.* **2008**, *42*, 395–407. [CrossRef]
30. Wehner, B.; Philippin, S.; Wiedensohler, A. Design and calibration of a thermodenuder with an improved heating unit to measure the size-dependent volatile fraction of aerosol particles. *J. Aerosol Sci.* **2002**, *33*, 1087–1093. [CrossRef]
31. Swanson, J.; Kittelson, D. Evaluation of thermal denuder and catalytic stripper methods for solid particle measurements. *J. Aerosol Sci.* **2010**, *41*, 1113–1122. [CrossRef]
32. Knox, A.; Evans, G.; Brook, J.; Yao, X.; Jeong, C.H.; Godri, K.; Sabaliauskas, K.; Slowik, J. Mass absorption cross-section of ambient black carbon aerosol in relation to chemical age. *Aerosol Sci. Technol.* **2009**, *43*, 522–532. [CrossRef]
33. Khalizov, A.F.; Lin, Y.; Qiu, C.; Guo, S.; Collins, D.; Zhang, R. Role of OH-initiated oxidation of isoprene in aging of combustion soot. *Environ. Sci. Technol.* **2013**, *47*, 2254–2263. [CrossRef] [PubMed]
34. Chow, J.C.; Watson, J.G.; Chen, L.W.A.; Arnott, W.P.; Moosmüller, H.; Fung, K. Equivalence of elemental carbon by thermal/optical reflectance and transmittance with different temperature protocols. *Environ. Sci. Technol.* **2004**, *38*, 4414–4422. [CrossRef] [PubMed]
35. Countess, R.J. Interlaboratory analyses of carbonaceous aerosol samples. *Aerosol Sci. Technol.* **1990**, *12*, 114–121. [CrossRef]
36. Cheng, Y.; Duan, F.K.; He, K.B.; Zheng, M.; Du, Z.Y.; Ma, Y.L.; Tan, J.H. Intercomparison of thermal–optical methods for the determination of organic and elemental carbon: Influences of aerosol composition and implications. *Environ. Sci. Technol.* **2011**, *45*, 10117–10123. [CrossRef] [PubMed]

37. Khan, B.; Hays, M.D.; Geron, C.; Jetter, J. Differences in the OC/EC ratios that characterize ambient and source aerosols due to thermal-optical analysis. *Aerosol Sci. Technol.* **2012**, *46*, 127–137. [CrossRef]

38. Stanmore, B.R.; Brilhac, J.F.; Gilot, P. The oxidation of soot: A review of experiments, mechanisms and models. *Carbon* **2001**, *39*, 2247–2268. [CrossRef]

39. Ebert, M.; Inerle-Hof, M.; Weinbruch, S. Environmental scanning electron microscopy as a new technique to determine the hygroscopic behaviour of individual aerosol particles. *Atmos. Environ.* **2002**, *36*, 5909–5916. [CrossRef]

40. Ma, X.; Zangmeister, C.D.; Gigault, J.; Mulholland, G.W.; Zachariah, M.R. Soot aggregate restructuring during water processing. *J. Aerosol Sci.* **2013**, *66*, 209–219. [CrossRef]

41. Tritscher, T.; Jurányi, Z.; Martin, M.; Chirico, R.; Gysel, M.; Heringa, M.F.; DeCarlo, P.F.; Sierau, B.; Prévôt, A.S.; Weingartner, E.; et al. Changes of hygroscopicity and morphology during ageing of diesel soot. *Environ. Res. Lett.* **2011**, *6*, 034026. [CrossRef]

42. Radney, J.G.; You, R.; Ma, X.; Conny, J.M.; Zachariah, M.R.; Hodges, J.T.; Zangmeister, C.D. Dependence of soot optical properties on particle morphology: Measurements and model comparisons. *Environ. Sci. Technol.* **2014**, *48*, 3169–3176. [CrossRef] [PubMed]

43. China, S.; Kulkarni, G.; Scarnato, B.V.; Sharma, N.; Pekour, M.; Shilling, J.E.; Wilson, J.; Zelenyuk, A.; Chand, D.; Liu, S.; et al. Morphology of diesel soot residuals from supercooled water droplets and ice crystals: Implications for optical properties. *Environ. Res. Lett.* **2015**, *10*, 114010. [CrossRef]

44. Popovicheva, O.; Persiantseva, N.; Kuznetsov, B.; Rakhmanova, T.; Shonija, N.; Suzanne, J.; Ferry, D. Microstructure and water adsorbability of aircraft combustor soots and kerosene flame soots: Toward an aircraft-generated soot laboratory surrogate. *J. Phys. Chem. A* **2003**, *107*, 10046–10054. [CrossRef]

45. Huang, P.F.; Turpin, B.J.; Pipho, M.J.; Kittelson, D.B.; McMurry, P.H. Effects of water condensation and evaporation on diesel chain-agglomerate morphology. *J. Aerosol Sci.* **1994**, *25*, 447–459. [CrossRef]

46. Schnitzler, E.G.; Gac, J.M.; Jäger, W. Coating surface tension dependence of soot aggregate restructuring. *J. Aerosol Sci.* **2017**, *106*, 43–55. [CrossRef]

47. Leung, K.; Schnitzler, E.G.; Jaeger, W.; Olfert, J.S. Relative humidity dependence of soot aggregate restructuring induced by secondary organic aerosol: Effects of water on coating viscosity and surface tension. *Environ. Sci. Technol. Lett.* **2017**, in press. [CrossRef]

48. Weber, A.; Baltensperger, U.; Gäggeler, H.; Schmidt-Ott, A. In situ characterization and structure modification of agglomerated aerosol particles. *J. Aerosol Sci.* **1996**, *27*, 915–929. [CrossRef]

49. Weber, A.P.; Friedlander, S.K. In situ determination of the activation energy for restructuring of nanometer aerosol agglomerates. *J. Aerosol Sci.* **1997**, *28*, 179–192. [CrossRef]

50. Jang, H.D.; Friedlander, S.K. Restructuring of chain aggregates of titania nanoparticles in the gas phase. *Aerosol Sci. Technol.* **1998**, *29*, 81–91. [CrossRef]

51. Schmidt-Ott, A. New approaches to in situ characterization of ultrafine agglomerates. *J. Aerosol Sci.* **1988**, *19*, 553–563. [CrossRef]

52. Chen, C.; Fan, X.; Shaltout, T.; Qiu, C.; Ma, Y.; Goldman, A.; Khalizov, A.F. An unexpected restructuring of combustion soot aggregates by subnanometer coatings of polycyclic aromatic hydrocarbons. *Geophys. Res. Lett.* **2016**, *43*, 11080–11088. [CrossRef]

53. Rothenbacher, S.; Messerer, A.; Kasper, G. Fragmentation and bond strength of airborne diesel soot agglomerates. *Part. Fibre Toxicol.* **2008**, *5*, 9. [CrossRef] [PubMed]

54. Lambe, A.; Ahern, A.; Williams, L.; Slowik, J.; Wong, J.; Abbatt, J.; Brune, W.; Ng, N.; Wright, J.; Croasdale, D.; et al. Characterization of aerosol photooxidation flow reactors: Heterogeneous oxidation, secondary organic aerosol formation and cloud condensation nuclei activity measurements. *Atmos. Meas. Tech.* **2011**, *4*, 445–461. [CrossRef]

55. Lambe, A.; Chhabra, P.; Onasch, T.; Brune, W.; Hunter, J.; Kroll, J.; Cummings, M.; Brogan, J.; Parmar, Y.; Worsnop, D.; et al. Effect of oxidant concentration, exposure time, and seed particles on secondary organic aerosol chemical composition and yield. *Atmos. Chem. Phys.* **2015**, *15*, 3063–3075. [CrossRef]

56. Lambe, A.; Ahern, A.; Wright, J.; Croasdale, D.; Davidovits, P.; Onasch, T. Oxidative aging and cloud condensation nuclei activation of laboratory combustion soot. *J. Aerosol Sci.* **2015**, *79*, 31–39. [CrossRef]

57. Schneider, C.A.; Rasband, W.S.; Eliceiri, K.W. NIH image to ImageJ: 25 years of image analysis. *Nat. Methods* **2012**, *9*, 671–675. [CrossRef] [PubMed]

58. Forrest, S.; Witten, T., Jr. Long-range correlations in smoke-particle aggregates. *J. Phys. A Math. Gen.* **1979**, *12*, L109. [CrossRef]

59. Sorensen, C.; Cai, J.; Lu, N. Light-scattering measurements of monomer size, monomers per aggregate, and fractal dimension for soot aggregates in flames. *Appl. Opt.* **1992**, *31*, 6547–6557. [CrossRef] [PubMed]

60. Klein, R.; Meakin, P. Universality in colloid aggregation. *Nature* **1989**, *339*, 360–392.

61. Köylü, Ü.Ö.; Faeth, G.; Farias, T.L.; Carvalho, M.d.G. Fractal and projected structure properties of soot aggregates. *Combust. Flame* **1995**, *100*, 621–633. [CrossRef]

62. Oh, C.; Sorensen, C. The effect of overlap between monomers on the determination of fractal cluster morphology. *J. Colloid Interface Sci.* **1997**, *193*, 17–25. [CrossRef] [PubMed]

63. Raj, A.; Tayouo, R.; Cha, D.; Li, L.; Ismail, M.A.; Chung, S.H. Thermal fragmentation and deactivation of combustion-generated soot particles. *Combust. Flame* **2014**, *161*, 2446–2457. [CrossRef]

64. Slowik, J.G.; Cross, E.S.; Han, J.-H.; Kolucki, J.; Davidovits, P.; Williams, L.R.; Onasch, T.B.; Jayne, J.T.; Kolb, C.E.; Worsnop, D.R. Measurements of morphology changes of fractal soot particles using coating and denuding experiments: Implications for optical absorption and atmospheric lifetime. *Aerosol Sci. Technol.* **2007**, *41*, 734–750. [CrossRef]

65. Dhaubhadel, R.; Pierce, F.; Chakrabarti, A.; Sorensen, C. Hybrid superaggregate morphology as a result of aggregation in a cluster-dense aerosol. *Phys. Rev. E* **2006**, *73*, 011404. [CrossRef] [PubMed]

66. Slowik, J.G.; Stainken, K.; Davidovits, P.; Williams, L.; Jayne, J.; Kolb, C.; Worsnop, D.R.; Rudich, Y.; DeCarlo, P.F.; Jimenez, J.L. Particle morphology and density characterization by combined mobility and aerodynamic diameter measurements. Part 2: Application to combustion-generated soot aerosols as a function of fuel equivalence ratio. *Aerosol Sci. Technol.* **2004**, *38*, 1206–1222. [CrossRef]

67. Wilks, D.S. *Statistical Methods in the Atmospheric Sciences*; Academic Press: Cambridge, MA, USA, 2011.

68. Chakrabarty, R.K.; Moosmüller, H.; Arnott, W.P.; Garro, M.A.; Walker, J. Structural and fractal properties of particles emitted from spark ignition engines. *Environ. Sci. Technol.* **2006**, *40*, 6647–6654. [CrossRef] [PubMed]

69. Virtanen, A.K.; Ristimäki, J.M.; Vaaraslahti, K.M.; Keskinen, J. Effect of engine load on diesel soot particles. *Environ. Sci. Technol.* **2004**, *38*, 2551–2556. [CrossRef] [PubMed]

70. Alfè, M.; Apicella, B.; Barbella, R.; Rouzaud, J.N.; Tregrossi, A.; Ciajolo, A. Structure–property relationship in nanostructures of young and mature soot in premixed flames. *Proc. Combust. Inst.* **2009**, *32*, 697–704. [CrossRef]

71. Vander Wal, R.L.; Tomasek, A.J. Soot oxidation: Dependence upon initial nanostructure. *Combust. Flame* **2003**, *134*, 1–9. [CrossRef]

72. Ishiguro, T.; Suzuki, N.; Fujitani, Y.; Morimoto, H. Microstructural changes of diesel soot during oxidation. *Combust. Flame* **1991**, *85*, 1–6. [CrossRef]

73. Song, J.; Alam, M.; Boehman, A.L.; Kim, U. Examination of the oxidation behavior of biodiesel soot. *Combust. Flame* **2006**, *146*, 589–604. [CrossRef]

74. Müller, J.-O.; Frank, B.; Jentoft, R.E.; Schlögl, R.; Su, D.S. The oxidation of soot particulate in the presence of NO_2. *Catal. Today* **2012**, *191*, 106–111. [CrossRef]

75. Kamm, S.; Möhler, O.; Naumann, K.-H.; Saathoff, H.; Schurath, U. The heterogeneous reaction of ozone with soot aerosol. *Atmos. Environ.* **1999**, *33*, 4651–4661. [CrossRef]

76. Disselkamp, R.; Carpenter, M.; Cowin, J.; Berkowitz, C.; Chapman, E.; Zaveri, R.; Laulainen, N. Ozone loss in soot aerosols. *J. Geophys. Res. Atmos.* **2000**, *105*, 9767–9771. [CrossRef]

77. Browne, E.C.; Franklin, J.P.; Canagaratna, M.R.; Massoli, P.; Kirchstetter, T.W.; Worsnop, D.R.; Wilson, K.R.; Kroll, J.H. Changes to the chemical composition of soot from heterogeneous oxidation reactions. *J. Phys. Chem. A* **2015**, *119*, 1154–1163. [CrossRef] [PubMed]

atmosphere

MDPI

Article

Immersion Freezing of Total Ambient Aerosols and Ice Residuals

Gourihar Kulkarni

Atmospheric Sciences and Global Change Division, Pacific Northwest National Laboratory,
Richland, WA 99352, USA; Gourihar.Kulkarni@pnnl.gov; Tel.: +1-509-375-3729

Received: 17 January 2018; Accepted: 6 February 2018; Published: 9 February 2018

Abstract: This laboratory study evaluates an experimental set-up to study the immersion freezing properties of ice residuals (IRs) at a temperature ranging from -26 to $-34\,^{\circ}\mathrm{C}$ using two continuous-flow diffusion chamber-style ice nucleation chambers coupled with a virtual impactor and heat exchanger. Ice was nucleated on the total ambient aerosol through an immersion freezing mechanism in an ice nucleation chamber (chamber 1). The larger ice crystals formed in chamber 1 were separated and sublimated to obtain IRs, and the frozen fraction of these IRs was investigated in a second ice nucleation chamber (chamber 2). The ambient aerosol was sampled from a sampling site located in the Columbia Plateau region, WA, USA, which is subjected to frequent windblown dust events, and only particles less than 1.5 µm in diameter were investigated. Single-particle elemental composition analyses of the total ambient aerosols showed that the majority of the particles are dust particles coated with organic matter. This study demonstrated a capability to investigate the ice nucleation properties of IRs to better understand the nature of Ice Nucleating Particles (INPs) in the ambient atmosphere.

Keywords: immersion freezing; ice nucleation; ice nucleation; aerosols; ice chamber

1. Introduction

Atmospheric INPs can be airborne particles such as pollen, biological spores, bacteria, plant debris, inorganic dust, volcanic ash, organics, salts, meteoritic particles, and also complex mixtures of organic and inorganic compounds [1,2]. Our understanding of the specific properties that govern the heterogeneous nucleation of ice is limited, and for the same reason, heterogeneous ice nucleation is very difficult to represent adequately in cloud models. Laboratory experiments have been conducted to investigate a wide range of potential INP sources [3–15]. While these studies have improved our understanding of INP number concentration and freezing temperatures across the range of ice nucleation mechanisms significantly, it is often assumed in cloud models that the history of an individual particle does not influence its ice-nucleating properties [16–20]. However, upon sublimation of ice crystals, IRs (ice residuals) are released and may affect cloud properties, leading to poorly constrained feedback processes such as interactions between clouds and ocean/sea ice surface in current cloud resolving models [21,22]. Recently, it has been demonstrated that these IRs may further induce ice crystal formation through recycling and maintain cloud production [23].

Previously, many ground- and airborne-based studies have collected and characterized the nature of IRs using real-time and offline analysis techniques [24–36]. For example, DeMott et al. (2003) [35] utilized a ice nucleation chamber to induce ice nucleation at cirrus cloud temperature conditions on ambient aerosols, and these ice crystals were separated to obtain IRs. The nature of IRs was further characterized using the mass spectrometer. Recently, an ice selective inlet was deployed at the Jungfraujoch research site located in the Swiss Alps to sample ice crystals from mixed-phase clouds to characterize the properties of IRs [29]. Previous aircraft studies have used a counterflow virtual impactor (CVI) to separate and evaporate ice crystals to characterize IRs using online mass spectrometry and electron microscopy techniques. Recently, Cziczo et al. (2013) [25] used a CVI inlet to sample ice crystals and IRs. Further, these IRs were characterized using cabin based mass spectrometer and various electron microscopy-based techniques.

These efforts greatly helped to understand the morphological and composition properties of IRs. The studies that investigate the ice nucleation behavior of IRs are also needed, particularly in the immersion freezing mode because it is likely the most common mechanism of formation of ice crystals in the atmosphere [2], to further improve our understanding of the ice nucleating ability of atmospheric aerosols. Such a study was undertaken here with an aim to demonstrate the experimental method that can be used to investigate the ice nucleating properties of IRs using two ice nucleation chambers connected in series via a pumped counterflow virtual impactor (PCVI) [37,38] and a heat exchanger. These experiments were conducted at temperatures between −26 and −34 °C, and each experiment was repeated twice. The chemical and morphological properties of the total aerosols were also analyzed.

2. Experiments

2.1. Sample Collection

The total aerosol particles were sampled from the air inlet located on the rooftop of the Atmospheric Measurements Laboratory building located at the Pacific Northwest National Laboratory (PNNL) site (Richland, WA, USA). The sampling site falls within the Columbia Plateau region, which was covered with basalt lava at one time, but now is covered with a layer of loess. Because of the dryland farming practice and dry climate, windblown dust events are commonly observed. These dust particles may or may not have an organic component depending upon the season and nearby farming practice. Infrequent regional wildfires can add biomass burning aerosol into the atmosphere, although this was not observed in this study. The ambient temperature and relative humidity (RH) conditions during measurements were 26 °C and 34%, respectively. The sampling port height was ~9 m above the ground, and the site was minimally influenced by the local vehicle pollution as measurements were performed on the weekend. The total particles drawn into the inlet are further passed through the cyclone inlet (model: URG-2000-30EH) operated at 30 LPM to sample only particles smaller than 1.5 μm in diameter. Further, this sample flow was passed through two diffusion driers connected in series to dry the ambient aerosol before they are transported to various laboratory instruments. The concentration of these size-selected particles was measured using a condensation particle counter (CPC) and in some cases by an ultra-high sensitivity aerosol spectrometer (UHSAS; Droplet Measurement Technology, Boulder, CO, USA) (see below). The CPC provides a number concentration of the total particles of size above ~10 nm per unit volume of air, whereas the UHSAS provides the number concentration of total particles of size above ~60 nm per unit volume of air.

2.2. INP Measurements of Total Aerosol and IRs

Measurements of INPs were investigated at various temperatures using two PNNL ice nucleation chambers (chamber 1 and chamber 2) as shown in Figure 1. The detailed procedures to operate the ice chambers have been previously described in Kulkarni et al. (2012) [19] and Friedman et al. (2011) [39], and the chamber geometry design is described in Stetzer at al. (2008) [40]. Briefly, the ice chamber consists of two vertical parallel plates with an evaporation section attached at the bottom of the chamber to remove supercooled water droplets. These plates are cooled using independent external cooling baths (Lauda Brinkmann Inc.; New York, NY, USA). To produce the desired ice-supersaturation conditions a linear temperature gradient between the plates is applied, and the corresponding temperature and relative humidities with respect to water (RH_w) and ice (RH_i) are calculated using the Murphy and Koop (2005) [41] vapor pressure formulations. The sheath and sample flow rates in chamber 1 are 9 and 1 LPM, respectively. These flow conditions limit the particle residence time within this chamber to ~12 s, whereas the flow rates within chamber 2 are 5 and 1 LPM, respectively, and particle residence time is also ~12 s. Although the dimensions of chamber 1 and 2 are slightly different (the length of chamber 2 is exactly half of the chamber 1), the experimental protocol to measure the immersion freezing efficiency of particles is similar. Dissimilarities in terms of dimensions only affect the water droplet breakthrough RH_w limit, which is the RH_w threshold above which water droplets can co-exist with ice crystals and phase discrimination is not possible. Such limits at temperature $-26\,°C$ for chamber 1 and 2 were observed at ~112% and >115%, respectively. The temperature dependent water droplet breakthrough limits at other temperatures are shown in Table S1. Each ice chamber has an in-built diffusion drier (maintained at room temperature) connected permanently to the chamber aerosol inlet port to dry the ambient aerosol before introducing into the chamber. The temperature and RH_i uncertainty limits for both chambers are $\pm 1.0\,°C$ and $\pm 3\%$, respectively. The uncertainties are based on aerosol lamina profile located between the two plates of the chamber at experimental conditions where sheath and sample flows are 9 LPM and 1 LPM, respectively, and the warm and cold plate temperatures are -24 and $-40\,°C$, respectively. The wall temperatures have uncertainties of ~± 0.2 K, and this temperature uncertainty translates into $RH_i = ~\pm 2.5\%$ (slightly higher $RH_i = \pm 3\%$ is used in the study). Also, the uncertainty in the aerosol lamina temperature is $\pm 0.5\,°C$ (the temperature difference across the aerosol lamina width is ~$1.0\,°C$). In this study slightly higher uncertainty $= \pm 1.0\,°C$ are used. The optical particle counter (OPC; CLiMET, model: CI-3100) was used to classify the particles as ice crystals if they were greater than 4 μm. The OPC and UHSAS results were further used to determine the frozen fractions using Equation (1):

$$\text{Frozen fraction} = \text{OPC/UHSAS} \qquad (1)$$

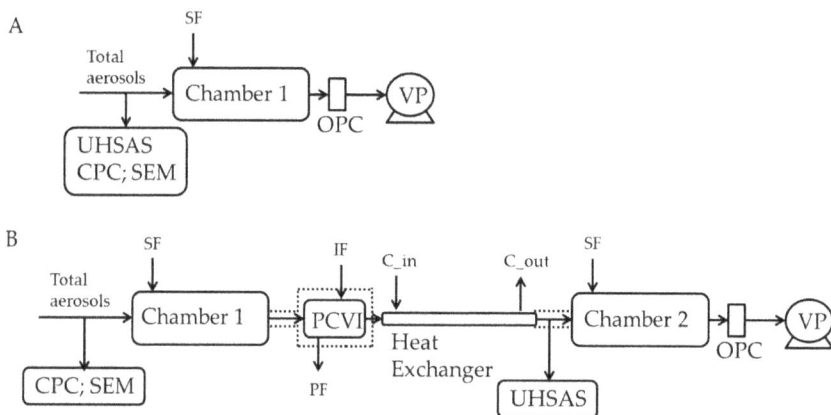

Figure 1. Experimental setup to investigate the ice nucleation ability of ambient aerosols. **A** and **B** show the experimental setup to determine the frozen fraction in immersion mode of only total aerosols and IRs, respectively. In experiment (**A,B**), a CPC was used to monitor ambient particle concentration. Dashed lines indicate the thermal insulation to prevent warming of the set up above ice melting temperature. The heat exchanger was thermally insulated and kept at a constant temperature ($-30\,°C$) using a liquid circulating bath. See text for more details. SF: sheath flow; IF: input flow; PF: pump flow; C_in: coolant in; C_out: coolant out; OPC: optical particle counter; UHSAS: ultra-high sensitivity aerosol spectrometer; PCVI: pumped counterflow virtual impactor; CPC: condensation particle counter; VP: vacuum pump. PCVI flows were: inlet flow = 10.0 LPM; IF = 2.8 LPM; PF = 11.8 LPM; and output flow was 1.0 LPM.

A blank experiment at the beginning and end of the experiment was performed, which included sampling only filtered air (i.e., particle-free) at each temperature. These experiments provided the background ice fraction (~0.01%), which was subtracted from the frozen fraction measured at each temperature. Both chambers were operated at RH_w = ~108%, and the evaporation section of both the chambers was maintained at aerosol lamina temperature. In addition, the PSL particles (2 μm) with known concentration are passed through the OPC, and the fraction of PSL particles that are detected by the OPC are measured. One standard deviation of the fraction of particles detected by the OPC is used as the uncertainty (~±0.1) within the frozen fraction.

Experiments were performed to understand the sensitivity of operational RH_w = 108% upon the frozen fraction. Size-selected 200 nm mobility diameter ammonium sulfate particles were transported to the ice nucleation chambers 1 and 2 in succession. The particle concentration was ~4000 (# per cm³), which is nearly similar to ambient particle concentration (see Table 1). The 1st chamber was operated at −20 °C and at either two RH_w 108 and 115% conditions. A fraction of sulfate particles would then activate to droplets. These droplets would move to evaporation section maintained at −42 °C to freeze the droplets through homogeneous ice nucleation. Those droplets which activated would freeze and those which did not would remain as solid sulfate particles. The idea here was to calculate the fraction of activated droplets as a function of RH_w. The frozen fraction of nearly 1 was observed when the chamber was operated at 115%, but the fraction was ~62% when chamber operated at 108%. This understanding was used to develop the correction factor = ~1.6 (=1/0.62) that was applied to the frozen fraction results (Experiments A and B; Figure 1).

Table 1. The average UHSAS concentrations (# per cm³) in experiment A and B at various temperatures. In A and B UHSAS measured particle concentration of total ambient aerosol and IRs, respectively.

Experiment	−26 °C	−28 °C	−30 °C	−32 °C	−34 °C
A	4000 ± 50	4200 ± 40	4100 ± 50	4150 ± 50	4100 ± 50
B	200 ± 20	250 ± 20	300 ± 30	400 ± 30	600 ± 30

Experiments A and B (see Figure 1) at various temperatures (−26, −28, −30, −32, and −34 °C) were performed on the same day. The aim of experiment A was to determine the immersion freezing efficiency of total ambient aerosols with chamber 1. In experiment B, the immersion freezing efficiency of IRs was investigated using both chambers 1 and 2. Specifically, chamber 1 was used to form ice crystals from INPs present in the total aerosol population. Next, the grown ice crystals were separated using PCVI and transported through the heat exchanger to obtain the IRs. The PCVI output flow and heat exchanger conditions produced efficient sublimation of ice crystals (see below). These IRs are further transported to the UHSAS and chamber 2. Table 1 shows the average UHSAS concentration at different temperature values.

The accuracy of experiment B was validated using well-characterized reference dust particles. K-feldspar (BCS-CRM 376/1; Bureau of Analysed Samples Ltd., Middlesbrough, UK) and illite (NX Nanopowder, Arginotec, Karlsruhe, Germany) dust particles size-selected at 250 nm by a differential mobility analyzer (DMA; TSI, 3080) were transported to the CPC and ice chamber experimental set-up (Figure 1). In this experiment chamber 1 and 2 were operated at RH_w = ~115%.

The PCVI cut size or counterflow was determined as follows (Figure S1). First, chamber 1 was operated at RH_i = 100% and the counterflow within the PCVI was set to zero, allowing transmission of all ambient particles through the output port of the PCVI. The concentration of these particles was continuously monitored using the CPC. Next, the counterflow was increased in a step pattern (0.2 LPM flow change every 5 min; see Figure S2) until no particles were transmitted, and threshold counterflow at these conditions was ~1.8 LPM, which was then used throughout all the experiments. The theoretical particle cut-size corresponding to these flow conditions was ~2.5 μm [37]. More performance and design details regarding PCVI including particle transmission efficiency and computational fluid dynamics (CFD) simulations were shown in a previous study [37], and the CFD model details are described in the supplementary material. The zero transmission efficiency test of the PCVI was verified once again at the end of the experiment to ensure that no particles are transmitted.

These flow settings should separate interstitial and small ice crystals less than ~2.5 μm, but under experimental conditions where larger ice crystals (>2.5 μm) co-exist and because of unknown artifacts (wake capture and flow fluctuations; [42]), interstitial particles may also transmit in addition to large ice crystals. However, based on previous studies using similar PCVI flow conditions and an ice nucleation chamber instrument (e.g., Baustian et al. 2012 [43]; China et al. 2015 [24]) interstitial particles could contribute up to 5% of the hydrometeors separated by the PCVI. Therefore, frozen fraction data from experiment B obtained after applying correction factor 1.6 (discussed above) is further corrected by subtracting 5% to account for the potential transmission of interstitial aerosol. A better estimate of these interstitial particles through CFD simulations was not possible because of unknown artifacts and partial imperfections [37] within the geometry that cannot be simulated.

CFD simulations using experimental flow conditions (Figure 1) were performed in this study to confirm that the output flow that carries large ice crystals separated by the PCVI is dry (RH = 0%) (Figure 2). These flow conditions led to the sublimation of ice crystals within the PCVI (output flow) and the heat exchanger. More information regarding velocity vector direction and flow conditions within the PCVI (labels A and B in Figure 2) obtained through simulations are shown in Figures S2 and S3. The notation RH represents RH_w henceforth.

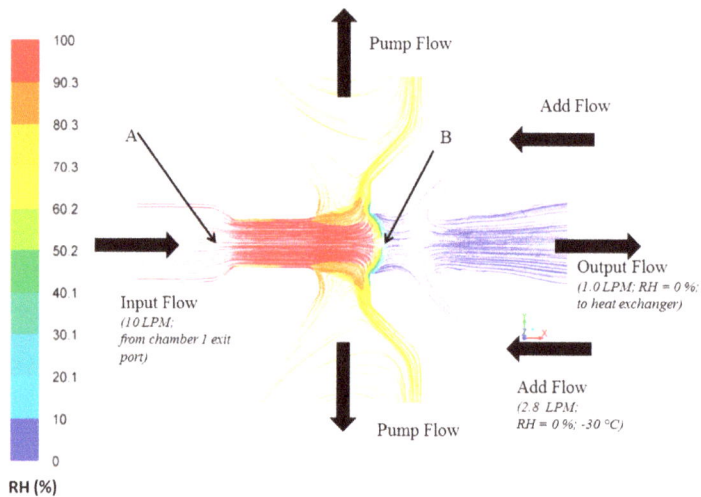

Figure 2. Modelled pathlines colored by RH within the PCVI indicating output flow that carries separated large ice crystals is dry (RH = 0%). The input flow (10 LPM) is the exit flow from the chamber 1. Add flow (2.8 LPM; RH = 0%; −30 °C) is divided into counterflow (not shown) and output flow (1 LPM) that joins as an inlet flow for the heat exchanger unit. The velocity vectors again colored by RH are shown in Figure S2. Labels A and B shows the location within the PCVI where input flow begins to increase and the first stagnation plane is observed, respectively. The RH and velocity magnitudes corresponding to these two locations are shown in Figure S3. The RH corresponds to RH_w.

The schematic of a heat exchanger is shown in Figure S4. It has two concentric cylinders made from stainless steel material. The ice crystals are sublimated inside the inner cylinder that is maintained at −30 °C through actively cooling the outer cylinder. The choice of this coolant temperature value was arbitrary, but caution was taken such that the coolant temperature is warmer than homogeneous freezing threshold temperature (~−38 °C) and colder than ice crystal melting temperature. The diameter and length of the inner cylinder are 0.007 m and 0.68 m, respectively, and the inner cylinder carries the dry flow of ~1 LPM which limits the residence of particles to ~1.6 s. High precision humidity and temperature inline probe (E+E Elektronik; model: EE08-PFT1V11D6/T02) was intermittently used to confirm the cold and dry air conditions (−30 °C; RH = 0%) at the inlet port of the heat exchanger (or the output flow port of the PCVI). Simulations also show that the temperature (−30 °C) at the output flow port is insensitive to the input flow temperature conditions (Figure S5). This was performed because the PCVI input flow temperature conditions changes as the temperature of evaporation section of the chamber 1 vary (−26 to −34 °C; see below Section 3.2). Theoretical ice crystal sublimation calculations [44] confirmed that ice crystals of 6 μm in diameter (the maximum size of ice crystals from chamber 1) are completely sublimated under 1 s at −30 °C and dry RH = 0% conditions. This ensures ice phase was completely sublimated within the heat exchanger and only IRs are available for downstream UHSAS and chamber 2 measurements. However, as hypothesized previously [45] that ice may persist within the pores of the IRs. Additional drying of IRs that would occur within the long in-built diffusion drier permanently installed to the chamber 2 will melt any ice and evaporate the liquid water.

UHSAS measured the IRs concentration, and the OPC from chamber 2 measured the ice crystal concentration of IRs. The particle losses from the exit of the heat exchanger to the chamber 2 are typically ~2%, and this loss correction factor was applied to the UHSAS measurements. Next, the frozen fraction as per Equation (1) was calculated.

The experimental procedure to determine the frozen fraction was as follows. Experiment A was carried out at a defined temperature for ~15 min. Next, only filtered air was sampled for ~3 min to clear out any particles and hydrometeors. Immediately after, at the same temperature, experiment B was carried out for again ~15 min. Next, this experimental procedure was repeated at various temperatures, and the frozen fraction results (for experiment A and B) as a function of temperature were obtained. This experimental procedure was repeated again, and the second set of frozen fraction results were obtained. Finally, these two sets of frozen fraction results were averaged and analyzed to investigate immersion freezing behavior of total ambient aerosols and IRs.

2.3. Microscopy Analysis

The morphology and elemental analysis of the total aerosols were investigated using computer-controlled scanning electron microscopy (SEM) with energy dispersive X-ray analysis (EDS). The acceleration voltage and magnification used were 20 kV and 20,000×, respectively.

The particles were collected in parallel with ongoing ice nucleation experiments (Figure 1) on a carbon type-B support film (Ted Pella, Inc., Redding, CA, USA; 01814-F). Particles were not coated with electrically-conducting metals (e.g., gold, platinum) to enhance the image quality in this analysis. The mesh grid was located on the C- and D-stage of the SKC Sioutas impactor that had a 50% cutoff diameter of 0.5 and 1.0 μm, respectively. Figure 3a shows the example of SEM images of representative individual particles collected from the C-stage. Approximately 1000 randomly selected particles were used to determine the elemental composition of total aerosol particles, and this atomic wt % data was further analyzed to categorize the particles into various subgroups: CNO, CNO_T1, CNO_T2, and others using the classification scheme (Table 2) that was formulated to understand the general composition of particles in a semi-qualitative manner. Figure 3b shows the size distribution of these selected particles. The particle diameter is based on area equivalent diameter reported by the SEM-EDS instrument.

Table 2. Single-particle elemental classification scheme implemented to understand the composition of total aerosol particle population.

Composition Category	Elemental Classification Scheme
CNO	Only C, N, O
CNO_T1	C, N, O and trace elements of Al, Si, Ca
CNO_T2	C, N, O and trace elements of Na, Mg, P, K, Cl
Others	Mixtures of the above including carbonaceous soot

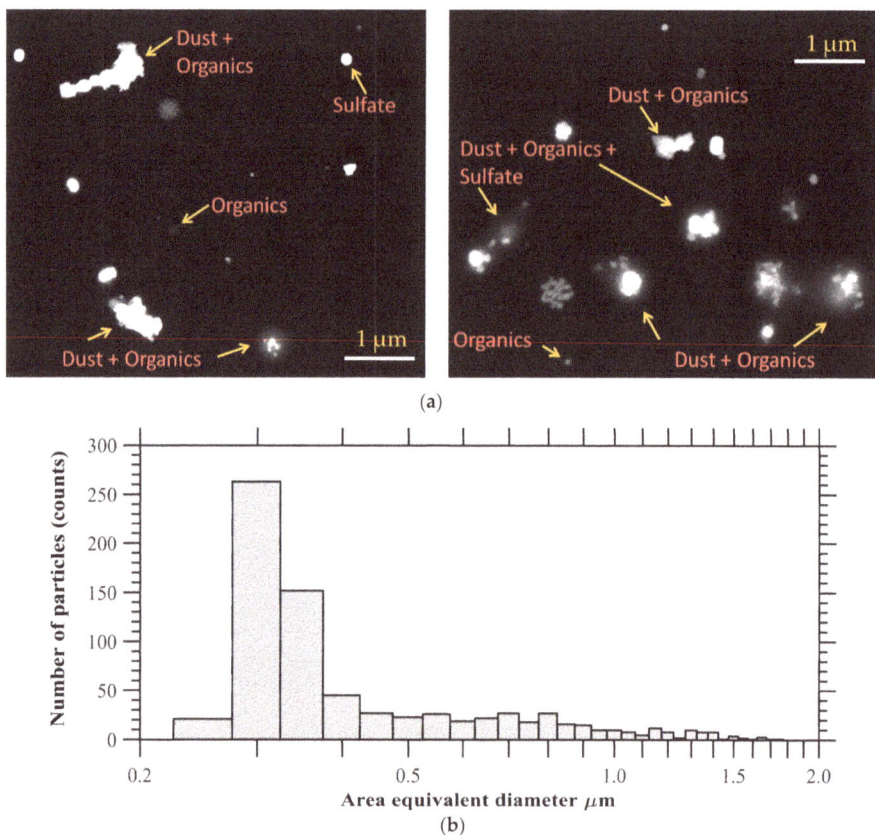

Figure 3. (a) SEM images showing various morphology and mixing state of total aerosol particles. (b) Size distribution of ambient particles. The particle diameter is based on area equivalent diameter reported by the SEM-EDS instrument.

3. Results and Discussion

3.1. Total Ambient Aerosol Particle Characterization

SEM with the EDS technique provided the elemental composition of the total aerosol particles. The composition results revealed that minerals coated with organics dominated the total aerosol particle population (Figure 4). Pure organic particles were also identified. It was not possible to discern the type of organic substances that were condensed on mineral dust particles, but they could represent a multi-component solution of various organic molecules.

The organic particles that were condensed on the dust particles may have formed in the atmosphere by photochemical oxidation reactions of pre-cursor gas species (e.g., α-pinene, isoprene) [13,46–48] or these organics could be soil organic matter that was originally contained with the dust particles. Organic particles without any inorganic inclusions (possibly homogeneously nucleated) were also observed, which could have been emitted directly from the combustion, vegetation, and biomass burning sources.

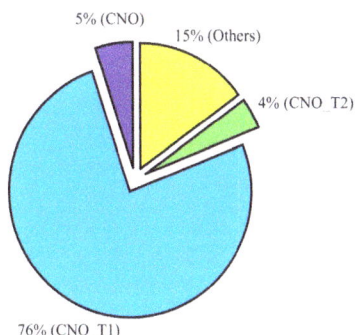

Figure 4. The chemical composition of total aerosol collected from the C-stage of the Sioutus impactor based on number fraction. EDS analysis was based on ~1000 particles. The various subdivisions of composition were characterized based on atomic weight percent intensity of various elements. See Table 2 for more details.

3.2. INP Measurements

Figure 5 shows the frozen fraction of total ambient aerosols and IRs in the immersion freezing mode as a function of temperature (-26, -28, -30, -32 and -34 °C). In water-saturated conditions, it is assumed that particles are activated in a droplet prior to freezing. The vertical and horizontal error bars show the uncertainty in calculating the frozen fraction and the temperature measurements, respectively. In general, the results show that the magnitude of frozen fraction increased with decreasing temperature, which is in agreement with many previous immersion freezing studies (e.g., [49,50]). The results that verify the experiment B are also shown in Figure 5. It was observed that the frozen fraction of total aerosol and ice residuals was ~0.2 and nearly one, respectively. This shows that all the IRs from chamber 1 induced nucleation of ice in the chamber 2 and suggest that local surface features that were responsible for nucleation of water ice on these particles were unaffected after first ice nucleation event (chamber 1). Recently, Kiselev et al. (2016) [51] through a combination of experimental observations and molecular-level model simulations concluded that surface defects such as steps, cracks, and cavities are the ice nucleating active-site features of the K-feldspar particle surface. This implies such features may be responsible for ice nucleation on the surface of dust particles in the chamber 1, and they were also responsible for ice nucleation in the chamber 2. The ice crystal sublimation and residual drying processes in experiment B (Figure 1) seem do not affect the ice nucleating ability of these surface features. In addition, the size distribution of ice residuals was obtained by collecting residuals after the heat exchanger (Figures S6 and S7). As DMA does not classify monodispersed dust particles, larger (multiple-charged) particles were also transmitted in addition to the 250 nm size particles (Figure S7).

Results from experiment A are compared with the previous studies [52–56] by calculating ice nucleation active surface site density (n_s), see Figure 6. The n_s approach has been widely used in previous studies to compare the ice nucleation efficiency of aerosol particles. This approach does not take into account the time dependence of the nucleation events but does describe the number of ice nucleation actives sites at a defined temperature and humidity conditions normalized by the particle surface area. The SEM-EDS reported area equivalent diameter is assumed to equal to the diameter of the spherical particle, which is used for the surface area calculations. The sampling site is often dominated by the windblown dust particles (see Section 2.1 for more details), and therefore results are compared against n_s values from natural deserts dusts reported by the previous studies [56,57]. n_s calculations are performed using total surface area and total number of ice nucleating particles based on 12 s particle residence time and averaged over 15 min, Equation (2),

$$n_s = \frac{N_{ice\_total}}{Area_{total}} \tag{2}$$

where "$Area_{total}$" is the total surface area of ambient particles in all size bins (Figure 3b) that enter the chamber 1 for ~15 min, and N_{ice_total} is the total number of ice particles detected by the OPC of the chamber 1 during these ~15 min. The measurement uncertainty of the n_s density was determined by the error propagation through N_{ice_total} and $Area_{total}$ uncertainties. Additionally, the n_s densities are compared with the n_s fits based on the various soil dust (Figure 6). Niemand et al. (2012) [56] used various soil dusts from various locations (desert from China and Egypt, soil samples from Canary Island and a dust storm in Israel, and commercially available Arizona Test Dust) to derive a fit. Tobo et al. (2014) [57] used surface agricultural soil dust collected from Wyoming, USA and loess soils collected from a dry area in China. The comparison (Figure 6) shows that present data is within an order of magnitude from the literature data. In general, these results indicate that INP efficiency of ambient particles (Section 3.1) is comparable with the previously reported INP efficiency of natural desert particles.

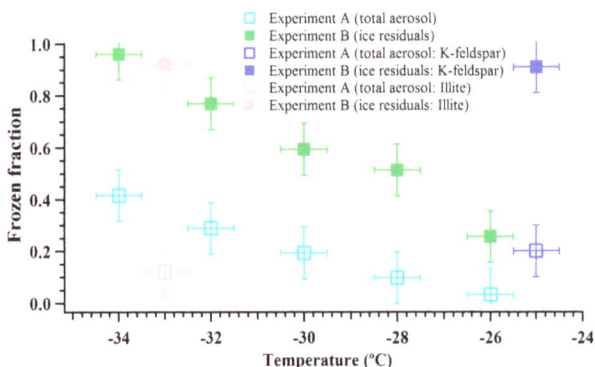

Figure 5. The corrected frozen fraction of total ambient aerosol particles in cyan and IRs in green obtained from the CFDC experiments. The frozen fraction of K-feldspar and illite particles are also shown.

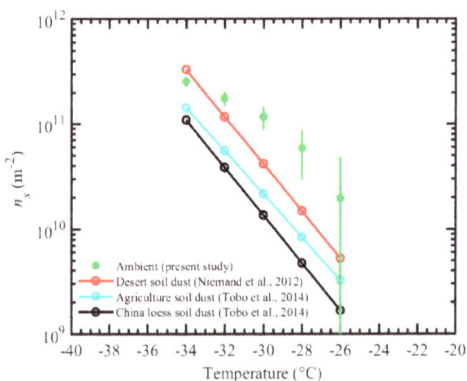

Figure 6. n_s for ambient particles from the present study (experiment A) for immersion freezing. Error bars show the uncertainties propagated through uncertainties within ice particle number concentration and total surface area of particles that enter the chamber 1. The n_s fits from previous studies for desert soil dust, agriculture soil dust, and China loess soil dust are also shown.

This study also suggests that dust particles coated with organic particles, which are observed within ~75% of the ambient particles (Figure 4) can nucleate ice (see experiment A results; Figure 5) at water saturation conditions, These findings are consistent with many previous studies which investigated the ice nucleation ability of pure and organic-coated particles [58–61]. For example, Tobo et al. (2012) [59] observed that at water-saturated conditions only sulfuric acid-coated kaolinite particles showed a reduction in ice nucleation efficiency and that the results were unaffected when these particles were coated with levoglucosan, a water-soluble organic compound. A similar conclusion was also derived from a later study by Wex et al. (2014) [60] who examined kaolinite particles coated with levoglucosan and succinic acid.

Results from experiment B also showed that IRs can induce ice nucleation obtained after sublimation. Similar to experiment A, immersion freezing efficiency of IRs increased with decreasing temperature. The results show that ~25% to 95% IRs induced nucleation of ice at temperatures varying from -26 to -34 °C. This also suggests that the frozen fraction values were not equal to one at temperatures warmer than -34 °C. This indicates some IRs do not induce ice nucleation and could be due to the stochastic nature of IRs that may have limited 100% nucleation of ice [62]. To understand the influence of stochastic freezing the upper limit for the heterogeneous ice nucleation rate coefficient (J_{het}^{up}) and maximum ice fraction (F_{ice}^{max}) was calculated [63,64], see Equations (3) and (4), respectively,

$$J_{het}^{up} = \frac{1}{\tau \cdot Area_{total}} \, ln \left[\frac{1}{1-x} \right] \tag{3}$$

where τ is the total observation time (15 min), $Area_{total}$ is the total surface area of ambient particles in all size bins (Figure 3b) that enter the chamber 1 for ~15 min, and x is the confidence level (99.9%). Next, based on J_{het}^{up}, the F_{ice}^{max} that can be produced during particle residence time period within the chamber was calculated as follows,

$$F_{ice}^{max} = 1 - exp\left(-J_{het}^{up} \cdot Area_{res} \cdot \tau_{res}\right) \tag{4}$$

where $Area_{res}$ is the total surface area of particles in all size bins that was available for nucleation of ice over a particle residence time period (τ_{res}) within the chamber. The τ_{res} was ~6 s. These calculations produced F_{ice}^{max} = 0.0003 or 0.03%, which is less than uncertainty in the frozen fraction ($\pm$0.1; see Section 2.2), and therefore stochastic freezing process did not influence the frozen fraction results reported in this study.

It is also possible that the composition and morphology of IRs could be different from total particles that induced nucleation of ice in the chamber 1. Previous studies support this premise; for example, DeMott et al. (2003) [35] showed that aerosol composition of a total ambient aerosol is different than IRs. Recently, Adler et al. (2014) [65] showed that morphology of organic coated particles can change due to phase separation after the freeze-drying process at low temperatures and suggested that such particles could modify the propensity of organic material towards ice nucleation. Future work needs to be carried out to investigate the morphological changes and physio-chemical properties of IRs to better understand the INP efficiency of IRs.

4. Conclusions

The objective of this study was to demonstrate a laboratory-based experimental method to study the immersion freezing properties of IRs. This objective is achieved by measuring the immersion freezing behavior of total ambient aerosol particles and IRs at a temperature ranging from -26 to -34 °C using the two ice nucleation chambers, a PCVI, and a heat exchanger. Total ambient aerosol particles of mostly <~1.5 μm in diameter were sampled from a site located in the Columbia Plateau region in WA, USA, where frequent windblown dust events are observed throughout the year. A UHSAS instrument confirmed that particle concentration remained nearly constant during the measurements. Chemical composition analysis revealed that total ambient aerosols are mostly mixtures containing dust and organic compounds. IRs were obtained by first separating the larger ice crystals using the PCVI and later sublimating them in a heat exchanger that was maintained at dry cold temperature conditions. Sublimation of ice phase was confirmed by performing theoretical ice sublimation calculations using experimental conditions (RH = 0% and −30 °C) and CFD simulations that confirmed RH = 0% conditions within the output flow port of the PCVI. Immersion freezing fractions of ambient aerosol and IRs were compared, and in general, the results indicate that the experimental setup can be utilized to study the INP properties of IRs. Future experiments are required to identify and characterize the physio-chemical properties and size distributions of IRs from chamber 1 and 2.

Supplementary Materials: The following are available online at http://www.mdpi.com/2073-4433/9/2/55/s1; Table S1 provides the droplet breakthrough limits. Figures S1–S9 shows the details of the experimental details including CFD simulation results. Figures S8 and S9 shows the images of the chamber 1 and 2.

Acknowledgments: I would like to thank three anonymous reviewers and editor for comments and discussions that further improved the quality of the manuscript. This work was partially supported by the Quickstarter 2016 program at the PNNL. The work was also supported by the Office of Science of the U.S. Department of Energy (DOE) as part of the Atmospheric System Research Program. The author would like to thank M. Petters and Y. Agalgaonkar for useful discussions. PNNL is operated by the U.S. DOE by the Battelle Memorial Institute under contract DEAC05-76RL0 1830.

Conflicts of Interest: Author declares no conflict of interest.

Abbreviations

The following abbreviations are used in this manuscript.

SF	sheath flow
IF	input flow
PF	pump flow
C_in	coolant in
C_out	coolant out
OPC	optical particle counter
PCVI	pumped counterflow virtual impactor
CPC	condensation particle counter
CFD	computational fluid dynamics
VP	vacuum pump
RH_i	relative humidity with respect to ice
RH_w	relative humidity with respect to water
INP	ice-nucleating particle
LPM	liters per minute

References

1. Hoose, C.; Mohler, O. Heterogeneous ice nucleation on atmospheric aerosols: A review of results from laboratory experiments. *Atmos. Chem. Phys.* **2012**, *12*, 9817–9854. [CrossRef]
2. Murray, B.J.; O'Sullivan, D.; Atkinson, J.D.; Webb, M.E. Ice nucleation by particles immersed in supercooled cloud droplets. *Chem. Soc. Rev.* **2012**, *41*, 6519–6554. [CrossRef] [PubMed]
3. Alpert, P.A.; Aller, J.Y.; Knopf, D.A. Ice nucleation from aqueous NaCl droplets with and without marine diatoms. *Atmos. Chem. Phys.* **2011**, *11*, 5539–5555. [CrossRef]
4. Baustian, K.J.; Wise, M.E.; Tolbert, M.A. Depositional ice nucleation on solid ammonium sulfate and glutaric acid particles. *Atmos. Chem. Phys.* **2010**, *10*, 2307–2317. [CrossRef]
5. Boose, Y.; Sierau, B.; García, M.I.; Rodríguez, S.; Alastuey, A.; Linke, C.; Schnaiter, M.; Kupiszewski, P.; Kanji, Z.A.; Lohmann, U. Ice nucleating particles in the Saharan Air Layer. *Atmos. Chem. Phys.* **2016**, *16*, 9067–9087. [CrossRef]
6. Cantrell, W.; Bunker, K.; Woodward, X.X. Ice Nucleation in the Contact Mode: Temperature and Size Dependence for Selected Dusts. *AIP Conf. Proc.* **2013**, *1527*, 926–929.
7. Chou, C.; Stetzer, O.; Weingartner, E.; Jurányi, Z.; Kanji, Z.A.; Lohmann, U. Ice nuclei properties within a Saharan dust event at the Jungfraujoch in the Swiss Alps. *Atmos. Chem. Phys.* **2011**, *11*, 4725–4738. [CrossRef]

8. Cziczo, D.J.; Stetzer, O.; Worringen, A.; Ebert, M.; Weinbruch, S.; Kamphus, M.; Gallavardin, S.J.; Curtius, J.; Borrmann, S.; Froyd, K.D.; et al. Inadvertent climate modification due to anthropogenic lead. *Nat. Geosci.* **2009**, *2*, 333–336. [CrossRef]

9. DeMott, P.J.; Petters, M.D.; Prenni, A.J.; Carrico, C.M.; Kreidenweis, S.M.; Collett, J.L.; Moosmuller, H. Ice nucleation behavior of biomass combustion particles at cirrus temperatures. *J. Geophys. Res. Atmos.* **2009**, *114*. [CrossRef]

10. DeMott, P.J.; Sassen, K.; Poellot, M.R.; Baumgardner, D.; Rogers, D.C.; Brooks, S.D.; Prenni, A.J.; Kreidenweis, S.M. African dust aerosols as atmospheric ice nuclei. *Geophys. Res. Lett.* **2003**, *30*. [CrossRef]

11. Hiranuma, N.; Mohler, O.; Yamashita, K.; Tajiri, T.; Saito, A.; Kiselev, A.; Hoffmann, N.; Hoose, C.; Jantsch, E.; Koop, T.; et al. Ice nucleation by cellulose and its potential contribution to ice formation in clouds. *Nat. Geosci.* **2015**, *8*, 273–277. [CrossRef]

12. Knopf, D.A.; Koop, T. Heterogeneous nucleation of ice on surrogates of mineral dust. *J. Geophys. Res. Atmos.* **2006**, *111*. [CrossRef]

13. Prenni, A.J.; Petters, M.D.; Faulhaber, A.; Carrico, C.M.; Ziemann, P.J.; Kreidenweis, S.M.; DeMott, P.J. Heterogeneous ice nucleation measurements of secondary organic aerosol generated from ozonolysis of alkenes. *Geophys. Res. Lett.* **2009**, *36*. [CrossRef]

14. Wang, B.B.; Laskin, A.; Roedel, T.; Gilles, M.K.; Moffet, R.C.; Tivanski, A.V.; Knopf, D.A. Heterogeneous ice nucleation and water uptake by field-collected atmospheric particles below 273 K. *J. Geophys. Res. Atmos.* **2012**, *117*. [CrossRef]

15. Welti, A.; Luond, F.; Stetzer, O.; Lohmann, U. Influence of particle size on the ice nucleating ability of mineral dusts. *Atmos. Chem. Phys.* **2009**, *9*, 6705–6715. [CrossRef]

16. DeMott, P.J.; Prenni, A.J.; Liu, X.; Kreidenweis, S.M.; Petters, M.D.; Twohy, C.H.; Richardson, M.S.; Eidhammer, T.; Rogers, D.C. Predicting global atmospheric ice nuclei distributions and their impacts on climate. *Proc. Natl. Acad. Sci. USA* **2010**, *107*, 11217–11222. [CrossRef] [PubMed]

17. Eidhammer, T.; DeMott, P.J.; Kreidenweis, S.M. A comparison of heterogeneous ice nucleation parameterizations using a parcel model framework. *J. Geophys. Res. Atmos.* **2009**, *114*. [CrossRef]

18. Fan, J.; Leung, L.R.; DeMott, P.J.; Comstock, J.M.; Singh, B.; Rosenfeld, D.; Tomlinson, J.M.; White, A.; Prather, K.A.; Minnis, P.; et al. Aerosol impacts on California winter clouds and precipitation during CalWater 2011: Local pollution versus long-range transported dust. *Atmos. Chem. Phys.* **2014**, *14*, 81–101. [CrossRef]

19. Kulkarni, G.; Fan, J.; Comstock, J.M.; Liu, X.; Ovchinnikov, M. Laboratory measurements and model sensitivity studies of dust deposition ice nucleation. *Atmos. Chem. Phys.* **2012**, *12*, 7295–7308. [CrossRef]

20. Seinfeld, J.H.; Bretherton, C.; Carslaw, K.S.; Coe, H.; DeMott, P.J.; Dunlea, E.J.; Feingold, G.; Ghan, S.; Guenther, A.B.; Kahn, R.; et al. Improving our fundamental understanding of the role of aerosol-cloud interactions in the climate system. *Proc. Natl. Acad. Sci. USA* **2016**, *113*, 5781–5790. [CrossRef] [PubMed]

21. Verlinde, J.; Harrington, J.Y.; McFarquhar, G.M.; Yannuzzi, V.T.; Avramov, A.; Greenberg, S.; Johnson, N.; Zhang, G.; Poellot, M.R.; Mather, J.H.; et al. The mixed-phase Arctic cloud experiment. *Bull. Am. Meteorol. Soc.* **2007**, *88*, 205–221. [CrossRef]

22. Fan, J.W.; Ovtchinnikov, M.; Comstock, J.M.; McFarlane, S.A.; Khain, A. Ice formation in Arctic mixed-phase clouds: Insights from a 3-D cloud-resolving model with size-resolved aerosol and cloud microphysics. *J. Geophys. Res. Atmos.* **2009**, *114*. [CrossRef]

23. Solomon, A.; Feingold, G.; Shupe, M.D. The role of ice nuclei recycling in the maintenance of cloud ice in Arctic mixed-phase stratocumulus. *Atmos. Chem. Phys.* **2015**, *15*, 10631–10643. [CrossRef]

24. China, S.; Kulkarni, G.; Scarnato, B.V.; Sharma, N.; Pekour, M.; Shilling, J.E.; Wilson, J.; Zelenyuk, A.; Chand, D.; Liu, S.; et al. Morphology of diesel soot residuals from supercooled water droplets and ice crystals: Implications for optical properties. *Environ. Res. Lett.* **2015**, *10*, 114010. [CrossRef]

25. Cziczo, D.J.; Froyd, K.D.; Hoose, C.; Jensen, E.J.; Diao, M.H.; Zondlo, M.A.; Smith, J.B.; Twohy, C.H.; Murphy, D.M. Clarifying the Dominant Sources and Mechanisms of Cirrus Cloud Formation. *Science* **2013**, *340*, 1320–1324. [CrossRef] [PubMed]

26. Cziczo, D.J.; Murphy, D.M.; Hudson, P.K.; Thomson, D.S. Single particle measurements of the chemical composition of cirrus ice residue during CRYSTAL-FACE. *J. Geophys. Res. Atmos.* **2004**, *109*. [CrossRef]

27. Froyd, K.D.; Cziczo, D.J.; Hoose, C.; Jensen, E.J.; Diao, M.H.; Zondlo, M.A.; Smith, J.B.; Twohy, C.H.; Murphy, D.M. Cirrus Cloud Formation and the Role of Heterogeneous Ice Nuclei. *AIP Conf. Proc.* **2013**, *1527*, 976–978.

28. Hiranuma, N.; Mohler, O.; Kulkarni, G.; Schnaiter, M.; Vogt, S.; Vochezer, P.; Jarvinen, E.; Wagner, R.; Bell, D.M.; Wilson, J.; et al. Development and characterization of an ice-selecting pumped counterflow virtual impactor (IS-PCVI) to study ice crystal residuals. *Atmos. Meas. Tech.* **2016**, *9*, 3817–3836. [CrossRef]

29. Kupiszewski, P.; Weingartner, E.; Vochezer, P.; Schnaiter, M.; Bigi, A.; Gysel, M.; Rosati, B.; Toprak, E.; Mertes, S.; Baltensperger, U. The Ice Selective Inlet: A novel technique for exclusive extraction of pristine ice crystals in mixed-phase clouds. *Atmos. Meas. Tech.* **2015**, *8*, 3087–3106. [CrossRef]

30. Mertes, S.; Verheggen, B.; Walter, S.; Connolly, P.; Ebert, M.; Schneider, J.; Bower, K.N.; Cozic, J.; Weinbruch, S.; Baltensperger, U.; et al. Counterflow virtual impact or based collection of small ice particles in mixed-phase clouds for the physico-chemical characterization of tropospheric ice nuclei: Sampler description and first case study. *Aerosol Sci. Technol.* **2007**, *41*, 848–864. [CrossRef]

31. Targino, A.C.; Krejci, R.; Noone, K.J.; Glantz, P. Single particle analysis of ice crystal residuals observed in orographic wave clouds over Scandinavia during INTACC experiment. *Atmos. Chem. Phys.* **2006**, *6*, 1977–1990. [CrossRef]

32. Twohy, C.H.; Poellot, M.R. Chemical characteristics of ice residual nuclei in anvil cirrus clouds: Evidence for homogeneous and heterogeneous ice formation. *Atmos. Chem. Phys.* **2005**, *5*, 2289–2297. [CrossRef]

33. Worringen, A.; Kandler, K.; Benker, N.; Dirsch, T.; Mertes, S.; Schenk, L.; Kastner, U.; Frank, F.; Nillius, B.; Bundke, U.; et al. Single-particle characterization of ice-nucleating particles and ice particle residuals sampled by three different techniques. *Atmos. Chem. Phys.* **2015**, *15*, 4161–4178. [CrossRef]

34. Pratt, K.A.; DeMott, P.J.; Twohy, C.H.; Prather, K.A. Insights into cloud ice nucleation from real-time, single-particle aircraft-based measurements of ice crystal residues. In Proceedings of the 242nd American Chemical Society National Meeting, Denver, CO, USA, 28 August–1 September 2011; Volume 242.

35. DeMott, P.J.; Cziczo, D.J.; Prenni, A.J.; Murphy, D.M.; Kreidenweis, S.M.; Thomson, D.S.; Borys, R.; Rogers, D.C. Measurements of the concentration and composition of nuclei for cirrus formation. *Proc. Natl. Acad. Sci. USA* **2003**, *100*, 14655–14660. [CrossRef] [PubMed]

36. Prenni, A.J.; DeMott, P.J.; Twohy, C.; Poellot, M.R.; Kreidenweis, S.M.; Rogers, D.C.; Brooks, S.D.; Richardson, M.S.; Heymsfield, A.J. Examinations of ice formation processes in Florida cumuli using ice nuclei measurements of anvil ice crystal particle residues. *J. Geophys. Res. Atmos.* **2007**, *112*. [CrossRef]

37. Kulkarni, G.; Pekour, M.; Afchine, A.; Murphy, D.M.; Cziczo, D.J. Comparison of Experimental and Numerical Studies of the Performance Characteristics of a Pumped Counterflow Virtual Impactor. *Aerosol Sci. Technol.* **2011**, *45*, 382–392. [CrossRef]

38. Boulter, J.E.; Cziczo, D.J.; Middlebrook, A.M.; Thomson, D.S.; Murphy, D.M. Design and performance of a pumped counterflow virtual impactor. *Aerosol Sci. Technol.* **2006**, *40*, 969–976. [CrossRef]

39. Friedman, B.; Kulkarni, G.; Beranek, J.; Zelenyuk, A.; Thornton, J.A.; Cziczo, D.J. Ice nucleation and droplet formation by bare and coated soot particles. *J. Geophys. Res. Atmos.* **2011**, *116*. [CrossRef]

40. Stetzer, O.; Baschek, B.; Luond, F.; Lohmann, U. The Zurich Ice Nucleation Chamber (ZINC)—A new instrument to investigate atmospheric ice formation. *Aerosol Sci. Technol.* **2008**, *42*, 64–74. [CrossRef]

41. Murphy, D.M.; Koop, T. Review of the vapour pressures of ice and supercooled water for atmospheric applications. *Quart. J. R. Meteorol. Soc.* **2005**, *131*, 1539–1565. [CrossRef]

42. Pekour, M.S.; Cziczo, D.J. Wake Capture, Particle Breakup, and Other Artifacts Associated with Counterflow Virtual Impaction. *Aerosol Sci. Technol.* **2011**, *45*, 758–764. [CrossRef]

43. Baustian, K.J.; Cziczo, D.J.; Wise, M.E.; Pratt, K.A.; Kulkarni, G.; Hallar, A.G.; Tolbert, M.A. Importance of aerosol composition, mixing state, and morphology for heterogeneous ice nucleation: A combined field and laboratory approach. *J. Geophys. Res. Atmos.* **2012**, *117*. [CrossRef]

44. Curry, J.A.; Webster, P.J. *Thermodynamics of Atmospheres and Oceans*; Holton, J.R., Ed.; Academic Press: Cambridge, MA, USA, 1999; Volume 65, pp. 129–159.

45. Marcolli, C. Pre-activation of aerosol particles by ice preserved in pores. *Atmos. Chem. Phys.* **2017**, *17*, 1596–1623. [CrossRef]

46. Ignatius, K.; Kristensen, T.B.; Jarvinen, E.; Nichman, L.; Fuchs, C.; Gordon, H.; Herenz, P.; Hoyle, C.R.; Duplissy, J.; Garimella, S.; et al. Heterogeneous ice nucleation of viscous secondary organic aerosol produced from ozonolysis of α-pinene. *Atmos. Chem. Phys.* **2016**, *16*, 6495–6509. [CrossRef]

47. Jarvinen, E.; Ignatius, K.; Nichman, L.; Kristensen, T.B.; Fuchs, C.; Hoyle, C.R.; Hoppel, N.; Corbin, J.C.; Craven, J.; Duplissy, J.; et al. Observation of viscosity transition in α-pinene secondary organic aerosol. *Atmos. Chem. Phys.* **2016**, *16*, 4423–4438. [CrossRef]

48. Price, H.C.; Mattsson, J.; Zhang, Y.; Bertram, A.K.; Davies, J.F.; Grayson, J.W.; Martin, S.T.; O'Sullivan, D.; Reid, J.P.; Rickards, A.M.J.; et al. Water diffusion in atmospherically relevant α-pinene secondary organic material. *Chem. Sci.* **2015**, *6*, 4876–4883. [CrossRef] [PubMed]

49. Kanji, Z.A.; Welti, A.; Chou, C.; Stetzer, O.; Lohmann, U. Laboratory studies of immersion and deposition mode ice nucleation of ozone aged mineral dust particles. *Atmos. Chem. Phys.* **2013**, *13*, 9097–9118. [CrossRef]

50. Kohn, M.; Lohmann, U.; Welti, A.; Kanji, Z.A. Immersion mode ice nucleation measurements with the new Portable Immersion Mode Cooling chAmber (PIMCA). *J. Geophys. Res. Atmos.* **2016**, *121*, 4713–4733. [CrossRef]

51. Kiselev, A.; Bachmann, F.; Pedevilla, P.; Cox, S.J.; Michaelides, A.; Gerthsen, D.; Leisner, T. Active sites in heterogeneous ice nucleation-the example of K-rich feldspars. *Science* **2017**, *355*, 367–371. [CrossRef] [PubMed]

52. Connolly, P.J.; Mohler, O.; Field, P.R.; Saathoff, H.; Burgess, R.; Choularton, T.; Gallagher, M. Studies of heterogeneous freezing by three different desert dust samples. *Atmos. Chem. Phys.* **2009**, *9*, 2805–2824. [CrossRef]

53. DeMott, P.J.; Mohler, O.; Stetzer, O.; Vali, G.; Levin, Z.; Petters, M.D.; Murakami, M.; Leisner, T.; Bundke, U.; Klein, H.; et al. Resurgence in Ice Nuclei Measurement Research. *Bull. Am. Meteorol. Soc.* **2011**, *92*. [CrossRef]

54. Kanji, Z.A.; DeMott, P.J.; Mohler, O.; Abbatt, J.P.D. Results from the University of Toronto continuous flow diffusion chamber at ICIS 2007: Instrument intercomparison and ice onsets for different aerosol types. *Atmos. Chem. Phys.* **2011**, *11*, 31–41. [CrossRef]

55. Koehler, K.A.; Kreidenweis, S.M.; DeMott, P.J.; Petters, M.D.; Prenni, A.J.; Mohler, O. Laboratory investigations of the impact of mineral dust aerosol on cold cloud formation. *Atmos. Chem. Phys.* **2010**, *10*, 11955–11968. [CrossRef]

56. Niemand, M.; Mohler, O.; Vogel, B.; Vogel, H.; Hoose, C.; Connolly, P.; Klein, H.; Bingemer, H.; DeMott, P.; Skrotzki, J.; et al. A Particle-Surface-Area-Based Parameterization of Immersion Freezing on Desert Dust Particles. *J. Atmos. Sci.* **2012**, *69*, 3077–3092. [CrossRef]

57. Tobo, Y.; DeMott, P.J.; Hill, T.C.J.; Prenni, A.J.; Swoboda-Colberg, N.G.; Franc, G.D.; Kreidenweis, S.M. Organic matter matters for ice nuclei of agricultural soil origin. *Atmos. Chem. Phys.* **2014**, *14*, 8521–8531. [CrossRef]

58. Niedermeier, D.; Hartmann, S.; Clauss, T.; Wex, H.; Kiselev, A.; Sullivan, R.C.; DeMott, P.J.; Petters, M.D.; Reitz, P.; Schneider, J.; et al. Experimental study of the role of physicochemical surface processing on the IN ability of mineral dust particles. *Atmos. Chem. Phys.* **2011**, *11*, 11131–11144. [CrossRef]

59. Tobo, Y.; DeMott, P.J.; Raddatz, M.; Niedermeier, D.; Hartmann, S.; Kreidenweis, S.M.; Stratmann, F.; Wex, H. Impacts of chemical reactivity on ice nucleation of kaolinite particles: A case study of levoglucosan and sulfuric acid. *Geophys. Res. Lett.* **2012**, *39*. [CrossRef]

60. Wex, H.; DeMott, P.J.; Tobo, Y.; Hartmann, S.; Rosch, M.; Clauss, T.; Tomsche, L.; Niedermeier, D.; Stratmann, F. Kaolinite particles as ice nuclei: Learning from the use of different kaolinite samples and different coatings. *Atmos. Chem. Phys.* **2014**, *14*, 5529–5546. [CrossRef]

61. China, S.; Alpert, P.A.; Zhang, B.; Schum, S.; Dzepina, K.; Wright, K.; Owen, R.C.; Fialho, P.; Mazzoleni, L.R.; Mazzoleni, C.; et al. Ice cloud formation potential by free tropospheric particles from long-range transport over the Northern Atlantic Ocean. *J. Geophys. Res. Atmos.* **2017**, *122*, 3065–3079. [CrossRef]

62. Pruppacher, H.R.; Klett, J.D.; Wang, P.K. *Microphysics of Clouds and Precipitation*; Springer Publications: New York, NY, USA, 1998; Volume 8, pp. 381–382.

63. Koop, T.; Luo, B.P.; Biermann, U.M.; Crutzen, P.J.; Peter, T. Freezing of $HNO_3/H_2SO_4/H_2O$ solutions at stratospheric temperatures: Nucleation statistics and experiments. *J. Phys. Chem. A* **1997**, *101*, 1117–1133. [CrossRef]

64. Dymarska, M.; Murray, B.J.; Sun, L.M.; Eastwood, M.L.; Knopf, D.A.; Bertram, A.K. Deposition ice nucleation on soot at temperatures relevant for the lower troposphere. *J. Geophys. Res. Atmos.* **2006**, *111*. [CrossRef]

65. Adler, G.; Koop, T.; Haspel, C.; Taraniuk, I.; Moise, T.; Koren, I.; Heiblum, R.H.; Rudich, Y. Formation of highly porous aerosol particles by atmospheric freeze-drying in ice clouds. *Proc. Natl. Acad. Sci. USA* **2013**, *110*, 20414–20419. [CrossRef] [PubMed]

![atmosphere logo] *atmosphere*

MDPI

Article

Temperature-Dependent Diffusion of H_2SO_4 in Air at Atmospherically Relevant Conditions: Laboratory Measurements Using Laminar Flow Technique

David Brus [1,2,*], **Lenka Škrabalová** [2,3], **Erik Herrmann** [4], **Tinja Olenius** [5,6], **Tereza Trávníčková** [7], **Ulla Makkonen** [1] and **Joonas Merikanto** [1]

[1] Finnish Meteorological Institute, Erik Palménin aukio 1, P.O. Box 503, FIN-00100 Helsinki, Finland; Ulla.Makkonen@fmi.fi (U.M.); joonas.merikanto@fmi.fi (J.M.)

[2] Laboratory of Aerosols Chemistry and Physics, Institute of Chemical Process Fundamentals Academy of Sciences of the Czech Republic, Rozvojova 135, CZ-165 02 Prague 6, Czech Republic; skrabalova@icpf.cas.cz

[3] Department of Physical Chemistry, Faculty of Science, Charles University in Prague, Hlavova 8, CZ-128 43 Prague, Czech Republic

[4] Laboratory of Atmospheric Chemistry, Paul Scherrer Institute, CH-5232 Villigen PSI, Switzerland; erik.herrmann@psi.ch

[5] Department of Physics, University of Helsinki, Gustaf Hällströmin katu 2 A, P.O. Box 64, FIN-00014 Helsinki, Finland; Tinja.Olenius@aces.su.se

[6] Department of Environmental Science and Analytical Chemistry (ACES) and Bolin Centre for Climate Research, Stockholm University, SE-10691 Stockholm, Sweden

[7] Department of Multiphase Reactors, Institute of Chemical Process Fundamentals Academy of Sciences of the Czech Republic, Rozvojova 135, CZ-165 02 Prague 6, Czech Republic; travnickovat@icpf.cas.cz

* Correspondence: david.brus@fmi.fi

Received: 5 June 2017 ; Accepted: 19 July 2017; Published: 22 July 2017

Abstract: We report flow tube measurements of the effective sulfuric acid diffusion coefficient at ranges of different relative humidities (from ∼4 to 70%), temperatures (278, 288 and 298 K) and initial H_2SO_4 concentrations (from 1×10^6 to 1×10^8 molecules·cm^{-3}). The measurements were carried out under laminar flow of humidified air containing trace amounts of impurities such as amines (few ppt), thus representing typical conditions met in Earth's continental boundary layer. The diffusion coefficients were calculated from the sulfuric acid wall loss rate coefficients that were obtained by measuring H_2SO_4 concentration continuously at seven different positions along the flow tube with a chemical ionization mass spectrometer (CIMS). The wall loss rate coefficients and laminar flow conditions were verified with additional computational fluid dynamics (CFD) model FLUENT simulations. The determined effective sulfuric acid diffusion coefficients decreased with increasing relative humidity, as also seen in previous experiments, and had a rather strong power dependence with respect to temperature, around $\propto T^{5.6}$, which is in disagreement with the expected temperature dependence of $\sim T^{1.75}$ for pure vapours. Further clustering kinetics simulations using quantum chemical data showed that the effective diffusion coefficient is lowered by the increased diffusion volume of H_2SO_4 molecules via a temperature-dependent attachment of base impurities like amines. Thus, the measurements and simulations suggest that in the atmosphere the attachment of sulfuric acid molecules with base molecules can lead to a lower than expected effective sulfuric acid diffusion coefficient with a higher than expected temperature dependence.

Keywords: diffusion coefficient; sulfuric acid; laminar flow tube; CFD; amines; clustering kinetics simulations

Atmosphere **2017**, *8*, 132

1. Introduction

Sulfate aerosols play a major role in atmospheric chemistry and significantly influence humans' health and Earth's climate. Particulate matter contributes to air pollution and acts as seeds for cloud droplets, thus affecting cloud properties and radiation budget.Gaseous sulfuric acid H_2SO_4, formed via oxidation of SO_2 by OH radicals, is the most important driver of new particle formation in the present-day atmosphere, and the concentration of H_2SO_4 vapour has a significant impact on atmospheric particle number [1]. While gas-phase H_2SO_4 is reduced by molecular cluster formation and nucleation, its main sink is condensation on newly formed and pre-existing aerosol particles. The condensation process affects both the growth dynamics of atmospheric particle populations, and the amount of H_2SO_4 available for new particle formation. A key parameter in atmospheric mass transport calculations is the gas-phase diffusion coefficient (D), to which the condensation rate of vapour onto particle surface is proportional and dependent on the particle size [2]. Assessments of vapour concentrations [3] and modelling of particle growth and composition [4] are thus dependent on the values used for the binary diffusion coefficients. Under certain circumstances, the gas phase diffusion can even limit the overall rates of condensation and reactions of trace gases with aerosol particles via influencing the uptake of gas molecules onto the surface. The extensive and detailed discussion on gas phase diffusion limitations is given in Introduction section of Tang et al. (2014) [5]. Hanson and Eisele (2000) [6] showed that hydration of sulfuric acid molecules in the atmosphere can significantly lower the sulfuric acid diffusion coefficient. However, gaseous sulfuric acid vapour can also undergo strong clustering in the presence of base impurities, as noted in several previous experiments (e.g., [7–10]). Ammonia is the most abundant atmospheric base, but it does not cluster very strongly with sulfuric acid, in contrast to amines [8]. Amines are strong organic bases, and the most common and abundant amines in the atmosphere being the low-molecular weight aliphatic amines with carbon numbers of 1–6. In this study we focus only on mono-methyl-amine (MMA), dimethyl-amine (DMA) , and tri-methyl-amine (TMA). The sources of amines can be various, both anthropogenic (e.g., animal husbandry, fish processing, food industry, sewage) and natural (e.g., biodegradation of organic matter, ocean, vegetation) [11]. As the oxidative lifetime of gas-phase amines is relatively short, amines are likely to be important mainly around their sources, including densely populated regions and ocean areas with high biological activity. While the measurements of atmospheric amine concentration are sparse, it appears that amine mixing ratios in Earth's continental boundary layer vary between less than 1 to few tens of ppt and often exhibit a day-time peak [11,12]. At few ppt amine concentrations the total atmospheric sulfuric acid likely exists as a mixture of partly hydrated pure sulfuric acid monomers and partly hydrated sulfuric acid monomers bound to amines, with some further contributions from clusters containing several sulfuric acid molecules [13]. Clustering kinetics simulations have suggested that at such few ppt amine concentrations, the effective diffusion coefficient of sulfuric acid is lowered due to amine binding [14], which on the other hand may be sensitive to temperature.

The Chapman-Enskog theory on gas kinetics predicts the binary diffusion coefficient to depend on the temperature as $D \propto T^{1.5}$ when approximating the gas molecules as hard spheres. Fuller et al. (1966) [15] used a semi-empirical method based on the best nonlinear least square fit for a compilation of 340 experimental diffusion coefficients, and obtained a temperature dependence of $T^{1.75}$. According to a compilation work of Marrero and Mason (1972) [16], the temperature dependence of diffusion coefficients in binary gas mixtures in most cases varies between $T^{1.5}$ and T^2. The Fuller et al. method is known to yield the smallest average error, hence it is still recommended for use [17].

In this paper we report the first laboratory measurements, to our knowledge, of the temperature dependence of the diffusion coefficient of sulfuric acid. The diffusion coefficient of H_2SO_4 was estimated from the first order rate coefficients of the wall losses of H_2SO_4 in a flow tube. All previous measurements of the sulfuric acid diffusion coefficient have been carried out also in a flow tube, but at fixed temperatures varying between 295 and 303 K depending on the measurement, and by using nitrogen as the carrier gas [6,18–20]. Here, we use air at atmospheric pressure as a carrier gas, and vary the relative humidities (from 4 to 70%), temperatures (278, 288 and 298 K) and initial H_2SO_4 concentrations (from 1×10^6 to 1×10^8 molecules· cm^{-3}). Base impurities such as di-methyl-amine (DMA) and tri-methyl-amine (TMA) were unavoidably present in our experiment in concentrations of few ppt (as shown in Supplementary Materials, Table ST1), and most probably they originated from the carrier gas and its humidification (e.g., [9,21,22]). Thus, our measurements represent a range of typical conditions met in many locations of the lower troposphere.

We use several approaches to verify the experimental method, and examine our results against the predictions of the semi-empirical Fuller formula as well as data from the previous experiments. In addition, we assess the effect of molecular cluster formation by cluster kinetics simulations with quantum chemical input data.

2. Methods

The experimental setup used in this study was described in detail in our previous work [23,24] and therefore only a brief description is given here. All experiments were conducted at Finnish Meteorological Institute in Helsinki, and thus all above mentioned major sources of atmospheric amines are in the vicinity of the laboratory. The whole experimental apparatus consists of four main parts: a saturator, a mixing unit, a flow tube and the sulfuric acid detection system—Chemical Ionization Mass Spectrometer (CIMS) [25], presented in Figure 1. The H_2SO_4 wall loss measurements were carried out in a laminar flow tube at three temperatures of 278.20 ($\pm$0.2), 288.79 ($\pm$0.2), and 298.2 ($\pm$0.2) K using purified, particle free and dry air as a carrier gas. The flow tube is a vertically mounted cylindrical tube made of stainless steel with an inner diameter (I.D.) of 6 cm and a total length of 200 cm. The whole flow tube was insulated and kept at a constant temperature with two liquid circulating baths (Lauda RK-20). The flow tube consists of two 1 meter long parts; one of them is equipped with 4 holes in the distance of 20 cm from each other, see Figure 1. Sulfuric acid vapour was produced by passing a stream of carrier gas through a saturator filled with 95–97% wt sulfuric acid (J.T. Baker analysed). As a saturator we used a horizontal iron cylinder with Teflon insert (I.D. 5 cm) and it was thermally controlled using a liquid circulating bath (LAUDA RC-6). The temperature inside the saturator was measured with a PT100 themperature sensor ($\pm$0.05 K). The carrier gas saturated with H_2SO_4 was then introduced with a flow rate from 0.1 to 1 L·min^{-1} into the mixing unit made of Teflon and turbulently mixed with a stream of humidified particle free air. The saturator temperature and the flow tube temperature were kept the same (isothermal conditions, see Figure 1), and the amount of sulfuric acid in the system was then governed by the amount of flow through the saturator and the subsequent mixing flow. The only part that was not temperature-controlled was the mixing unit. The mixing unit, which was -18 cm from the beginning of the flow tube (see Figure 1), was kept slightly warmer to avoid any condensation and particle formation in it. However, with the experiments taking relatively long (hours, one profile) the mixing unit cooled down a bit (at 278 and 288 K), and thus the temperature gradient was not prominent. The mixing unit had following dimensions: O.D. = 10 cm, I.D. = 7 cm and height = 6 cm. The flow rate of the mixing air varied in most of the experiments from about 7 to 10 L·min^{-1}. The mixing air was humidified with one pair of Nafion humidifiers (Perma Pure MH-110-12) connected in parallel, where the flow of the mixing air was split into half for longer residence time and better humidification in both humidifiers. Ultrapure water (Millipore, TOC less than 10 ppb, resistivity 18.2 MΩ·cm @25 °C) circulating in both humidifiers was temperature controlled with liquid circulating bath (Lauda RC-6 CS). Both lines of the carrier gas (saturator and mixing air) were controlled by a mass flow controller to within $\pm$3% (MKS type 250). The relative humidity was measured at the centre and far end of the flow tube with two humidity and temperature probes (Vaisala HMP37E and humidity data processor Vaisala HMI38) within accuracy of $\pm$3%. The sulfuric acid diffusion coefficients were estimated as a function of relative humidity from the H_2SO_4 loss measured by CIMS along the flow tube. The detailed information regarding the operational principles and calibration of CIMS is given in [25–27] and therefore will not be given here again. The charging and detection efficiency of CIMS in the presence of trace concentrations of base impurities is discussed thoroughly theoretically (e.g., [13,28]) and also in recent experimental reports (e.g., [9,10]). Possible attachment of base and/or water molecules to single H_2SO_4 molecules (at levels of impurities of this work) is not expected to have a notable effect on their detection efficiency. However, both free H_2SO_4 molecules and those bound to base and/or water molecules are detected as single H_2SO_4 molecules by CIMS, since the ligands are quickly lost upon the chemical ionization (e.g., [28]). In this study the actual H_2SO_4 concentrations are not of particular interest, we focus here only on relative loss of H_2SO_4 along the flow tube. The concentration of sulfuric acid in gas phase was measured as 97 *m/z* Da using CIMS along the flow tube (see Figure 1) at the beginning (0 cm), in the middle (100 cm) and at the lower part in distances of 120, 140, 160, and 180 cm from the beginning and at the outlet (200 cm) of the flow tube in a wide range of relative humidities from 4 to 70%. The concentrations of H_2SO_4 are reported here as measured values. The CIMS sampling flow rate was set to 7 L·min^{-1}. In order to measure the H_2SO_4 concentration along the whole flow tube, an additional CIMS inlet sampling tube was used—a stainless steel tube with I.D. 10 mm and whole length of 122 cm (100 cm straight + 22 cm elbow-pipe). Considering the H_2SO_4 losses, the sampling tube behaves in the same way as the flow tube, that is, for constant steady conditions the relative losses are constant (independent of H_2SO_4 concentration). The detailed separate experiments on H_2SO_4 losses within the sampling tube are discussed in detail by Brus et al. (2011) [29]. The inlet of the CIMS was about the mid-height of the flow tube (110 cm from ground), the sampling tube thus covered whole necessary length for sampling from top to bottom of the flow tube. The CIMS' rack was equipped with wheels allowing movement, and the CIMS was moved in horizontal direction of maximum distance of about 1 m from the flow tube when the sampling tube was connected to mid-height port of the flow tube. The experimental measurement proceeded in the following way: First, all the experimental conditions (temperature of saturator and flow tube, flow rates, relative humidity) were adjusted. Then, when the steady state was reached, the CIMS' inlet was connected to the lowest hole at 200 cm and concentration of sulfuric acid was recorded for at least 20 min. Afterwards the CIMS' inlet was moved up to the hole at 180 cm, and the same procedure was repeated until the last hole at the top of the tube was reached. To confirm the reproducibility of the experimental data the H_2SO_4 concentration at any arbitrary distance along the flow tube was measured again. Moreover, the reproducibility was checked by exchanging the flow tube parts, so that the part with 4 holes was moved up, and H_2SO_4 losses were measured in the distances 0, 20, 40, 60, 80, 100 and 200 cm, respectively. The flow tube parts were exchanged only for experiments conducted at 298 K. At lower temperatures (288 and 278 K), we only used a setup where the flow tube part with sampling ports was at the lower position (100–200 cm). Further details on the flow tube operation tests are provided in the Supplementary Materials.

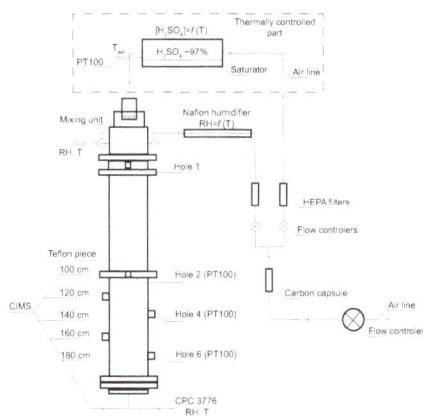

Figure 1. The schematic figure of the FMI flow tube.

2.1. The CFD Simulations

To verify the assumption of laminar flow inside the tube, we applied the computational fluid dynamics (CFD) model FLUENT (Fluent version 16.2, ANSYS Inc., Canonsburg, PA, USA) which simulates flow based on the equations for mass and momentum conservation. These equations and the general operating principles of FLUENT are described in detail in [30–33]. It has to be noted that unlike the earlier studies, this work did not include the Fine Particle Model (FPM). Particle formation within the tube was thus not taken into account, and only sulfuric acid and water vapours are considered in the CFD FLUENT simulations. The simulations only considered the flow tube part of the experimental setup described in Section 2. Methods; the flow tube can be set up as an axisymmetric 2D problem. For the calculations presented here, we chose a resolution of 50×1000 cells. The same geometry has been used previously [32]. Boundary conditions (volumetric flow, wall temperatures, relative humidity, and initial sulfuric acid concentration) were set to match the experimental conditions , please see Supplementary Materials for details on boundary conditions and example profiles of H_2SO_4 concentration and velocity inside the flow tube. Generally, the wall was assumed to be an infinite sink for sulfuric acid, which means that the H_2SO_4 concentration at the walls was set to 0 in the simulations. However, we also carried out simulations where the wall was considered as a source of sulfuric acid; please see the discussion in the next section. The assigned properties for the sulfuric acid vapour were identical to the ones described in our earlier work [32]. Differing from Herrmann et al. (2010) [32], the setup was considered isothermal, i.e., there was no temperature gradient or buoyancy phenomena disturbing parabolic radial flow profile. An initial parabolic flow profile was used as the default flow profile for the all nominal simulations. To verify the proper operation of the experimental setup we applied the diffusion coefficient derived experimentally in this work in FLUENT simulations. The simulations yielded a profile of sulfuric acid concentration inside the flow tube which we compared back to the experimental results. This comparison was done to verify that the same assumptions used in the simulations and experiments (laminar flow and wall acting as an infinite sink) produced the same loss profiles.

2.2. Experimental Determination of the Diffusion Coefficient

We assume that the loss of H_2SO_4 to the walls of the flow tube is a diffusion controlled first-order rate process, which can be described by a simple equation:

$$[H_2SO_4]_t = [H_2SO_4]_0 e^{-kt} \tag{1}$$

where $[H_2SO_4]_0$ is the initial concentration of H_2SO_4, $[H_2SO_4]_t$ is the concentration after time t and k (s^{-1}) is the rate constant, which is given by the equation:

$$k = 3.65 \frac{D}{r^2} \tag{2}$$

where r is the radius of the flow tube and D is the diffusion coefficient of H_2SO_4. Equation (2) is valid for diffusion in a cylindrical tube under laminar flow conditions and when the axial diffusion of the species investigated is negligible [34]. The slopes obtained from linear fits to the experimental data of $\ln([H_2SO_4])$ as a function of the distance in the flow tube stand for the loss rate coefficient, k_{obs} (cm^{-1}), assuming that the first order loss to the flow tube wall is the only sink for the gas phase H_2SO_4. Multiplying the loss rate coefficient k_{obs} with mean flow velocity in the flow tube $(cm \cdot s^{-1})$ yields the experimental first-order wall loss rate coefficient k_w (s^{-1}), from which the diffusion coefficients of H_2SO_4 were determined using Equation (2). The averaged values of mean flow velocities were 5.35 ($\pm$0.3), 5.27 ($\pm$0.2) and 4.5 ($\pm$1.1) for experiments at temperatures 278, 288 and 298 K, respectively. Most of the initial tests were carried out at 298 K, at which the spread of the mean flow velocities is wider. Hanson and Eisele (2000) [6] reported that the wall of the flow tube can act as a source of H_2SO_4 vapour after exposure in long lasting experiments and under very low relative humidity

(RH $\leq \sim 0.5\%$). The accuracy of our RH measurements is $\pm 3\%$ RH, so to avoid any influence of H_2SO_4 evaporation from the flow tube wall we only used data measured at RH $\geq 4\%$ in the final analysis. Furthermore, we performed CFD FLUENT simulations at RH = 5% and T = 298 K, where we do not assume an infinite sink of H_2SO_4 on the wall, but instead apply several boundary conditions for the CFD model with increased H_2SO_4 concentrations on the flow tube wall (0–100% of $[H_2SO_4]_0$), as shown in Figure 2. The comparison suggests that in our measurements the concentration on the flow tube wall is below 6% of the $[H_2SO_4]_0$ under all conditions. When the H_2SO_4 concentration on the wall is $\leq 6\%$ of $[H_2SO_4]_0$, the resulting difference in the obtained diffusion coefficient is within 10% when compared to diffusion coefficient obtained with infinite sink boundary condition on the wall, as indicated by the shaded box in bottom left corner in Figure 2b. In other words, when the flow tube wall is emitting up to 6% of $[H_2SO_4]_0$, we are not able to recognize it in our experiment, since the change in the obtained diffusion coefficient is smaller than our 10% experimental uncertainty. Any higher H_2SO_4 concentration at the wall than 6% of $[H_2SO_4]_0$ would lead to a larger than 10% decrease in the obtained diffusion coefficient. Another experimental method to examine the assumption that the wall of the flow tube acts as an infinite sink is to measure the diffusion coefficient as a function of pressure (e.g., [35,36]). Since in systems including easily nucleating substances like H_2SO_4 a small change in the pressure can initiate strong new particle formation, i.e., secondary losses in the system, this method was not suitable for our study.

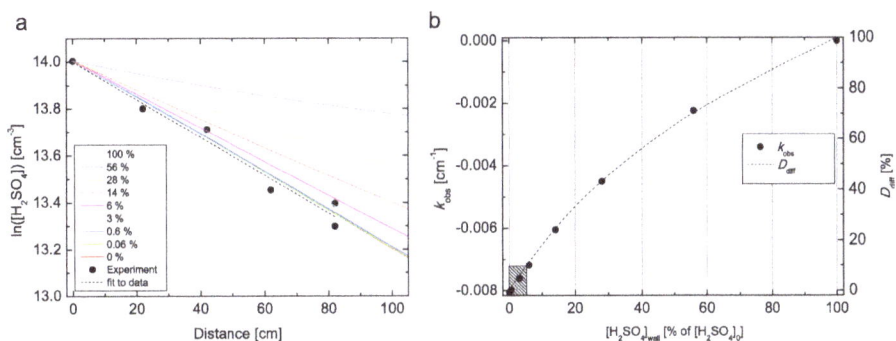

Figure 2. Results from computational fluid dynamics (CFD) FLUENT simulations on the influence of increased H_2SO_4 concentration on the flow tube wall. (**a**) $\ln[H_2SO_4]$ as function of distance in the flow tube. (**b**) k_{obs} and diffusion coefficient difference from the infinite sink boundary condition (D_{diff}) as a function of the H_2SO_4 concentration on the wall expressed as % of the initial H_2SO_4 concentration, $[H_2SO_4]_0$. The simulations conditions were RH = 5%, T = 298 K and Q_{tot} = 7.6 L·min^{-1}. When the H_2SO_4 concentration on the wall is $\leq 6\%$ of $[H_2SO_4]_0$, the resulting difference in the obtained diffusion coefficient is within 10% when compared to diffusion coefficient obtained with infinite sink boundary condition on the wall, as indicated by the shaded box in bottom left corner in Figure 2b. The H_2SO_4 concentrations are low, ranging from 6.5×10^5 to 1.2×10^6 molecules· cm^{-3}, and the points at each distance are averaged over 20-min with a standard deviation of less than 10% of the average value.

2.3. Quantum Chemical Data and Cluster Kinetics Modelling

To assess the effects of base impurities on the measurement results, we performed clustering kinetics simulations using quantum chemical input data for the stabilities of H_2SO_4—base clusters as described by Olenius et al. (2014) [14]. Since recent theoretical studies (e.g., [13,37,38]) suggest and experiments (e.g., [8,10,23,39,40]) confirm that amines are more effective in stabilizing sulfuric acid clusters than ammonia, even taking into account the much higher abundance of ammonia, we focus only on the clustering of sulfuric acid with dimethylamine (DMA) and trimethylamine (TMA) and their hydrates (see also Section 3.2 Effect of Base Contaminants in Olenius et al. 2014) [14]. The cluster kinetics approach does not consider the 2D flow profile, but only the central stream line of the flow, from which clusters and molecules are lost by diffusion. This is considered a reasonable assumption for a laminar flow, as also indicated by the CFD FLUENT modelling results (see the Section 3). Detailed information on the simulations, as well as extensive discussion on the effects of clustering on the apparent diffusion coefficient can be found in the study by Olenius et al. (2014) [14], and the details of the present simulations are in the Supplementary Materials. The effective diffusion coefficient corresponding to the experimental approach was determined by simulating the time evolution of the molecular cluster concentrations using cluster evaporation rates based on quantum chemical calculations at the B3LYP/CBSB7//RICC2/aug-cc-pV(T+d)Z level of theory. The initial base concentration was considered to be a) constant during the experiment, or b) RH dependent, i.e., base molecules enter the system with the water vapour; such a scenario seems to be reasonable since it was observed in several experimental set-ups (e.g., [21,22,29] and also confirmed in our experiments, see Supplementary Materials. In the second case we set the initial base concentration $[base]_{init}$ to be linearly proportional to RH as

$$[base]_{init}(ppt) = [base]_{dry}(ppt) + 0.1 \times RH(\%) \tag{3}$$

where the linear relationship was based on a fit to DMA concentrations measured at various RH, in a separate experiment, where only ultrapure water was used and no sulfuric acid involved. The DMA slope was chosen, since it was found steeper than TMA slope, thus providing higher estimate, see Supplementary Materials for a measurement method description. The measured dry values $[base]_{dry}$ were based on another separate experiment where only sulfuric acid at dry conditions was used, values at all three temperatures are tabulated in Supplementary Materials (Table ST1). The $[base]_{dry}$ values were found higher in experiment with sulfuric acid than ultrapure water, again higher estimates are used in Equation (3). The resulting initial base concentrations $[base]_{init}$ for DMA and TMA were 4 and 2 ppt, respectively, at dry conditions (RH = 0%), and 10 and 8 ppt at RH = 60%. However, we have to point out, that this RH dependence of amines is specific for the present set-up [41,42].

3. Results and Discussion

Figure 3 shows the diffusion coefficients of H_2SO_4, determined from the loss rate coefficients k_w (s^{-1}) using Equation (2) as a function of RH at the three temperatures of 278, 288 and 298 K. The measured points are accompanied with the fit and H_2SO_4–N_2 data at 298 K reported by Hanson and Eisele (2000) [6], and Hanson (2005) [20] at 295 K. Initially, the increase in RH decreased the obtained diffusion coefficient, and the dependence of diffusion coefficient on RH flattens in the range of RH between 20 and 70%. These results indicate slower diffusion to the wall due to strong hydration of H_2SO_4 molecules [43] and possibly H_2SO_4 clustering with base impurities. There are four previously reported experimental values of the H_2SO_4 diffusion coefficient in nitrogen. Pöschl et al. (1998) [19] reported a value of 0.088 cm^2·s^{-1} at T = 303 K and RH ≤ 3%, Lovejoy and Hanson (1996) [18] reported a value of 0.11 cm^2·s^{-1} at T = 295 K and RH ≤ 1%, the study of Hanson and Eisele (2000) [6] yielded a value of 0.094 cm^2·s^{-1} at T = 298 K, and Hanson (2005) [20] reported value of 0.088 cm^2·s^{-1} at T = 295 K, both for RH ≤ 1%. The value of the diffusion coefficient of H_2SO_4 in air at T = 298 K and RH = 4% determined in this study is 0.08 cm^2·s^{-1}, which is in reasonable agreement with previously reported values, although the comparison is complicated because of slightly different experimental conditions and different carrier gases. The relatively larger scatter in our dataset compared to Hanson and Eisele (2000) [6] and Hanson (2005) [20] is because our experiments are carried out at close to atmospheric concentration of H_2SO_4 i.e., about two orders lower concentrations. In Figure 4 the H_2SO_4 losses simulated with the CFD FLUENT model described in Section 2.1 are compared with experimental values, which were measured in a separate set of experiments conducted at T = 278, 288 and 298 K. The linear fit to the experimental data represents the loss rate coefficients (k_{obs}, cm^{-1}).

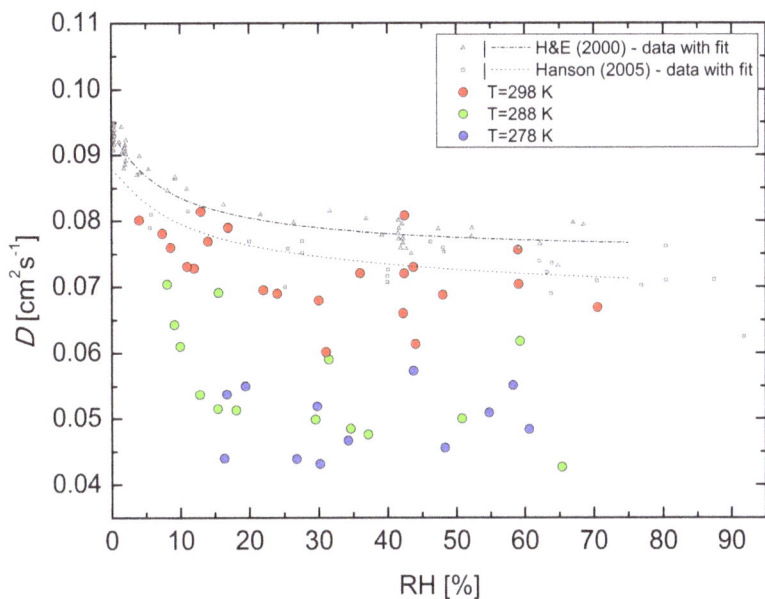

Figure 3. Experimental diffusion coefficient of H_2SO_4 in air as a function of relative humidity at different temperatures compared with the data of Hanson and Eisele (2000) [6] at 298 K and Hanson (2005) [20] at 295 K of H_2SO_4 diffusion in N_2.

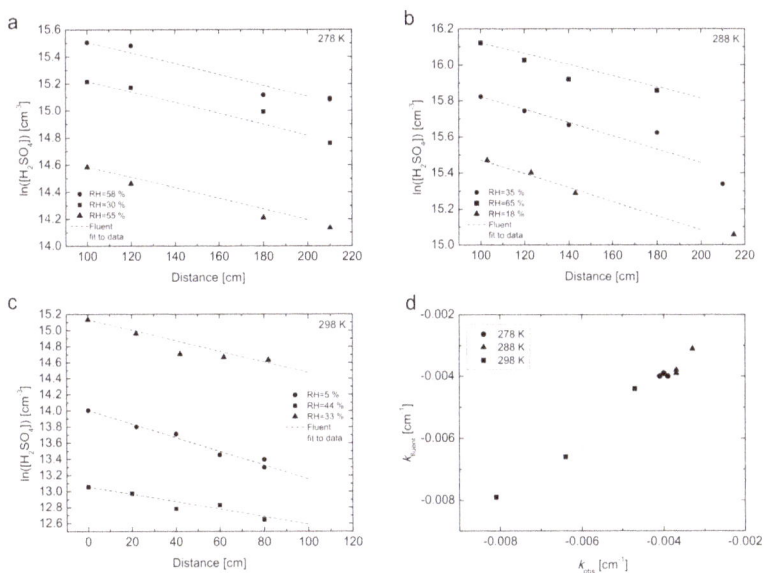

Figure 4. The sulfuric acid losses simulated with CFD FLUENT model when the experimentally obtained diffusion coefficients are used as an input at (**a**) $T = 278$ K; (**b**) $T = 288$ K and (**c**) $T = 298$ K. (**d**) simulated losses rate coefficients compared with experimental values of k_{obs} (cm^{-1}) at $T = 278$, 288 and 298 K. The instances of each symbol in panel (**d**) represent loss rate coefficient at different relative humidities, the lowest relative humidity having the steepest slope (the highest absolute value of loss rate coefficient).

As can be seen from Figure 4, the model describes the behaviour of H_2SO_4 in the flow tube very well and confirms the validity of laminar flow approximation for all three temperatures. The maximum difference between the experimental and simulated values of the loss rate coefficient (k_{obs}) was found 7%, see Figure 4d for details. Additionally, we made the following set of CFD simulations for all three temperatures and a constant mid-range RH $\approx$ 30%: (a) we used the diffusion coefficient obtained from the experiment, but with a constant (flat-plug type) flow profile as initial condition; (b) we used diffusion coefficients obtained from Equation (3) in Hanson and Eisele (2000) [6] with the temperature dependence of $\sim T^{1.75}$ obtained from literature, and an initial parabolic flow profile. The simulations are summarized in Figure S4 in Supplementary Materials. Clearly, the two alternative approaches were not able to reproduce the measured loss profile. The losses are underestimated for case (a) and overestimated for case (b) compared to our above mentioned approach, where we use the experimentally obtained diffusion coefficient and initial parabolic flow profile.

Semi-empirical predictions for binary diffusion coefficients can be calculated from the Fuller et al. equation which is based on fits to experimental data of various gases as described by Fuller et al. (1966) [15] and Reid et al. (1987) [17]:

$$D_{AB} = \frac{0.00143 T^{1.75}}{P\sqrt{M_{AB}} \times [\sqrt[3]{(\Sigma_v)_A} + \sqrt[3]{(\Sigma_v)_B}]^2} \tag{4}$$

where D_{AB} is the binary diffusion coefficient of species A and B (cm$^2\cdot$s^{-1}), T is the temperature (K), P is the pressure (bar), M_{AB} is $2((1/M_A) + (1/M_B))^{-1}$ (g$\cdot$mol^{-1}), where M_A and M_B are the molecular weights of species A and B (g$\cdot$mol^{-1}), and Σ_v is calculated for each component by summing its atomic diffusion volumes [17]. The functional form of Equation (4) is based on the kinetic gas theory (the Chapman-Enskog theory), and the temperature dependence is obtained from a fit to a large set of experimental diffusion coefficients. A purely theoretical approach based on the kinetic gas theory with the hard-spheres approximation would yield a dependence of $T^{1.5}$. The calculated values of the diffusion coefficients of H_2SO_4, dimethylamine- and trimethylamine-sulfate in dry air at 298 K using the Fuller method are 0.11, 0.08 and 0.074 cm$^2\cdot$s^{-1}, respectively, which is in a reasonable agreement with our experimental data—the measured diffusion coefficient of H_2SO_4 at T = 298 K under close to dry conditions (RH 4%) is 0.08 (cm$^2\cdot$s^{-1}). However, when calculating the diffusion coefficients of H_2SO_4 in dry air at lower temperatures (278 and 288 K) with the Fuller method, the agreement of the experimental values with the predictions deteriorates. The formula predicts significantly higher diffusion coefficients than those observed in the experiments. The calculated values of D_{AB} for H_2SO_4 are 0.104 cm$^2\cdot$s^{-1} at T = 288 K and 0.098 cm$^2\cdot$s^{-1} at T = 278 K, and the measured values are 0.07 cm$^2\cdot$s^{-1} at (T = 288 K and RH = 8%) and 0.054 cm$^2\cdot$s^{-1} at (T = 278 K and RH = 16%), respectively. The temperature dependence of the experimental diffusion coefficients was found to be a power of 5.6 for the whole dataset and temperature range, when averaged over the RH between 15–70%. Since the data show a clear stepwise temperature dependence we provide also two separate fits to data from 278 to 288 K and from 288 to 298 K, with power dependencies of 1.9 and 9.4, respectively. These numbers are striking when compared to the empirical method of Fuller et al. (1966) [15], who obtained the best fit to 340 experimental diffusion coefficients with the power dependence of $T^{1.75}$. In Figure 5 we show the temperature dependence of the experimental data obtained from literature [6,18,19], predictions of the Fuller method [15] for the diffusion coefficients of sulfuric acid, dimethylamine- and trimethylamine-sulfates in dry air, results of the clustering kinetics simulations using quantum chemical data for several simulated systems (discussed below), and the experimental data of this work. The data collected from literature, all obtained using laminar flow technique and N_2 as the carrier gas, show a temperature dependence opposite to the one expected from theory. However, the range of temperatures at which the measurements were carried out is quite narrow (only 8 K) and different experimental set-ups could explain such behaviour.

In order to explain the experimental observation of the temperature dependence of the diffusion coefficient, the dimer to monomer ratio at different temperatures was investigated. The CIMS was used to measure the concentrations of H_2SO_4 gas phase monomers and dimers during the experiments; larger clusters were outside the mass range of the CIMS used. The H_2SO_4 dimer formation is a result of H_2SO_4 monomer collisions, and thus the observed H_2SO_4 dimer CIMS signal depends on the H_2SO_4 monomer concentration and also on the residence time, which determinates the time available for the clustering to take place [7]. No significant temperature dependence of the dimer to monomer concentration ratio was observed in our experiments, which is in agreement with Eisele and Hanson (2000) [6], who reported a relatively constant H_2SO_4 dimer to monomer concentration ratio with lowering temperature (the temperature range investigated in their study was 235–250 K). On the other hand, they reported a substantial increase in the larger clusters' (trimer and tetramer) concentration with decreasing temperature while the monomer concentration was almost constant. There are only a very few previously reported values of the sulfuric acid dimer to monomer ratio from laboratory experiments. Petäjä et al. (2011) [7] studied the close to collision-limited sulfuric acid dimer formation under experimental conditions similar to our study ($T = 293$ K, RH = 22%, initial H_2SO_4 concentrations from 10^6 to 10^8 molecule cm^{-3} with saturator containing liquid H_2SO_4 and in-situ H_2SO_4 using O_3-photolysis as methods for producing gas phase H_2SO_4). They reported H_2SO_4 dimer to monomer concentration ratios ranging from 0.05 to 0.1 at RH = 22% and a residence time of 32 s. Petäjä et al. (2011) [7] speculate about the presence of a third stabilizing compound, and their experimental dimer formation rates correspond well to modelled rates at a DMA concentration of about 5 ppt. Almeida et al. (2013) [8] reported the dimer to monomer concentration ratios from 0.01 to 0.06 for the experiments in CLOUD chamber with addition of DMA (3–140 ppt, with the effect saturated for addition >5 ppt) and the dimer to monomer concentration ratios from 1×10^{-4} to 0.003 for pure binary H_2SO_4-water system, both at RH = 38% and $T = 278$ K. In our measurements the H_2SO_4 dimer to monomer concentration ratio under conditions $T = 298$ K, RH = 24% and a residence time of ~37 s, spans the range from 0.03 to 0.11, which is in reasonable agreement with reported values [7,8], when trace impurity levels of DMA are present in the system.

The formation of particles inside the flow tube during the experiments was measured regularly using Ultrafine Condensation Particle Counter (UCPC model 3776, TSI Inc., Shoreview, MN, USA USA) with the lower detection limit of 3 nm. The highest determined concentration of particles was approximately 2×10^4 yielding the maximum nucleation rate J of ~500 particles cm$^3 \cdot$s^{-1} at $T = 278$ K and RH = 60%. Since the nucleation rate was increasing with decreasing temperature and elevated RH in the flow tube, the loss of gas phase sulfuric acid to the particles was more pronounced at temperatures of 288 and 278 K. The losses of H_2SO_4 to particles were minimal—units of percent [9,29] and cannot explain our experimental observation of increased H_2SO_4 diffusion coefficient temperature dependence. The additional losses of H_2SO_4 would lead to increased values of observed loss rate coefficient (k_{obs}) and subsequently to higher diffusion coefficient. The origin of the discrepancy in the temperature dependence of the diffusion coefficient in our experiment remains unclear; however, a plausible explanation is the increased clustering of H_2SO_4 at lower temperatures (see explanation below) with base molecules such as amines. The cluster population simulations using quantum chemical data (see Figures 5 and 6, Table 1) show that the presence of base impurities can decrease the effective H_2SO_4 diffusion coefficient via the attachment of base molecules to the acid molecule. Simulations considering only hydrated H_2SO_4 molecules and no bases give higher values for the diffusion coefficient, and also a notably less steep temperature dependence (Table 1). Also, the stepwise behaviour of temperature dependence can be found in cluster population simulations when fits are performed separately for temperatures 278–288 K and 288–298 K, see Table 1.

This stepwise behaviour corresponds to that obtained in the experiments. Overall, cluster population simulations demonstrate that temperature-dependent clustering can change the magnitude and temperature behaviour of the effective diffusion coefficient. Results obtained by simulating clusters containing H_2SO_4 and DMA are closer to the experimental diffusion coefficient values than those obtained using H_2SO_4 and TMA. On the other hand, the power dependence shows a better agreement for the H_2SO_4–TMA system (see Table 1). The best agreement between the simulations and the experiment was found for the temperature 298 K. Also, changing the base concentration according to Equation (3) shows a better performance than keeping the base concentration constant. Allowing the formation of clusters containing up to two H_2SO_4 and two base molecules has no significant effect. In principle, the larger clusters bind H_2SO_4 molecules and may thus increase the apparent diffusion coefficient, but here their effect is minor due to the relatively low initial H_2SO_4 concentration of 5×10^6 cm^{-3} used in the simulations. More analysis on the effects of the amines can be found in the work by Olenius et al. (2014) [14].

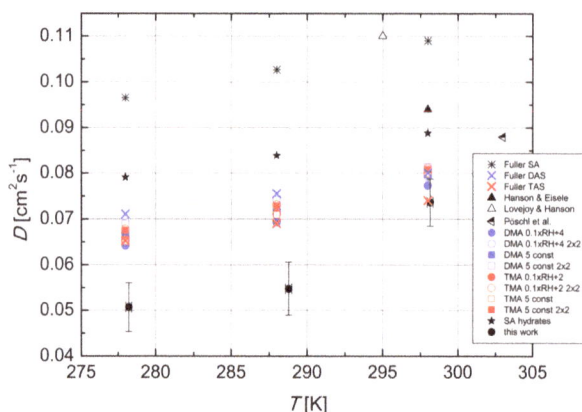

Figure 5. The temperature dependence of the effective H_2SO_4 diffusion coefficient, calculated using the Fuller method for dry H_2SO_4 (SA), dimethylamine-(DAS) and trimethylamine-sulfate (TAS), both in dry air, data from literature, several assemblies of cluster population simulations (see text for details) and data measured experimentally in this work. The temperature dependence of the experimental diffusion coefficients was found to be a power of 5.6.

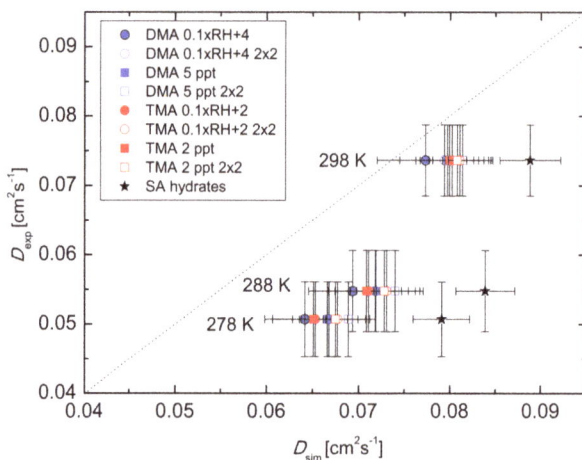

Figure 6. Comparison of the experiment and the cluster population simulations at different temperatures, considered are also different levels and sources of impurities in the system. The formation of clusters containing up to two H_2SO_4 and two base molecules is denoted as "2 × 2" in the legend.

Table 1. Summary of simulated and experimental averages (unweighted over the range of RH that is covered at all three temperatures (15–70%)) of the effective H_2SO_4 diffusion coefficients D $(cm^2 \cdot s^{-1})$ with standard deviations in parenthesis for three temperatures 278, 288 and 298 K. The initial base concentration $[base]_{init}$ is set to be either RH-dependent according to Equation (3) or RH-independent, and the simulations consider clusters containing up to one acid and one base molecule (1×1) or two acid and two base molecules (2×2) as well as hydrates of the clusters. Power dependencies with respect to the temperature, obtained as linear fits to the data, are also listed.

Base	$[Base]_{init}$	Simed Clusters	D (T = 278 K)	D (T = 288 K)	D (T = 298 K)	Power Dep. *
DMA	($0.1 \times$ RH+4) ppt	1×1	0.064 (7%)	0.069 (7%)	0.077 (7%)	2.20/3.18/2.68
		2×2	0.067 (6%)	0.072 (6%)	0.079 (6%)	2.06/2.92/2.48
	Constant 5 ppt	1×1	0.067 (4%)	0.072 (4%)	0.080 (4%)	2.16/3.01/2.58
		2×2	0.069 (3%)	0.074 (4%)	0.081 (4%)	2.02/2.77/2.39
TMA	($0.1 \times$ RH+2) ppt	1×1	0.065 (7%)	0.071 (6%)	0.080 (4%)	2.40/3.53/2.95
		2×2	0.068 (6%)	0.073 (5%)	0.081 (4%)	2.16/3.04/2.59
	Constant 2 ppt	1×1	0.065 (2%)	0.071 (2%)	0.080 (1%)	2.41/3.52/2.95
		2×2	0.067 (2%)	0.073 (2%)	0.081 (1%)	2.15/3.04/2.58
	Only SA hydrates		0.079 (4%)	0.084 (4%)	0.089 (4%)	1.67/1.67/1.67
	This experiment		0.051 (11%)	0.055 (11%)	0.074 (7%)	1.9/9.4/5.56

* power dependence given separately for the temperature ranges 278–288 K/288–298 K/the whole dataset temperature range (278–298 K), the same RH range is used for both simulations and experiment.

4. Conclusions

We have presented measurements of sulfuric acid diffusion coefficient in air derived from the first-order rate coefficients of wall loss of H_2SO_4. The experiments were performed in a laminar flow tube at temperatures of 278, 288 and 298 K, relative humidities from 4 to 70%, under atmospheric pressure and at initial H_2SO_4 concentrations from 10^6 to 10^8 molecules $\cdot$ cm^{-3}. The carrier gas also contained trace amounts of impurities such as amines (few ppt). The chemical ionization mass spectrometer (CIMS) was used to measure H_2SO_4 gas phase concentration at seven different positions along the flow tube. The wall losses were determined from the linear fits to experimental $\ln[H_2SO_4]$ as a function of axial distance in the flow tube. The losses of H_2SO_4 inside the flow tube were also simulated using a computational fluid dynamics model (CFD FLUENT), in which the wall is assumed to be an infinite sink for H_2SO_4. The experimentally determined H_2SO_4 losses along the flow tube were in a very good agreement with profiles calculated using the FLUENT model, where experimentally obtained diffusion coefficients were used as an input. A maximum difference of 7% for experiments conducted at T = 278, 288 and 298 K and in the whole RH range was found when compared to model. The results of the fluid dynamics model (CFD FLUENT) also satisfactory confirm the assumption of fully developed laminar profile inside the flow tube and infinite sink boundary conditions on the wall for H_2SO_4 loss.

To explain an unexpectedly high power dependence of the H_2SO_4 diffusion coefficient on temperature observed in our system we accounted in our calculations for involvement of base impurities: dimethyl- (DMA) and trimethyl-amine (TMA). The semi-empirical Fuller formula [15] was used to calculate the diffusion coefficients at dry conditions for solely H_2SO_4, and H_2SO_4 neutralized with amine bases, namely dimethyamine- and trimethylamine-sulfate. Further, a molecular cluster kinetics model [14] with quantum chemical input data was used to simulate acid-base cluster formation that may lead to the observed behaviour. With the simulations we obtained an effective diffusion coefficient determined in the same way as in the experiments.

The experimental H_2SO_4 diffusion coefficients were found to be independent of different initial $[H_2SO_4]$ and a wide range of total flow rates. The values of the diffusion coefficient were found to decrease with increasing relative humidity owing to stronger hydration of H_2SO_4 molecules. The observed power dependence of the experimental diffusion coefficients as a function of temperature was found to be of the order of 5.6 when the range of RH that is covered at all three temperatures

(15–70%) is accounted for which is in a clear disagreement with predictions from the Fuller method [15] having a power dependence of 1.75. Since the experimental diffusion coefficients deviate more from the theory towards the lower temperatures of 278 and 288 K, we suggest that a plausible explanation for this discrepancy is involvement of impurities such as amines, capable of binding to acid molecules with the binding strength increasing with decreasing temperature. This hypothesis is qualitatively supported by clustering kinetics simulations performed using quantum chemical input data for H_2SO_4–dimethylamine and H_2SO_4–trimethylamine clusters. Our results indicate that the effective diffusion coefficient of H_2SO_4 in air exhibits a stronger temperature dependence than predicted from a theory that does not consider cluster formation. Neglecting this dependence might result in incorrect determination of residual H_2SO_4 concentration in laboratory experiments and lead to biases in atmospheric aerosol models in the presence of amines due to effects on the effective diffusion coefficient and thus on the condensation rate. However, more measurements are needed to gain a better understanding on the role of clustering and the magnitude of the temperature dependence of the H_2SO_4 diffusion coefficient.

Supplementary Materials: Tests of flow tube proper operation, impurity measurements and additional CFD simulations are available online at www.mdpi.com/2073-4433/8/7/132/s1.

Acknowledgments: Authors would like to acknowledge KONE foundation, project CSF No. P209/11/1342, ERC project 257360-MOCAPAF, the Academy of Finland Centre of Excellence (project number: 272041), Formas project 2015-749 and Academy of Finland (project number: 288440) for their financial support.

Author Contributions: D.B. and L.S. designed and performed the experiments; E.H. and T.T. run the CDF-FLUENT simulations, analyzed and interpreted the simulation data; T.O. performed the clustering kinetics simulations using quantum chemical data; U.M. performed the MS-MARGA chemical composition analysis, all authors contributed to synthesis of data and the article writting.

Conflicts of Interest: The authors declare no conflict of interest.The founding sponsors had no role in the design of the study; in the collection, analyses, or interpretation of data; in the writing of the manuscript, and in the decision to publish the results.

Abbreviations

The following abbreviations are used in this manuscript:

CIMS	Chemical Ionization Mass Spectrometer
CFD	Computational Fluid Dynamics
FPM	Fine Particle Model
TOC	Total Organic Carbon
UCPC	Ultrafine Condensation Particle Counter
SA	Sulfuric Acid
MMA	Monomethylamine
DMA	Dimethylamine
TMA	Trimethylamine
DAS	Dimethylamine-sulfate
TAS	Trimethylamine-sulfate

References

1. Dunne, E.M.; Gordon, H.; Kürten, A.; Almeida, J.; Duplissy, J.; Williamson, C.; Ortega, I.K.; Pringle, K.J.; Adamov, A.; Baltensperger, U.; et al. Global atmospheric particle formation from CERN CLOUD measurements. *Science* **2016**, *354*, 1119–1124.

2. Seinfeld, J.H.; Pandis, S.N. *Atmospheric Chemistry and Physics: From Air Pollution to Climate Change*; John Wiley: New York, NY, USA, 1998.

3. Pierce, J.R.; Adams, P.J. A Computationally Efficient Aerosol Nucleation/Condensation Method: Pseudo-Steady-State Sulfuric Acid. *Aerosol Sci. Tech.* **2009**, *43*, 216–226, doi:10.1080/02786820802587896.

4. Hodshire, A.L.; Lawler, M.J.; Zhao, J.; Ortega, J.; Jen, C.; Yli-Juuti, T.; Brewer, J.F.; Kodros, J.K.; Barsanti, K.C.; Hanson, D.R.; et al. Multiple new-particle growth pathways observed at the US DOE Southern Great Plains field site. *Atmos. Chem. Phys.* **2016**, *16*, 9321–9348, doi:10.5194/acp-16-9321-2016.

5. Tang, M.J.; Cox, R.A.; Kalberer, M. Compilation and evaluation of gas phase diffusion coefficients of reactive trace gases in the atmosphere: Volume 1. Inorganic compounds. *Atmos. Chem. Phys.* **2014**, *14*, 9233–9247, doi:10.5194/acp-14-9233-2014.

6. Hanson, D.R.; Eisele, F.L. Diffusion of H_2SO_4 in humidified nitrogen: Hydrated H_2SO_4. *J. Phys. Chem. A* **2000**, *104*, 1715–1719.

7. Petäjä, T.; Sipilä, M.; Paasonen, P.; Nieminen, T.; Kurtén, T.; Ortega, I.K.; Stratmann, F.; Hanna Vehkamäki, H.; Berndt, T.; Kulmala, M. Experimental Observation of Strongly Bound Dimers of Sulphuric Acid: Application to Nucleation in the Atmosphere. *Phys. Rev. Lett.* **2011**, *106*, 228302.

8. Almeida, J.; Schobesberger, S.; Kürten, A.; Ortega, I.K.; Kupiainen-Määttä, O.; Praplan, A.P.; Adamov, A.; Amorim, A.; Bianchi, F.; Breitenlechner, M.; et al. Molecular understanding of sulphuric acid-amine particle nucleation in the atmosphere. *Nature* **2013**, *502*, 359–363.

9. Neitola, K.; Brus, D.; Makkonen, U.; Sipilä, M.; Mauldin R.L., III; Sarnela, N.; Jokinen, T.; Lihavainen, H.; Kulmala, M. Total sulfate vs. sulfuric acid monomer concenterations in nucleation studies. *Atmos. Chem. Phys.* **2015**, *15*, 3429–3443, doi:10.5194/acp-15-3429-2015.

10. Rondo, L.; Ehrhart, S.; Kürten, A.; Adamov, A.; Bianchi, F.; Breitenlechner, M.; Duplissy, J.; Franchin, A.; Dommen, J.; Donahue, N.M.; et al. Effect of dimethylamine on the gas phase sulfuric acid concentration measured by Chemical Ionization Mass Spectrometry (CIMS). *J. Geophys. Res. Atmos.* **2016**, *120*, 3036–3049, doi: 10.1002/2015JD023868.

11. Ge, X.; Wexler, A. S.; Clegg, S.L. Atmospheric amines—Part I. A review. *Atmos. Environ.* **2011**, *45*, 524–546.

12. Hanson, D.R.; McMurry, P.H.; Jiang, J.; Tanner, D.; Huey, L.G. Ambient pressure proton transfer mass spectrometry: Detection of amines and ammonia. *Environ. Sci. Technol.* **2011**, *45*, 8881–8888.

13. Kupiainen-Määttä, O.; Olenius, T.; Kurtén, T.; Vehkamäki, H. CIMS Sulfuric Acid Detection Efficiency Enhanced by Amines Due to Higher Dipole Moments: A Computational Study. *J. Phys. Chem. A* **2013**, *117*, 14109–14119, doi:10.1021/jp4049764.

14. Olenius, T.; Kurtén, T.; Kupiainen-Määttä, O.; Henschel, H.; Ortega, I.K.; Vehkamäki, H. Effect of Hydration and Base Contaminants on Sulfuric Acid Diffusion Measurement: A Computational Study. *Aerosol Sci. Technol.* **2014**, *48*, 593–603, doi:10.1080/02786826.2014.903556.

15. Fuller, E.N.; Schettler, P.D.; Giddings, J.C. New method for prediction of binary gas-phase diffusion coefficients. *Ind. Eng. Chem.* **1966**, *58*, 18–27.

16. Marrero, T.R.; Mason, E.A. Gaseous Diffusion Coefficients. *J. Phys. Chem. Ref. Data* **1972**, *1*, 3–118, doi:10.1063/1.3253094.

17. Reid, R.C.; Prausnitz, J.M.; Polling B.E. *The Properties of Gases and Liquids*, 4th ed.; Mc Graw-Hill Inc.: New York, NY, USA, 1987.

18. Lovejoy, E.R.; Hanson, D.R. Kinetics and products of the reaction SO_3 + NH_3 + N_2. *J. Phys. Chem.* **1996**, *100*, 4459–4465.

19. Pöschl, U.; Canagaratna, M.; Jayne, J.T.; Molina, L.T.; Worsnop, D.R.; Kolb, C.E.; Molina, M.J. Mass accommodation coefficient of H_2SO_4 vapor on aqueous sulphuric acid surfaces and gaseous diffusion coefficient of H_2SO_4 in N_2H_2O. *J. Phys. Chem. A* **1998**, *102*, 10082–10089.

20. Hanson, D.R. Mass accommodation of H_2SO_4 and CH_3SO_3H on water-sulfuric acid solutions from 6% to 97% RH. *J. Phys. Chem. A* **2005**, *109*, 6919–6927.

21. Benson, D.R.; Yu, J.H.; Markovich, A.; Lee, S.-H. Ternary homogeneous nucleation of H_2SO_4, NH_3, and H_2O under conditions relevant to the lower troposphere. *Atmos. Chem. Phys.* **2011**, *11*, 4755–4766, doi:10.5194/acp-11-4755-2011.

22. Kirkby, J.; Curtius, J.; Almeida, J.; Dunne, E.; Duplissy, J.; Ehrhart, S.; Franchin, A.; Gagné, S.; Ickes, L.; Kürten, A.; et al. Role of sulphuric acid, ammonia and galactic cosmic rays in atmospheric aerosol nucleation. *Nature* **2011**, *476*, 429–433, doi:10.1038/nature10343.

23. Neitola, K.; Brus, D.; Makkonen, U.; Sipilä, M.; Lihavainen, H.; Kulmala, M. Effect of addition of four base compounds on sulphuric-acid–water new-particle formation: A laboratory study. *Boreal Env. Res.* **2014**, *19* (Suppl. B), 257–274.

24. Skrabalova, L.; Brus, D.; Antilla, T.; Zdimal, V.; Lihavainen, H. Growth of sulphuric acid nanoparticles under wet and dry conditions. *Atmos. Chem. Phys.* **2014**, *14*, 1–15.

25. Eisele, F.L.; Tanner, D. Measurement of the gas phase concentration of H_2SO_4 and methane sulphonic acid and estimates of H_2SO_4 production and loss in the atmosphere. *J. Geophys. Res.* **1993**, *98*, 9001–9010.

26. Mauldin, R.L., III; Frost, G.; Chen, G.; Tanner, D.; Prevot, A.; Davis, D.; Eisele, F. OH measurements during the First Aerosol Characterization Experiment (ACE 1): Observations and model comparisons. *J. Geophys. Res.* **1998**, *103*, 16713–16729.

27. Petäjä, T.; Mauldin, R.L., III; Kosciuch, E.; McGrath, J.; Nieminen, T.; Paasonen, P.; Boy, M.; Adamov, A.; Kotiaho, T.; Kulmala, M. Sulphuric acid and OH concentrations in a boreal forest site. *Atmos. Chem. Phys.* **2009**, *9*, 7435–7448, doi:10.5194/acp-9-7435-2009.

28. Ortega, I.; Olenius, T.; Kupiainen-Määttä, O.; Loukonen, V.; Kurtén, T.; Vehkamäki, H. Electrical charging changes the composition of sulfuric acid–ammonia/dimethylamine clusters. *Atmos. Chem. Phys.* **2014**, *14*, 7995–8007, doi: 10.5194/acp-14-7995-2014.

29. Brus, D.; Neitola, K.; Hyvärinen, A.-P.; Petäjä, T.; Vanhanen, J.; Sipilä, M.; Paasonen, P.; Kulmala, M.; Lihavainen, H. Homogenous nucleation of sulfuric acid and water at close to atmospherically relevant conditions. *Atmos. Chem. Phys.* **2011**, *11*, 5277– 5287, doi:10.5194/acp-11-5277-2011.

30. Herrmann, E.; Lihavainen, H.; Hyvärinen, A. P.; Riipinen, I.; Wilck, M.; Stratmann, F.; Kulmala, M. Nucleation simulations using the fluid dynamics software FLUENT with the fine particle model FPM. *J. Phys. Chem. A* **2006**, *110*, 12448–12455.

31. Herrmann, E.; Hyvärinen, A.P.; Brus, D.; Lihavainen, H.; Kulmala, M. Re-evaluation of the pressure effect for nucleation in laminar flow diffusion chamber experiments with fluent and the fine particle model. *J. Phys. Chem. A.* **2009**, *113*, 1434–1439.

32. Herrmann, E.; Brus, D.; Hyvärinen, A.-P.; Stratmann, F.; Wilck, M.; Lihavainen, H. and Kulmala, M. A computational fluid dynamics approach to nucleation in the water-sulphuric acid system. *J. Phys. Chem. A* **2010**, *114*, 8033–8042.

33. Travnickova, T.; Havlica, J.; Zdimal, V. Description of fluid dynamics and coupled transports in models of a laminar flow diffusion chamber. *J. Chem. Phys.* **2013**, *139*, 14, doi: 10.1063/1.4816963.

34. Brown, R.L. Tubular flow reactor with first-order kinetics. *J. Res. Natl. Bur. Stand.* **1978**, *83*, 1–6.

35. Fickert, S.; Adams, J.W.; Crowley, J.N. Activation of Br_2 and BrCl via uptake of HOBr onto aqueous salt solutions. *J. Geophys. Res. Atmos.* **1999**, *104*, 23719–23727.

36. Liu, Y.; Ivanov, A.V.; Molina, M.J. Temperature dependence of OH diffusion in air and He. *Geophys. Res. Lett.* **2009**, *36*, L03816, doi: 10.1029/2008gl036170.

37. Ortega, I.K.; Kupiainen, O.; Kurtén, T.; Olenius, T.; Wilkman, O.; McGrath, M.J.; Loukonen, V.; Vehkamäki, H. From quantum chemical formation free energies to evaporation rates. *Atmos. Chem. Phys.* **2012**, *12*, 225–235, doi:10.5194/acp-12-225-2012.

38. Loukonen, V.; Kuo, I.-F.W.; McGrath, M.J.; Vehkamäki, H. On the stability and dynamics of (sulfuric acid) (ammonia) and (sulfuric acid) (dimethylamine) clusters: A first-principles molecular dynamics investigation. *Chem. Phys.* **2014**, *428*, 164–174, doi:10.1016/j.chemphys.2013.11.014.

39. Zollner, J.H.; Glasoe, W.A.; Panta, B.; Carlson, K.K.; McMurry, P.H.; Hanson, D.R. Sulfuric acid nucleation: Power dependencies, variation with relative humidity, and effect of bases. *Atmos. Chem. Phys.* **2012**, *12*, 4399–4411, doi:10.5194/acp-12-4399-2012.

40. Kürten, A.; Jokinen, T.; Simon, M.; Sipilä, M.; Sarnela, N.; Junninen, H.; Adamov, A.; Almeida, J.; Amorim, A.; Bianchi, F.; et al. Neutral molecular cluster formation of sulfuric acid-dimethylamine observed in real-time under atmospheric conditions. *Proc. Natl. Acad. Sci. USA* **2014**, *111*, 15019–15024, doi:10.1073/pnas.1404853111.

41. Yu, H.; McGraw R.; Lee S.H. Effects of amines on formation of sub-3 nm particles and their subsequent growth. *Geophys. Res. Lett.* **2012**, *39*, doi: 10.1029/2011gl050099.

42. Erupe, M.E.; Viggiano, A.A.; Lee, S.H. The effect of trimethylamine on atmospheric nucleation involving H_2SO_4. *Atmos. Chem. Phys.* **2011**, *11*, 4767–4775.

43. Jaecker-Voirol, A.; Mirabel, P. Nucleation rate in binary mixture of sulphuric acid-water vapour: A re-examination. *J. Phys. Chem.* **1988**, *92*, 3518–3521.

atmosphere

MDPI

Article

Monthly and Diurnal Variation of the Concentrations of Aerosol Surface Area in Fukuoka, Japan, Measured by Diffusion Charging Method

Miho Kiriya [1], Tomoaki Okuda [1,*], Hana Yamazaki [1], Kazuki Hatoya [1], Naoki Kaneyasu [2], Itsushi Uno [3], Chiharu Nishita [4], Keiichiro Hara [4], Masahiko Hayashi [4], Koji Funato [5], Kozo Inoue [5], Shigekazu Yamamoto [6], Ayako Yoshino [7] and Akinori Takami [7]

[1]   Faculty of Science and Technology, Keio University, 3-14-1 Hiyoshi, Yokohama 223-8522, Japan; miear.k.117@gmail.com (M.K.); hanayamazaki@z5.keio.jp (H.Y.); k.hatoya.ec@gmail.com (K.H.)
[2]   Atmospheric Environment Research Group, National Institute of Advanced Industrial Science and Technology, Tsukuba 305-8569, Japan; kane.n@aist.go.jp
[3]   Research Institute for Applied Mechanics, Kyushu University, Fukuoka 816-8580, Japan; uno@riam.kyushu-u.ac.jp
[4]   Faculty of Science, Fukuoka University, Fukuoka 814-0180, Japan; cnishita@fukuoka-u.ac.jp (C.N.); harakei@fukuoka-u.ac.jp (K.H.); mhayashi@fukuoka-u.ac.jp (M.H.)
[5]   Tokyo Dylec Corporation, Tokyo 160-0014, Japan; k-funato@tokyo-dylec.co.jp (K.F.); inoue@tokyo-dylec.co.jp (K.I.)
[6]   Department of Environmental Science, Fukuoka Institute of Health and Environmental Sciences, Dazaifu 818-0135, Japan; yamamoto@fihes.pref.fukuoka.jp
[7]   Center for Regional Environmental Research, National Institute for Environmental Studies, Tsukuba 305-8506, Japan; yoshino.ayako@nies.go.jp (A.Y.); takamia@nies.go.jp (A.T.)
*   Correspondence: okuda@applc.keio.ac.jp; Tel./Fax: +81-45-566-1578

Received: 6 May 2017; Accepted: 24 June 2017; Published: 28 June 2017

Abstract: Observation of the ambient aerosol surface area concentrations is important to understand the aerosol toxicity because an increased surface area may be able to act as an enhanced reaction interface for certain reactions between aerosol particles and biological cells, as well as an extended surface for adsorbing and carrying co-pollutants that are originally in gas phase. In this study, the concentration of aerosol surface area was measured from April 2015 to March 2016 in Fukuoka, Japan. We investigated the monthly and diurnal variations in the correlations between the aerosol surface area and black carbon (BC) and sulfate concentrations. Throughout the year, aerosol surface area concentration was strongly correlated with the concentrations of BC, which has a relatively large surface area since BC particles are usually submicron agglomerates consisting of much smaller (tens of nanometers) sized primary soot particles. The slopes of the regression between the aerosol surface area and BC concentrations was highest in August and September 2015. We presented evidence that this was caused by an increase in the proportion of airmasses that originated on the main islands of Japan. This may enhance the introduction of the BC to Fukuoka from the main islands of Japan which we hypothesize to be relatively fresh or "uncoated", thereby maintaining its larger surface area.

Keywords: Asian monsoon; black carbon; long-range transport; land and sea breeze; sulfate; surface area; surface coating of particles

---

1. Introduction

Many studies investigating the adverse effects, such as respiratory and cardiovascular diseases, of exposure to ambient aerosols on human health have been conducted. This is based on the globally acknowledged possibility that these aerosols are hazardous to humans [1–3]. In particular, ultrafine particles with a diameter less than 100 nm (nanoparticles) are considered to be more harmful than larger particles because these exogenous nanoparticles can be inhaled and deposited in the respiratory tract, enter the blood stream, and translocate to other organs [4–7]. Surface area is considered to be an appropriate indicator as opposed to mass for evaluating pulmonary inflammatory responses for rats and mice caused by exposure to manufactured nanomaterials, such as TiO_2, fullerenes, and carbon nanotubes [8–11]. The surface area measurement is also important to understand the aerosol toxicity because an increased surface area may be able to act as an enhanced reaction interface for certain reactions between aerosol particles and biological cells, as well as an extended surface for adsorbing and carrying co-pollutants that are originally in gas phase [12–14].

The most popular method to measure the particle specific surface area is the Brunauer-Emmett-teller (BET) method [15]. The toxicity led by exposure to manufactured materials related to the specific surface area has been discussed on the basis of the BET specific surface area values [8–10]. However, it is challenging to experimentally measure the actual surface area of atmospheric aerosol particles due to low quantity of aerosols that can be collected using ordinary filter sampling methods [16–18].

Other practical and continuous techniques are required to measure the concentrations of ambient aerosol surface area. A nanoparticle surface area monitor (NSAM) using the diffusion charging method has been developed for continuous measurement for the concentration of particle surface area [19–23]. Particles introduced in the NSAM first become charged with positive ions emitted by a corona discharger in a mixing chamber. Particle charge is measured by an electrometer installed downstream from the chamber. Surface area is calculated assuming that it is ideally proportional to the particle charge [19]. The actual NSAM output is the lung-deposited surface area (LDSA, the product of particle surface area and lung deposition efficiency) concentration of particles that can be converted to the concentration of ambient aerosol surface area [18,23–28].

These previous reports noted that the concentration of aerosol surface area is strongly correlated with the concentrations of black carbon (BC) and polycyclic aromatic hydrocarbons in particulate phase. These facts are reasonable since BC (or BC-like) particles are usually submicron agglomerates consisting of much smaller (tens of nanometers) sized primary soot particles [27,29]. However, the ways in which chemical species other than BC and meteorological conditions contribute to the variation in the concentrations of ambient aerosol surface area is not well understood. Particularly, sulfate aerosols may occasionally have a significant effect on aerosol surface area concentration [23].

In this study, we measured the concentration of aerosol surface area together with BC and sulfate, as well as meteorological data for one year at Fukuoka, Japan, and investigated the effect of the aerosol chemical compositions and meteorological conditions on the aerosol surface area. This study is a follow-up experiment of Okuda et al. (2016) [23] in order to perform a year-round investigation of aerosol surface area and some related parameters.

2. Experiments

2.1. Observation Site and Period

The monitoring site for the aerosol surface area, BC, and particle number concentrations was the fourth floor of a building of the Fukuoka Institute for Atmospheric Environment and Health (33.55° N, 130.36° E) at Fukuoka University, Japan [23,30–32]. Another monitoring site for the sulfate concentration was the Fukuoka Institute of Health and Environmental Science (33.51° N, 130.48° E). These two sites are ~15 km from each other. Fukuoka City is one of the largest cities in northern Kyushu district, which faces the Asian continent (Figure 1). Generally, sulfate concentrations in this area (northern Kyushu district) are mainly affected by a long-range transport, and exhibit very similar variation over the scale of hundreds of kilometers [30]. For example, sulfate in Fukue and Fukuoka showed very similar concentrations and variations even though these two sites are approximately 200 km from each other [30]. Therefore, we assume that the distance between these observational sites can be ignored. The area and population of Fukuoka City were approximately 340 km^2 and

1.5 million, respectively. Aerosols observed in Fukuoka originate on the continent, the ocean, and in the local area. This site is therefore well situated for an examination of differences in aerosol surface area and chemical composition between aerosol sources. $PM_{2.5}$ concentrations over this site are generally affected by the long-range transport process from the Asian continent [30,33]. The observation period was from 1 April 2015 to 31 March 2016.

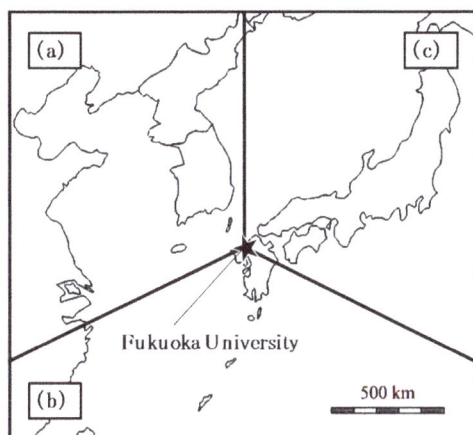

Figure 1. Map of East Asia showing each sector classified by airmass backward trajectories from the monitoring site (Fukuoka University, Japan): (**a**) Asian continent, (**b**) southern part of Kyushu, and (**c**) main islands of Japan.

2.2. Aerosol Surface Area

The surface area concentration was automatically and continuously measured using an NSAM (Model 3550, TSI Inc., Shoreview, MN, USA). The flow rate of the NSAM was 2.5 L·min^{-1}, and the time resolution was set to 10 min. The NSAM has a cyclone with a 50% cut-off of 1 μm at the inlet. The NSAM can measure reliable LDSA of the particles between the ranges of 20 and 400 nm [26]. The procedure for the conversion from LDSA (the actual NSAM reading) to aerosol surface area has already been described elsewhere [18], but is briefly summarized here. The calibration constant is determined with passing monodisperse aerosols simultaneously through the scanning mobility particle sizer (SMPS) and the NSAM by the manufacturer (TSI Inc.). Specifically, the total surface area of the 80-nm NaCl particles determined by the SMPS is multiplied by the lung deposition efficiency of 80-nm particles, which is determined using the lung deposition curve for a reference worker reported by the International Commission on Radiological Protection [34]. In this study, we measured tracheobronchial-deposited surface area values using the NSAM, and then converted them into aerosol surface area by dividing them by the ICRP deposition efficiency of 80-nm particles. In order to check the validity of the calibration constant, we measured the total surface area of polydisperse SiO_2 particles (sicastar, micromod Partikeltechnologie GmbH, Rostock, Germany) using SMPS and NSAM simultaneously. The geometric surface area concentrations were calculated using particle diameters of the spherical particle. The experiments were conducted for several levels of the particle concentrations by changing the mixing volume of dilution air. The results are shown in Figure 2. The surface area measured by NSAM matched very well with that measured by SMPS. Therefore, we applied this calibration constant to the field measurement results obtained in this study.

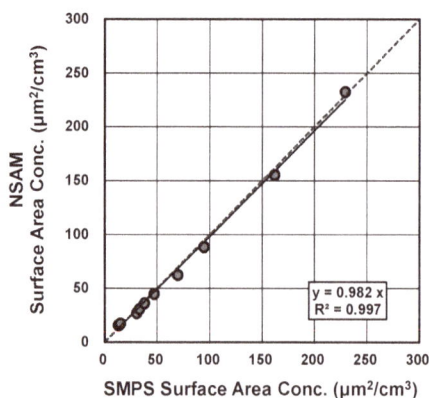

Figure 2. Total surface area of SiO_2 using scanning mobility particle sizer (SMPS) and nanoparticle surface area monitor (NSAM) simultaneously. The geometric surface area concentrations were calculated using particle diameters of the spherical particle.

2.3. Black Carbon, Aerosol Number and Mass, Sulfate Ion Concentration, and Wind Direction and Speed

The mass concentration of BC in $PM_{2.5}$ was automatically and continuously measured using an aethalometer (AE-16U, Magee Scientific Corp., Berkeley, CA, USA) [33]. A Sharp-Cut Cyclone SCC1.829 (BGI Inc., Butler, NJ, USA) was used as the $PM_{2.5}$ inlet. The BC concentration measured using an aethalometer based on the rate of absorption of incident light (880 nm) by BC shows reasonable agreement with elemental carbon (EC) measured using the thermal-optical method [35,36]. The flow rate of the aethalometer was 5.0 L·min^{-1}, and the time resolution was set to 15 min.

Sulfate concentrations were automatically measured and analyzed using a continuous dichotomous aerosol chemical speciation analyzer (ACSA-12, Kimoto Electric Co., Ltd., Osaka, Japan). The ACSA-12 determines the sulfate ion concentration using the $BaSO_4$-based turbidimetric method after the addition of $BaCl_2$ dissolved in polyvinyl pyrrolidone solution, and the results correspond closely with the values determined using the denuder-filter/ion chromatography method [37]. According to the previous study, sulfate concentration values measured by ACSA were approximately 10% higher than those measured by the denuder-filter/ion chromatography [37]; this might be caused by the effect of organosulfates.

The aerosol number concentrations and particle size distributions (optical equivalent diameter: >0.3, >0.5, >1.0, >2.0, >3.0, and >5.0 µm) were measured using an optical particle counter (OPC, TD100; Sigma Tech., Yokohama, Japan), which was installed on the rooftop of the building in Fukuoka University, and operated in ambient conditions without using a heating drier. The size discriminator of the OPC was calibrated using polystyrene latex spheres with a refractive index of 1.59-0i. The OPC data were corrected for coincidence loss. The flow rate of the OPC was 1.0 L·min^{-1}, and the time resolution was set to 1 min.

$PM_{2.5}$ mass concentration, wind direction, and wind speed data were obtained from websites operated by national and local governmental offices [38,39].

The time resolution varied according to the instrument or the sources of the downloaded data. As a result, we used 1-h average values for all further analyses in this study. All times are expressed according to the local time zone (JST:UTC + 9 h).

2.4. Airmass Backward Trajectory Analysis

Airmass backward trajectories were calculated for each day of the measurement campaign using the NOAA HYSPLIT model [40,41]. The trajectories were calculated based on the following conditions:

start latitude and longitude: 33.55° N, 130.36° E; start altitude: 1500 m above sea level; and calculation time: 72 h (three days). The trajectories were calculated every 3 h (0:00 to 21:00 of local time) and were allocated to one of three sectors ((a) Asian continent, (b) southern Kyushu, and (c) main islands of Japan, see Figure 1) if the trajectory positions for every 6-h interval were within the sector at least 36 h (50% of the time) and if the trajectory positions were within other sectors less than 24 h (33% of the time).

3. Results and Discussion

3.1. Diurnal Variations

Table 1 shows the mean values of the concentrations of aerosol surface area, PM$_{2.5}$ mass, BC, sulfate, and particle number measured in this study. The mean concentrations of surface area and other variables measured in this study were not so different from those presented in several previous papers [18,23–25,30].

Table 1. Mean, standard deviation (SD), and number of samples (*n*) of each variable measured in this study in Fukuoka, Japan (April 2015 to March 2016, 1-h average value). BC, black carbon.

Variables	Unit	Mean	SD	*n*
Surface Area	$\mu m^2 \cdot cm^{-3}$	127	62	8032
PM$_{2.5}$ mass	$\mu g \cdot m^{-3}$	17.3	10.7	8064
BC	$ng \cdot m^{-3}$	579	430	7698
SO$_4^{2-}$	$\mu g \cdot m^{-3}$	4.92	3.22	7511
Particle Number (>0.3 μm)	# cm^{-3}	419	469	7281

The diurnal variations of each variable were analyzed to investigate the effect of diurnal human activity on the concentration of aerosol surface area (Figure 3). The error bars indicate the standard deviation of the data and reflect the high variability of day to day measurements. In order to establish statistically significant trends, we performed *t*-tests and found that the concentrations of aerosol surface area and BC in the morning (8:00–10:00) were significantly higher than at other times ($p < 0.01$). Apparently, concentrated automotive traffic near the monitoring site resulted in the morning peak of aerosol surface area and BC [23–25,27,30].

The other variables did not exhibit morning peaks. This finding is supported by a previous study that suggested that the elemental carbon in Fukuoka City originated mainly from local emission sources rather than from long-range transport [30]. Of all the variables, BC was consistently and more strongly correlated with aerosol surface area over the observation site. This fact means that we should pay much more attention to BC when considering the aerosol surface area as a metric of adverse health effects caused by exposure to aerosols.

A detailed investigation of the diurnal variation of the aerosol surface area and BC concentrations are shown in Figure 4. The aerosol surface area and BC concentrations exhibited a clear daily morning peak in winter (October 2015 to March 2016). On the other hand, the morning peak of BC concentration was unclear in summer (April to September). Furthermore, the aerosol surface area concentration in the afternoon was higher than that in the morning in summer. Correlation plots of the aerosol surface area vs. BC concentrations are shown in Figure 5. The correlation between the aerosol surface area and BC concentrations was much stronger in winter than summer. The high-surface area-low-BC type plots were found frequently in summer, but not much in winter. Apparently, there are some reasons that the aerosol surface area becomes high other than the BC concentration in the afternoon in summer. The seasonal variations are discussed in the following sections.

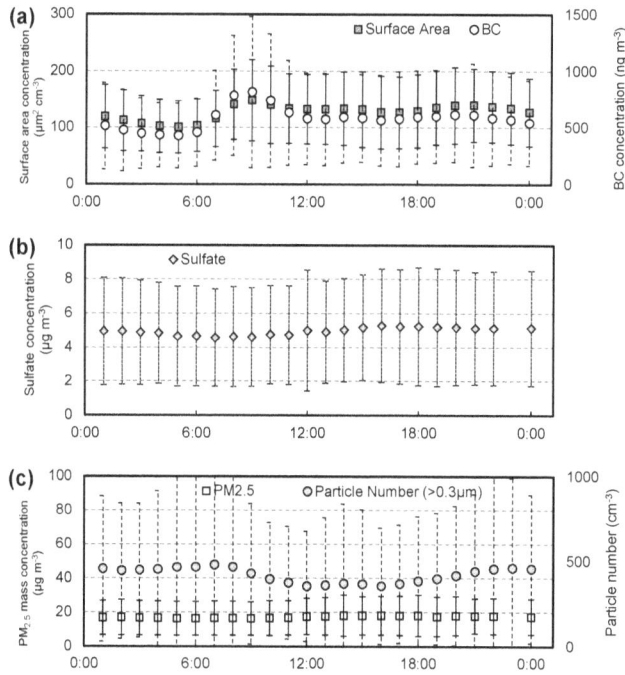

Figure 3. Diurnal variation in (**a**) aerosol surface area and BC concentrations; (**b**) sulfate concentrations; (**c**) PM$_{2.5}$ mass concentration and particle number (>0.3 µm), measured from April 2015 to March 2016 in Fukuoka, Japan.

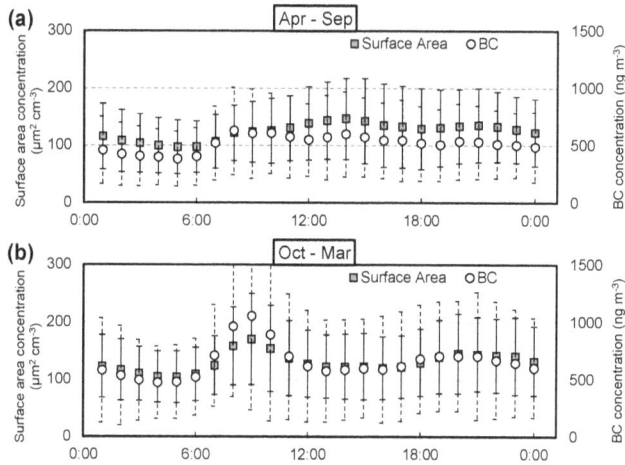

Figure 4. Diurnal variation in the aerosol surface area and BC concentrations in Fukuoka, Japan, measured from (**a**) April to September 2015, and (**b**) October 2015 to March 2016.

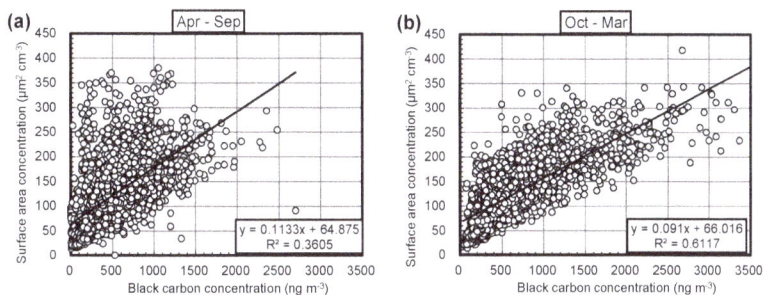

Figure 5. Correlation plots of the aerosol surface area vs. BC concentrations in Fukuoka, Japan, measured from (**a**) April to September 2015, and (**b**) October 2015 to March 2016.

3.2. Monthly Correlations between Aerosol Surface Area and BC Concentrations

The correlation between the aerosol surface area and BC concentrations was analyzed on a monthly basis to investigate this relationship in more detail. Table 2 shows the slopes and intercepts of regression lines, the coefficients of determination, and the number of data in the correlation between aerosol surface area and BC or sulfate concentrations for each month through the year. The correlation between the aerosol surface area and BC concentrations were relatively strong ($R^2 = 0.41-0.76$) for all months except June and July, for which the coefficient of determination was low ($R^2 = 0.23$). This may have been the result of increasing sulfate concentrations [23], due to the fact that the correlation between the correlation between the aerosol surface area and sulfate concentrations was higher in June ($R^2 = 0.33$) than in other months ($R^2 = 0.030-0.23$; Table 2). The correlations between the aerosol surface area and sulfate concentrations were shown in Figure 6, and that in June (Figure 6a) actually shows that the high concentrations of sulfate increased aerosol surface area concentrations. These events did not appear in other months. For June, the number of data points for which the hourly sulfate concentration was ≥ 15 $\mu g \cdot m^{-3}$ and the aerosol surface area concentration was ≥ 150 $\mu m^2 \cdot cm^{-3}$ was extremely large (Figure 7). That is, high sulfate concentrations increased aerosol surface area concentrations, which resulted in a weakening of the correlation between the aerosol surface area and BC concentrations. According to a previous study, high-sulfate-high-surface area events were possibly caused by volcanic SO_2 [23]. However, in this study, SO_2 concentrations did not increase in June at Fukuoka, thus the high sulfate concentrations were likely not due to volcanic emissions. Furthermore, the coefficient of determination in the correlation between the aerosol surface area and sulfate concentrations was not high in July ($R^2 = 0.030$; Figure 6b); on the contrary, it was the lowest of the year (Table 2). We therefore surmise that other chemical compounds, such as organic carbon, may have contributed to the July aerosol surface area concentration in Fukuoka.

The slopes of the regression between the aerosol surface area and BC concentrations, and thus the ratio of the former to the latter, in August and September was larger than in the other months (Table 2). This appears to have been the result of different airmass sources in summer from the rest of the year. The proportion of the airmass originating in continental Asia was smaller in July, August, and September than in the rest of the year, and the proportion coming from the main islands of Japan was greater in August and September (Figure 8). In Japan, meteorological conditions are generally dominated by the Asian monsoon. As a result, the prevailing wind direction was easterly in August and September 2015 in Fukuoka City. From our results, therefore, we can conclude that the BC from the Asian continent did not have a large surface area but had a high mass concentration, whereas, the mass concentrations of BC transported by easterly wind was low, but that BC had a larger surface area. A possible explanation for this is that some chemical compounds were adsorbed by (or "coated") the surface of the BC during its long-range transport [42–45], thereby reducing its surface area. Conversely, the BC from the main islands of Japan was relatively fresh (or "uncoated"), thereby maintaining its larger surface area. For example, China et al. (2015) clearly showed the morphology of soot particles that are thinly and heavily coated [42], and Moffet et al. (2016) reported that all soot particles in urban air were associated with organic carbon, and that soot was frequently at the center of particles as inclusions with thin or thick organic coatings [45]. Apparently this coating would cause the reduction of the surface area of soot particles. This caused the slopes of the regression between the aerosol surface area and BC concentrations for August and September to increase, but did not reduce the coefficient of determination.

Table 2. Linear regressions between black carbon concentration or sulfate concentration and the aerosol surface area concentration for each month from continuous measurements conducted from April 2015 to March 2016.

		Surface Area vs. BC			
Year	Month	Slope $(\mu m^2 \cdot cm^{-3})/(ng \cdot m^{-3})$	Intercept $\mu m^2 \cdot cm^{-3}$	Coefficient of Determination	n
2015	April	0.12	40	0.45	489
	May	0.093	81	0.41	659
	June	0.11	73	0.23	696
	July	0.10	100	0.23	625
	August	0.15	44	0.42	511
	September	0.17	39	0.62	678
	October	0.088	77	0.54	623
	November	0.10	51	0.59	546
	December	0.089	55	0.76	374
2016	January	0.079	64	0.71	535
	February	0.12	57	0.66	713
	March	0.13	46	0.64	632
		Surface Area vs. SO_4^{2-}			
Year	Month	Slope $(\mu m^2 \cdot cm^{-3})/(ng \cdot m^{-3})$	Intercept $\mu m^2 \cdot cm^{-3}$	Coefficient of Determination	n
2015	April	0.0072	72	0.21	508
	May	0.0068	99	0.10	651
	June	0.0084	70	0.33	676
	July	0.0034	112	0.030	570
	August	0.0056	87	0.15	461
	September	0.014	65	0.22	641
	October	0.0073	110	0.15	642
	November	0.017	79	0.12	428
	December	0.013	78	0.18	343
2016	January	0.012	68	0.23	679
	February	0.0066	100	0.084	655
	March	0.0084	80	0.19	683

(a)

(b)

(c)

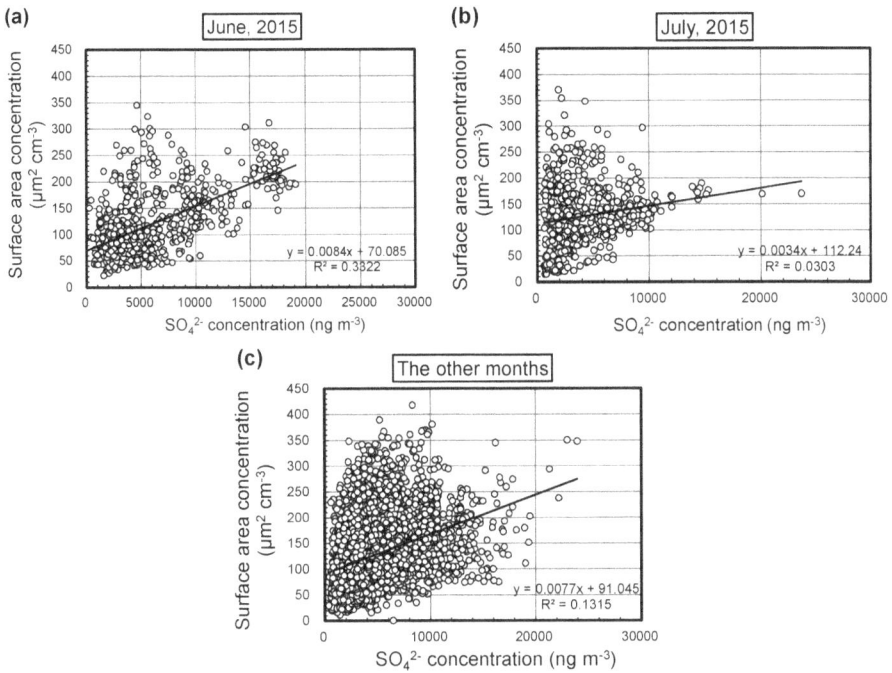

Figure 6. Comparison between the aerosol surface area and sulfate concentrations recorded in continuous measurements for (**a**) June, (**b**) July, and (**c**) the other months of the study.

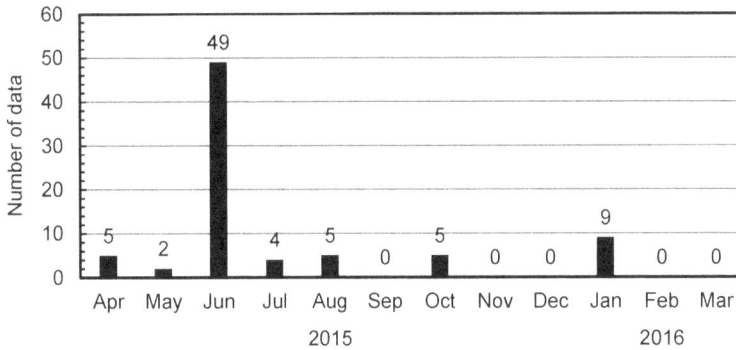

Figure 7. The monthly number of data points for which the sulfate concentration was ≥ 15 $\mu g \cdot m^{-3}$ and the surface area concentration was ≥ 150 $\mu m^2 \cdot cm^{-3}$ in Fukuoka.

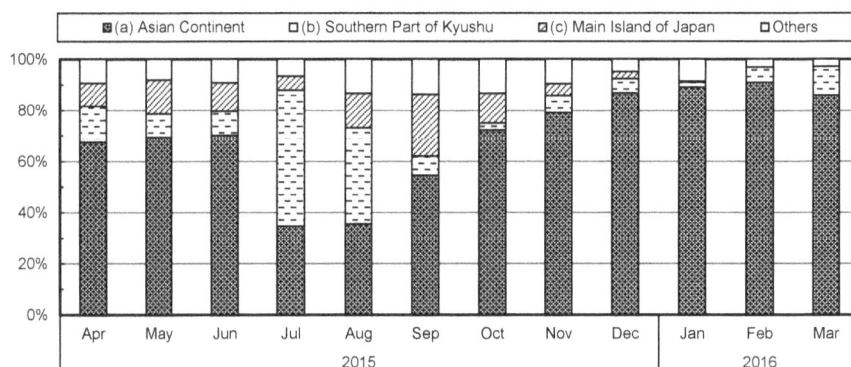

Figure 8. Monthly airmass backward trajectory analysis, starting at Fukuoka, from April 2015 to March 2016. The sector classification is shown in Figure 1.

3.3. Effect of Land and Sea Breeze on the Correlation between Aerosol Surface Area and Black Carbon Concentrations

As we have seen, the correlation between the aerosol surface area and BC concentrations varied from month to month as a result of sulfate input or coating of BC surfaces. In this section, we discuss how this correlation changed with time of day. We divided the day into four segments: (a) night (0:00–6:00), (b) morning (6:00–12:00), (c) afternoon (12:00–18:00), and (d) evening (18:00–24:00). Table 3 shows the correlation between the aerosol surface area and BC concentrations in the four time segments. The coefficients of determination for the afternoon (R^2 = 0.07–0.71) in particular were lower than that for the other times (night: 0.41–0.87, morning: 0.09–0.84, and evening: 0.29–0.81). This diurnal variation can be related to wind direction and speed. Fukuoka lies on the coast of the Sea of Japan, and is thus often subject to the land and sea breeze. Figure 9 shows a diurnal variation in wind speed and wind direction in Fukuoka from April 2015 to March 2016. In Fukuoka, the direction of sea breezes is northerly ($0°$) and that of land breezes is southerly ($180°$). We generally recorded a sea breeze during the afternoon (12:00 to 18:00), which was strongest from 13:00 to 16:00 (~4 m·s^{-1}), and a gentler land breeze (~2 m·s^{-1}) at night and the early morning. This may have caused that BC concentrations to decrease in the afternoon due to the inflow of a clean airmass via the sea breeze. The nocturnal reversal of the wind direction back to the land breeze would then prevent any further decrease in BC concentrations. Figure 10 shows the BC concentrations corresponding to the wind direction and wind speed. These plots clearly show that stronger northerly winds make the BC concentration low. In addition, sulfate concentrations were higher in the afternoon (12:00–18:00) than at other times (*t*-test, $p < 0.01$; Figure 3b). Generally, sulfate is formed by the oxidation of SO_2, and this reaction is promoted by solar radiation (i.e., during the daytime). The additional sulfate particles formed would thus contribute to the aerosol surface area concentration. This explains the lower coefficients of determination between the aerosol surface area and BC concentrations in the afternoon and the reversal of this pattern in the evening (Table 3).

Table 3. Linear regressions between the aerosol surface area and black carbon (BC) concentrations in the four time segments: (**a**) night (0:00–6:00), (**b**) morning (6:00–12:00), (**c**) afternoon (12:00–18:00), and (**d**) evening (18:00–24:00). The measurements were conducted from April 2015 to March 2016.

		(a) Night (0:00–6:00)				(b) Morning (6:00–12:00)			
Year	Month	Slope $(\mu m^2 \cdot cm^{-3})/(ng \cdot m^{-3})$	Intercept $\mu m^2 \cdot cm^{-3}$	Coefficient of Determination	n	Slope $(\mu m^2 \cdot cm^{-3})/(ng \cdot m^{-3})$	Intercept $\mu m^2 \cdot cm^{-3}$	Coefficient of Determination	n
2015	Apr	0.12	30	0.87	119	0.068	70	0.17	112
	May	0.12	62	0.47	172	0.070	83	0.42	160
	Jun	0.17	52	0.41	174	0.095	72	0.27	173
	Jul	0.15	36	0.63	163	0.051	102	0.09	149
	Aug	0.20	24	0.70	136	0.094	68	0.25	124
	Sep	0.19	28	0.70	164	0.12	55	0.48	160
	Oct	0.10	69	0.60	161	0.085	74	0.70	153
	Nov	0.13	40	0.63	138	0.088	51	0.74	135
	Dec	0.091	45	0.78	92	0.085	47	0.84	89
2016	Jan	0.080	51	0.68	132	0.083	65	0.76	132
	Feb	0.11	59	0.64	174	0.12	60	0.74	169
	Mar	0.13	47	0.50	156	0.11	57	0.61	156

		(c) Afternoon (12:00–18:00)				(d) Evening (18:00–24:00)			
Year	Month	Slope $(\mu m^2 \cdot cm^{-3})/(ng \cdot m^{-3})$	Intercept $\mu m^2 \cdot cm^{-3}$	Coefficient of Determination	n	Slope $(\mu m^2 \cdot cm^{-3})/(ng \cdot m^{-3})$	Intercept $\mu m^2 \cdot cm^{-3}$	Coefficient of Determination	n
2015	Apr	0.12	50	0.31	128	0.16	21	0.68	130
	May	0.084	95	0.30	160	0.12	74	0.62	167
	Jun	0.072	99	0.13	178	0.15	78	0.29	171
	Jul	0.060	119	0.07	155	0.14	53	0.45	157
	Aug	0.15	48	0.34	122	0.22	13	0.59	123
	Sep	0.18	34	0.62	174	0.21	27	0.71	168
	Oct	0.087	79	0.32	151	0.091	82	0.53	158
	Nov	0.096	61	0.36	140	0.13	43	0.72	133
	Dec	0.089	56	0.71	95	0.10	64	0.81	98
2016	Jan	0.063	73	0.63	133	0.083	73	0.71	138
	Feb	0.099	61	0.49	172	0.15	49	0.69	174
	Mar	0.12	46	0.56	159	0.17	29	0.81	161

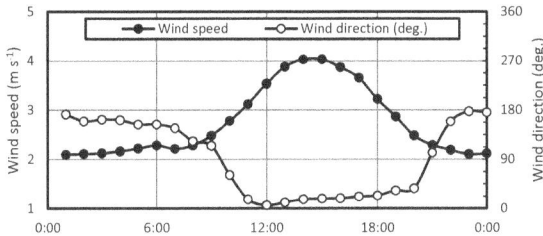

Figure 9. Diurnal variation in wind speed and wind direction in Fukuoka, measured from April 2015 to March 2016. Wind directions are as follows: $0°$: Northerly, $90°$; Easterly, $180°$; Southerly; and $270°$: Westerly.

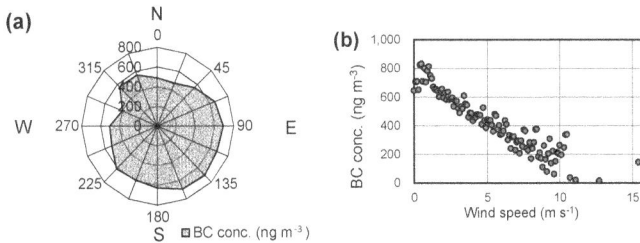

Figure 10. BC concentrations corresponding to (**a**) the wind direction, and (**b**) the wind speed, in Fukuoka, measured from April 2015 to March 2016.

Atmosphere **2017**, *8*, 114

4. Conclusions

We measured the aerosol surface area, black carbon (BC), and sulfate concentrations for one year in Fukuoka, Japan, and investigated the monthly and diurnal variations in the correlation between the aerosol surface area and BC concentrations. Throughout the year, the aerosol surface area concentration was strongly correlated with BC concentration. In June 2015, the coefficient of determination for this correlation was lower than in other months, which was evidently due to high sulfate concentrations. In August and September 2015, the slopes of the regression between the aerosol surface area and BC concentrations was highest. This appears to have been the result of an increase in the proportion of the airmass that originated on the main islands of Japan. This may enhance the introduction of the BC from the main islands of Japan which is relatively fresh (or "uncoated"), thereby maintaining its larger surface area. In addition, the correlation between the aerosol surface area and BC concentrations was weakest in the afternoon, and this could be because certain secondary formed aerosols increase. This may also be because Fukuoka is generally dominated by the land and sea breeze, and BC concentrations decrease under those conditions due to the afternoon inflow of a clean airmass.

Acknowledgments: This study was partly supported by JSPS/MEXT KAKENHI Grant Numbers 26340010, JP25220101, 17H01864 and 17H04480, Ministry of Environment, Environment Research and Technology Development Fund, Japan (Grant Nos. 5-1452 and 5-1651), Grant for Environmental Research Projects from The Sumitomo Foundation, Steel Foundation for Environmental Protection Technology, and Keio Gijuku Academic Development Funds.

Author Contributions: T.O. conceived and designed the experiments, and wrote the paper; M.K. analyzed the data and wrote the draft of the paper, H.Y. analyzed the data; K.H. (Keio University), K.F., and K.I. installed and operated the NSAM, N.K. performed the BC measurements; I.U. and S.Y. installed and operated the ACSA; A.Y., A.T., C.N., K.H. (Fukuoka University), and M.H. administrated the entire experiments. All authors contributed to revising the manuscript.

Conflicts of Interest: The authors declare no conflict of interest. The founding sponsors had no role in the design of the study; in the collection, analyses, or interpretation of data; in the writing of the manuscript, and in the decision to publish the results.

References

1. Dockery, D.W.; Pope, C.A., III; Xu, X.; Spengler, J.D.; Ware, J.H.; Fay, M.E.; Ferris, B.G., Jr.; Speizer, F.E. An association between air pollution and mortality in six U.S. cities. *N. Engl. J. Med.* **1993**, *329*, 1753–1759. [CrossRef] [PubMed]

2. Pope, C.A., III; Thun, M.J.; Namboodiri, M.M.; Dockery, D.W.; Evans, J.S.; Speizer, F.E.; Heath, C.W., Jr. Particulate air pollution as a predictor of mortality in a prospective study of U.S. adults. *Am. J. Respir. Crit. Care Med.* **1995**, *151*, 669–674. [CrossRef] [PubMed]

3. International Agency for Research on Cancer. *Air Pollution and Cancer*; IARC Scientific Publication: Lyon, France, 2013; No. 161.

4. Ferin, J.; Oberdörster, G.; Penney, D.P. Pulmonary retention of ultrafine and fine particles in rats. *Am. J. Respir. Cell Mol. Biol.* **1992**, *6*, 535–542. [CrossRef] [PubMed]

5. Oberdörster, G.; Gelein, R.M.; Ferin, J.; Weiss, B. Association of particulate air pollution and acute mortality: Involvement of ultrafine particles? *Inhal. Toxicol.* **1995**, *7*, 111–124. [CrossRef] [PubMed]

6. Donaldson, K.; Li, X.Y.; MacNee, W. Ultrafine (nanometre) particle mediated lung injury. *J. Aerosol Sci.* **1998**, *29*, 553–560. [CrossRef]

7. The National Institute for Occupational Safety and Health (NIOSH). *The National Institute for Occupational Safety and Health (NIOSH): Approaches to Safe Nanotechnology*; DHHS (NIOSH) Publication No. 2009-125; The National Institute for Occupational Safety and Health (NIOSH): Washington, DC, USA, 2009.

8. Oberdörster, G.; Finkelstein, J.N.; Johnston, C.; Gelein, R.; Cox, C.; Baggs, R.; Elder, A.C.P. Acute pulmonary effects of ultrafine particles in rats and mice. *Res. Rep. Health Effects Inst.* **2000**, *96*, 5–74.

9. Oberdörster, G.; Oberdörster, E.; Oberdörster, J. Nanotoxicology: An emerging discipline evolving from studies of ultrafine particles. *Environ. Health Perspect.* **2005**, *113*, 823–839. [CrossRef] [PubMed]

10. Nakanishi, J. *Risk Assessment of Manufactured Nanomaterials: Carbon Nanotubes (CNT)*; Final Report Issued on 12 August 2011, Executive Summary; NEDO Project "Research and Development of Nanoparticle Characterization Methods." (P06041); New Energy and Industrial Technology Development Organization: Kawasaki, Japan, 2011.

11. Schmid, O.; Stoeger, T. Surface area is the biologically most effective dose metric for acute nanoparticle toxicity in the lung. *J. Aerosol Sci.* **2016**, *99*, 133–143. [CrossRef]

12. Oberdörster, G. Pulmonary effects of inhaled ultrafine particles. *Int. Arch. Occup. Environ. Health* **2001**, *74*, 1–8. [CrossRef] [PubMed]

13. Giechaskiel, B.; Alföldy, B.; Drossinos, Y. A metric for health effects studies of diesel exhaust particles. *J. Aerosol Sci.* **2009**, *40*, 639–651. [CrossRef]

14. Skuland, T.; Øvrevik, J.; Låg, M.; Refsnes, M. Role of size and surface area for pro-inflammatory responses to silica nanoparticles in epithelial lung cells: Importance of exposure conditions. *Toxicol. In Vitro* **2014**, *28*, 146–155. [CrossRef] [PubMed]

15. Brunauer, S.; Emmett, P.H.; Teller, E. Adsorption of gases in multimolecular layers. *J. Am. Chem. Soc.* **1938**, *60*, 309–319. [CrossRef]

16. Okuda, T. Measurement of the specific surface area and particle size distribution of atmospheric aerosol reference materials. *Atmos. Environ.* **2013**, *75*, 1–5. [CrossRef]

17. Okuda, T.; Isobe, R.; Nagai, Y.; Okahisa, S.; Funato, K.; Inoue, K. Development of a high-volume $PM_{2.5}$ particle sampler using impactor and cyclone techniques. *Aerosol Air Qual. Res.* **2015**, *15*, 759–767. [CrossRef]

18. Hatoya, K.; Okuda, T.; Funato, K.; Inoue, K. On-line measurement of the surface area concentration of aerosols in Yokohama, Japan, using the diffusion charging method. *Asian J. Atmos. Environ.* **2016**, *10*, 1–12. [CrossRef]

19. Jung, H.; Kittelson, D.B. Characterization of aerosol surface instruments in transition regime. *Aerosol Sci. Technol.* **2005**, *39*, 902–911. [CrossRef]

20. Fissan, H.; Neumann, S.; Trampe, A.; Pui, D.Y.H.; Shin, W.G. Rationale and principle of an instrument measuring lung deposited nanoparticle surface area. *J. Nanopart. Res.* **2007**, *9*, 53–59. [CrossRef]

21. Shin, W.G.; Pui, D.Y.H.; Fissan, H.; Neumann, S.; Trampe, A. Calibration and numerical simulation of Nanoparticle Surface Area Monitor (TSI Model 3550 NSAM). *J. Nanopart. Res.* **2007**, *9*, 61–69. [CrossRef]

22. Heitbrink, W.A.; Evans, D.E.; Ku, B.K.; Maynard, A.D.; Slavin, T.J.; Peters, T.M. Relationships among particle number, surface area, and respirable mass concentrations in automotive engine manufacturing. *J. Occup. Environ. Hyg.* **2009**, *6*, 19–31. [CrossRef] [PubMed]

23. Okuda, T.; Yamazaki, H.; Hatoya, K.; Kaneyasu, N.; Yoshino, A.; Takami, A.; Funato, K.; Inoue, K.; Nishita, C.; Hara, K.; et al. Factors Controlling the Variation of Aerosol Surface Area Concentrations Measured by a Diffusion Charger in Fukuoka, Japan. *Atmosphere* **2016**, *7*, 33. [CrossRef]

24. Velasco, E.; Siegmann, P.; Siegmann, H.C. Exploratory study of particle-bound polycyclic aromatic hydrocarbons in different environments of Mexico City. *Atmos. Environ.* **2004**, *38*, 4957–4968. [CrossRef]

25. Ntziachristos, L.; Polidori, A.; Phuleria, H.; Geller, M.D.; Sioutas, C. Application of a diffusion charger for the measurement of particle surface concentration in different environments. *Aerosol Sci. Technol.* **2007**, *41*, 571–580. [CrossRef]

26. Asbach, C.; Fissan, H.; Stahlmecke, B.; Kuhlbusch, T.A.J.; Pui, D.Y.H. Conceptual limitations and extensions of lung-deposited Nanoparticle Surface Area Monitor (NSAM). *J. Nanopart. Res.* **2009**, *11*, 101–109. [CrossRef]

27. Albuquerque, P.C.; Gomes, J.F.; Bordado, J.C. Assessment of exposure to airborne ultrafine particles in the urban environment of Lisbon, Portugal. *J. Air Waste Manag. Assoc.* **2012**, *62*, 373–380. [CrossRef] [PubMed]

28. Gomes, J.F.P.; Albuquerque, P.C.S.; Esteves, H.M.D.S.; Carvalho, P.A. Notice on a methodology for characterizing emissions of ultrafine particles/nanoparticles in microenvironments. *Energy Emiss. Cont. Technol.* **2013**, *1*, 15–27. [CrossRef]

29. China, S.; Salvadori, N.; Mazzoleni, C. Effect of traffic and driving characteristics on morphology of atmospheric soot particles at freeway on-ramps. *Environ. Sci. Technol.* **2014**, *48*, 3128–3135. [CrossRef] [PubMed]

30. Kaneyasu, N.; Yamamoto, S.; Sato, K.; Takami, A.; Hayashi, M.; Hara, K.; Kawamoto, K.; Okuda, T.; Hatakeyama, S. Impact of long-range transport of aerosols on the $PM_{2.5}$ composition at a major metropolitan area in the northern Kyushu area of Japan. *Atmos. Environ.* **2014**, *97*, 416–425. [CrossRef]

31. Takami, A.; Miyoshi, T.; Irei, S.; Yoshino, A.; Sato, K.; Shimizu, A.; Hayashi, M.; Hara, K.; Kaneyasu, N.; Hatakeyama, S. Analysis of organic aerosol in Fukuoka, Japan using a PMF method. *Aerosol Air Qual. Res.* **2016**, *16*, 314–322. [CrossRef]

32. Yoshino, A.; Takami, A.; Sato, K.; Shimizu, A.; Kaneyasu, N.; Hatakeyama, S.; Hara, K.; Hayashi, M. Influence of Trans-Boundary Air Pollution on the Urban Atmosphere in Fukuoka, Japan. *Atmosphere* **2016**, *7*, 51. [CrossRef]

33. Kaneyasu, N.; Takami, A.; Sato, K.; Hatakeyama, S.; Hara, S.; Kawamoto, K.; Yamamoto, S. Year-round behavior of $PM_{2.5}$ in a remote island and urban site in the northern Kyushu area, Japan. *J. Jpn. Soc. Atmos. Environ.* **2011**, *46*, 111–118. (In Japanese)

34. International Commission on Radiological Protection (ICRP). Human respiratory tract model for radiological protection. *Ann. ICRP* **1994**, *24*, 1–482.

35. Venkatachari, P.; Zhou, L.; Hopke, P.K.; Schwab, J.J.; Demerjian, K.L.; Weimer, S.; Hogrefe, O.; Felton, D.; Rattigan, O. An intercomparison of measurement methods for carbonaceous aerosol in the ambient air in New York City. *Aerosol Sci. Technol.* **2006**, *40*, 788–795. [CrossRef]

36. Ng, I.P.; Ma, H.; Kittelson, D.B.; Miller, A.L. *Comparing Measurements of Carbon in Diesel Exhaust Aerosols Using the Aethalometer, NIOSH Method 5040, and SMPS*; SAE Technical Paper Series 2007-01-0334; University of Minnesota: Minneapolis, MN, USA, 2007.

37. Osada, K.; Kamiguchi, Y.; Yamamoto, S.; Kuwahara, S.; Pan, X.; Hara, Y.; Uno, I. Comparison of ionic concentrations on size-segregated atmospheric aerosol particles based on a denuder-filter method and a Continuous Dichotomous Aerosol Chemical Speciation Analyzer (ACSA-12). *Earozoru Kenkyu* **2016**, *31*, 203–209. (In Japanese)

38. Fukuoka Prefecture Website. Available online: http://www.fihes.pref.fukuoka.jp/taiki-new/Nipo/OyWbNpKm0151.htm (accessed on 6 May 2017).

39. Japan Meteorological Agency Website. Available online: http://www.data.jma.go.jp/obd/stats/etrn/index.php (accessed on 6 May 2017).

40. Stein, A.F.; Draxler, R.R.; Rolph, G.D.; Stunder, B.J.B.; Cohen, M.D.; Ngan, F. NOAA's HYSPLIT atmospheric transport and dispersion modeling system. *Bull. Am. Meteor. Soc.* **2015**, *96*, 2059–2077. [CrossRef]

41. Rolph, G.D. Real-Time Environmental Applications and Display sYstem (READY) Website. NOAA Air Resources Laboratory: Silver Spring, MD, USA, 2017. Available online: http://ready.arl.noaa.gov/HYSPLIT.php (accessed on 26 June 2017).

42. China, S.; Scarnato, B.; Owen, R.C.; Zhang, B.; Ampadu, M.T.; Kumar, S.; Dzepina, K.; Dziobak, M.P.; Fialho, P.; Perlinger, J.A.; et al. Morphology and mixing state of aged soot particles at a remote marine free troposphere site: Implications for optical properties. *Geophys. Res. Lett.* **2015**, *42*, 1243–1250. [CrossRef]

43. Shiraiwa, M.; Kondo, Y.; Moteki, N.; Takegawa, N.; Sahu, L.K.; Takami, A.; Hatakeyama, S.; Yonemura, S.; Blake, D.R. Radiative impact of mixing state of black carbon aerosol in Asian outflow. *J. Geophys. Res.* **2008**, *113*, D24210. [CrossRef]

44. Takami, A.; Mayama, N.; Sakamoto, T.; Ohishi, K.; Irei, S.; Yoshino, A.; Hatakeyama, S.; Murano, K.; Sadanaga, Y.; Bandow, H.; et al. Structural analysis of aerosol particles by microscopic observation using a time-of-flight secondary ion mass spectrometer. *J. Geophys. Res. Atmos.* **2013**, *118*, 6726–6737. [CrossRef]

45. Moffet, R.C.; O'Brien, R.E.; Alpert, P.A.; Kelly, S.T.; Pham, D.Q.; Gilles, M.K.; Knopf, D.A.; Laskin, A. Morphology and mixing of black carbon particles collected in central California during the CARES field study. *Atmos. Chem. Phys.* **2016**, *16*, 14515–14525. [CrossRef]

atmosphere

MDPI

Article

Influence of Common Assumptions Regarding Aerosol Composition and Mixing State on Predicted CCN Concentration

Manasi Mahish [1], Anne Jefferson [2,3] and Don R. Collins [1,*]

[1] Department of Atmospheric Sciences, Texas A&M University, College Station, TX 77843, USA; manasi.mahish@gmail.com

[2] Cooperative Institute for Research in Environmental Science (CIRES), University of Colorado, Boulder, CO 80309, USA; anne.jefferson@noaa.gov

[3] NOAA Earth System Research Laboratory, Boulder, CO 80305, USA

[*] Correspondence: dcollins@tamu.edu; Tel.: +1-979-862-4401

Received: 26 October 2017; Accepted: 6 February 2018; Published: 8 February 2018

Abstract: A 4-year record of aerosol size and hygroscopic growth factor distributions measured at the Department of Energy's Southern Great Plains (SGP) site in Oklahoma, U.S. were used to estimate supersaturation (S)-dependent cloud condensation nuclei concentrations (N_{CCN}). Baseline or reference $N_{CCN}(S)$ spectra were estimated using κ-Köhler Theory without any averaging of the measured distributions by creating matrices of size- and hygroscopicity-dependent number concentration (N) and then integrating for S > critical supersaturation (S_c) calculated for the same size and hygroscopicity pairs. Those estimates were first compared with directly measured N_{CCN} at the same site. Subsequently, N_{CCN} was calculated using the same dataset but with an array of simplified treatments in which the aerosol was assumed to be either an internal or an external mixture and the hygroscopicity either assumed or based on averages derived from the growth factor distributions. The CCN spectra calculated using the simplified treatments were compared with those calculated using the baseline approach to evaluate the error introduced with commonly used approximations.

Keywords: aerosol; CCN; hygroscopicity; mixing state

1. Introduction

Indirect forcing by aerosols involves complex interactions between the aerosol and clouds that affect the droplet number concentration, albedo, precipitation efficiency, and lifetime of clouds and the chemical processing and scavenging of aerosol particles. Mediating this interaction are cloud condensation nuclei (CCN), which are those particles that activate to form droplets at a given supersaturation (S), known as the critical supersaturation (S_c). In part because of the high temporal and spatial variability of aerosol and clouds, climate model predictions of indirect forcing have a high uncertainty. Estimates of CCN concentration (N_{CCN}) using detailed size-dependent aerosol chemical composition are computationally intensive and simplifications used to minimize computation introduce additional uncertainty.

Methods of determining aerosol activation have evolved to consider complex aerosol composition and structure. Particle S_c can be determined for inorganic species using Köhler Theory, provided the physico-chemical properties of the solutes are known [1]. However, atmospheric aerosols frequently contain a significant amount of organic material as well [2,3]. Compared with soluble inorganic particles, those composed entirely of organic species have higher S_c and are often comparatively inefficient in droplet formation [4,5]. But the solubility and surface tension-reducing properties of the organic component can sometimes have significant influence on S_c [6–13]. Extended Köhler Theory can predict CCN concentration of a multi-component aerosol reasonably well from a description of its size distribution and chemical composition [9,10,14–16]. Introduction of a single hygroscopicity parameter by Petters and Kreidenweis [17] has simplified description and comparison of hygroscopicity and CCN activity for particles composed of single or multiple inorganic and organic species [18].

CCN concentration is also dependent upon aerosol mixing state. Previous closure studies have shown that assumption of an internal mixture generally results in an overestimate of the CCN concentration and assumption of an external mixture in an underestimate [19–34]. However, assumption of either mixing state leads to reasonable results for aged aerosols [35]. While inclusion of mixing state [36–40] and chemical composition [37,38,41–44] can increase the accuracy with which CCN concentration can be estimated, both can be highly variable with time and with particle size, and are often unavailable with current measurement techniques and not easily incorporated into aerosol descriptions used in models. Moreover, chemical composition and mixing state are greatly simplified in large scale models, with simple categorizations such as inorganic/organic and internal/external mixture. Supporting such treatment, several studies have shown that N_{CCN} is most sensitive to the aerosol size distribution [45,46], and the assumption of internal mixing has resulted in fairly accurate predictions [42,47–52]. Ervens et al. [35,53] reported that description of the mixing state is relatively more important than that of the size-resolved chemical composition. Other studies suggest detailed information of size distribution, chemical composition, and mixing state is important for achieving closure among aerosol and CCN measurements [38,43,44,54–58].

Here we use size-resolved aerosol concentration and subsaturated hygroscopicity measurements to estimate CCN concentration using an array of assumptions for composition and mixing state. The data were collected by the U.S. Department of Energy's (DOE) Atmospheric Radiation Measurement (ARM) program at the Southern Great Plains (SGP) site in rural Oklahoma. The study evaluates N_{CCN} estimation methods of varying degrees of complexity with respect to size, chemical composition, and mixing state. Baseline CCN spectra (N_{CCN} vs. S) were first derived by using (i) each combined set of size and hygroscopicity distributions and (ii) κ-Köhler Theory to create pairs of matrices describing (i) N and (ii) S_c, respectively, as a function of particle dry diameter, D_d, and hygroscopic growth factor, GF. The $N_{CCN}(S)$ for each measurement set was calculated by integrating the N matrix over all elements for which $S > S_c$. Those results were compared with direct measurements of N_{CCN} made at the site to confirm that the calculated spectra were reasonably accurate. The objective was not to perform a rigorous closure analysis with consideration of measurement uncertainties but rather to simply evaluate the consistency of the datasets. Spectra calculated using a number of alternate approaches and assumptions were then compared to those from the baseline approach and the results were used to consider the scatter and bias introduced with simplifications commonly employed in large scale models.

2. Experiments

2.1. Site Description and Measurements

The data were recorded at the Southern Great Plains (SGP) central facility (CF1) (36°36'18.0" N, 97°29'6.0" W), located in a mixed land use area of cattle pastures and agricultural fields (mainly wheat, hay and corn) near Lamont, OK, USA. The climate at the site is continental with hot and humid summers and cool winters. The site is impacted by air masses originating from several regions, with accompanying diversity in aerosol concentration and properties. The chemical composition of the aerosol found at the site is complex and highly variable with time (evident in Figures S1 and S2) and with particle size. The size dependence of aerosol composition as reflected in that of GF was described by Mahish and Collins [59]. Table 1 lists the routine aerosol measurements at the site that were used for the analysis presented here. All datasets used for this analysis are available for download from the ARM archive. Data from the scanning mobility particle sizer (SMPS)/hygroscopicity tandem differential mobility analyzer (HTDMA) system were used for most of the analyses described here.

That instrument sequentially measures a size distribution and then a set of hygroscopic growth factor distributions at 7 dry particle sizes every ~45 min. Nafion tube bundles are used to dry the sample flow prior to entering the upstream DMA and to humidify it to 90% RH between the two DMAs when measuring hygroscopicity. The sheath and excess flows are recirculated using a variable speed blower. The sample flow rate is controlled by varying the amount of dilution flow that is introduced upstream of the 3.0 L min^{-1} TSI 3762 condensation particle counter (CPC). Details of the SMPS/HTDMA system and processing of the data it generates are available in the Tandem Differential Mobility Analyzer/Aerodynamic Particle Sizer handbook posted on the ARM archive site [60] and in the work of Gasparini et al. [61]. As noted above, the calculated CCN spectra were compared with direct measurements at the site made with a DMT CCN-100 counter (CCNc) [62]. The CCNc cycles through 7 fixed supersaturations every half hour. A description of the instrument and data processing is available in the Aerosol Observing System (AOS) handbook [63]. Non-refractory chemical composition and black carbon content measured with an Aerosol Chemical Speciation Monitor [64] and a Particle Soot Absorption Photometer [65] were used in the selection of a representative hygroscopicity of the organic component for use in some of the calculations. Details of those instruments are available in the ACSM [66] and AOS [63] handbooks.

Table 1. List of instruments, measured quantities, manufacturer, and year installed.

Instrument	Measurement	Manufacturer/Model	Installation Date
Scanning Mobility Particle Sizer (SMPS; part of the "TDMA" system)	Size distribution from 0.012 to 0.74 µm dry diameter (D_d)	Fabricated, Texas A&M University, TX, U.S.	2005
Hygroscopic Tandem Differential Mobility Analyzer (HTDMA; part of the "TDMA" system)	Hygroscopic growth factor distributions of 0.013, 0.025, 0.05, 0.1, 0.2, 0.4, and 0.6µm D_d particles at 90% RH	Fabricated, Texas A&M University, TX, U.S.	2005
Cloud Condensation Nuclei counter (CCNc)	CCN concentration at a fixed set of supersaturations	CCN-100 Droplet Measurement Technologies, Longmont, CO, U.S.	2009
Condensation Particle Counter (CPC)	Concentration of $D_d > 0.01$ µm particles	Model 3010 TSI, Inc., Shoreview, MN, U.S.	1996
Aerosol Chemical Speciation Monitor (ACSM)	Sub 1-µm chemical composition (organics, sulfate, nitrate, ammonium, and chloride)	Aerodyne Research, Inc., Billerica, MA, U.S.	2010
Particle Soot Absorption Photometer (PSAP)	Sub 1-µm absorption coefficient (used to infer black carbon concentration)	Radiance Research, Seattle, WA, U.S.	1992

2.2. Screening and Time Interval Selection

Data from each instrument were validated separately and periods having erroneous data or no data were excluded from analysis. Data from time periods during which instrument problems or failure was evident or when one of the following occurred were not used:

a. The total particle concentration (N_{CN}) calculated by integrating the SMPS size distribution differed significantly from that directly measured with the CPC

b. The N_{CCN} measured with the CCNc exceeded the N_{CN} measured by the CPC, possibly due to malfunction of the CCNc

c. The sample flow entering the upstream (1st) DMA had an RH > 30%, or

d. The sample flow entering the downstream (2nd) DMA had an RH < 85%.

The categorized data quality during the period of analysis is shown in Figure 1.

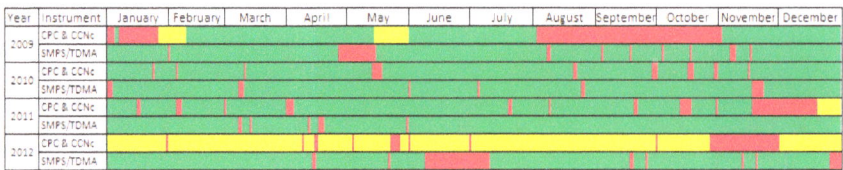

Figure 1. Data quality during the analysis period. Periods during which data are available and no significant problems were identified are colored green, those during which confidence in at least some subset of the data is low are yellow, and those during which data are unavailable or thought to be erroneous are red. Data used to generate this graph are available at the site identified in the Acknowledgments section below.

2.3. Use of K-Köhler Theory

For each of the approaches used to estimate N_{CCN}, the calculation of S_c from measured particle size and measured or assumed hygroscopicity employed κ-Köhler Theory [17]. The κ values of aerosol constituents were assumed to be the same in the aqueous particles as measured in the HTDMA at 90% RH and in the more dilute solutions at the point of activation to form cloud droplets. This has the same effect as assuming that the water activity coefficient in the solutions is constant and that all soluble species are fully dissolved at the RH in the HTMDA. A more accurate treatment for which activity coefficients or κ are calculated as a function of RH (and solution concentration) is not justified because size-resolved composition measurements are not available and the bulk submicron measurements made with the ACSM show the aerosol composition is complex and varies considerably during the year. Specifically, there is a strong seasonality in the soluble inorganic content, with sulfate dominant from roughly April through October and nitrate dominant from November through March, as is evident in the sulfate:nitrate ratio shown in Figure 2. Thus, the choice of soluble inorganic component(s) needed to model the extent and effect of variation in κ between 90% RH and at S_c would vary by month over the 4-year period of this analysis, as well as over shorter periods of days or even hours accompanying changes in the origin and processing of the sampled aerosol. Furthermore, any attempt to estimate RH-dependent changes in κ would require consideration of the influence of the significant organic content at the site. As shown in Table 2, averaged throughout the year organics contribute over 50% to the total submicron mass concentration.

Table 2. Mass concentration fraction from ACSM measurements at SGP.

Year	Mass Concentration Fraction (%)		
	Total Organics	Ammonium Sulfate	Ammonium Nitrate
2011	57	17	26
2012	56	18	26
2013	56	24	20

For all CCN concentration estimate approaches for which the HTDMA data were used, GF was related to κ using the following expression from Petters and Kreidenweis [17].

$$\kappa = \left([GF(RH)]^3 - 1 \right) \frac{\left(\exp\left(\frac{A}{GF(RH) \cdot D_d} \right) - RH \right)}{RH}$$
$$A = \frac{4 \cdot \sigma_{sol} \cdot M_w}{R \cdot T \cdot \rho_w}$$

(1)

where M_w and ρ_w are the molecular weight and liquid density of water, respectively. The solution surface tension, σ_{sol}, was assumed to be that of pure water, 0.072 J m^{-2}. Because it is assumed that κ values were the same for the more concentrated and the more dilute aqueous solutions in the HTDMA and at activation, respectively, it is not necessary to know the contributions of different aerosol components to the overall hygroscopicity for the internal mixture calculations. But for assumed external mixtures the hygroscopicity of two or more particle types must be determined or assumed. For this analysis, particle types assumed to be present in external mixtures were (i) particles composed of soluble inorganics; (ii) particles composed of soluble organics; and (iii) particles composed of insoluble (and non-hygroscopic) components. The hygroscopicity parameter of the soluble inorganic particles, κ_{inorg}, was assumed to be 0.6, which is similar to that of ammonium sulfate and ammonium nitrate. As aerosol organic components at the SGP site are not well characterized, direct derivation of κ_{org} is not possible. Here, κ_{org} was estimated using the mixing rule [17].

$$\kappa_{org} = \frac{V_{overall}\kappa_{overall} - \left\{V_{overall} - \left(\frac{m_{orgNR}}{\rho_{orgNR}} + \frac{\frac{b_{absBC}}{\beta_{BC}}}{\rho_{BC}}\right)\kappa_{inorg}\right\}}{\frac{m_{orgNR}}{\rho_{orgNR}}} \tag{2}$$

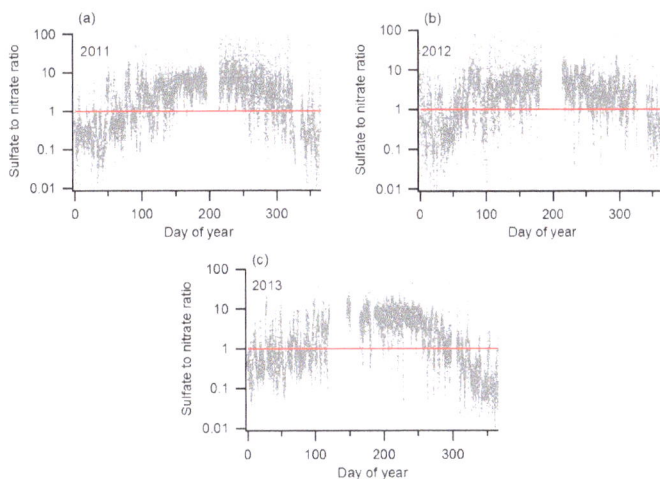

Figure 2. Sulfate to nitrate equivalent concentration ratio based on ACSM chemical composition data from (**a**) 2011, (**b**) 2012, and (**c**) 2013. Equivalent concentration = $\frac{molar\ concentration}{equivalence\ factor}$, where the equivalence factor for sulfate and nitrate are 0.5 and 1, respectively.

For these calculations the total particle volume concentration ($V_{overall}$) was determined from the measured size distribution. The submicron average hygroscopicity parameter ($\kappa_{overall}$) was calculated from the measured size and hygroscopicity distributions as the volume concentration-weighted average κ. The mass concentration of non-refractory organics (m_{orgNR}) was measured by the ACSM and that of black carbon, BC, was estimated as the ratio of measured submicron particle light absorbance (b_{absBC}) and an assumed absorption efficiency at 0.55 μm wavelength of 7.5 m^2 g^{-1} [67]. The density, ρ, of non-refractory organics and BC were both assumed to be 1.3 g cm^{-3} [68].

The resulting seasonal profiles of $\kappa_{overall}$ and κ_{org} in 2011 are shown in Figure 3. Unlike the κ_{org} profile, $\kappa_{overall}$ was highest in the winter and lowest in the summer. The high wintertime $\kappa_{overall}$ is a result of high concentrations of inorganic compounds, especially nitrate, while the relatively low summertime $\kappa_{overall}$ is caused by higher organic mass concentrations, as shown in Figure 4. Positive Matrix Factorization (PMF) analysis used for the Organic Aerosol Component (OACOMP) ARM Value Added Product (VAP) [69] indicates that less-hygroscopic biomass burning organic aerosol (BBOA) was prevalent from February through April in 2011, thus lowering κ_{org} in winter and spring. Aged SOA (MO-OOA), which is moderately hygroscopic, was more abundant in summer, and thus raised κ_{org}. The κ_{org} in the spring and fall lies between that of the winter and summer. The 2011 average of 0.1 was used for this study, while the sensitivity of the results to the selected value was assessed by repeating the calculations for the seasonal minimum ($\kappa_{org} = 0.06$, winter) and maximum ($\kappa_{org} = 0.16$, summer).

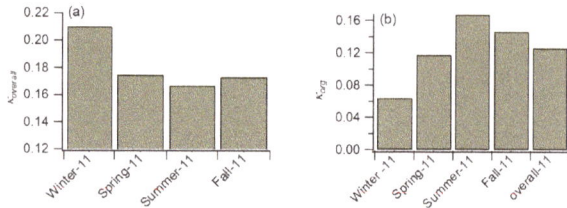

Figure 3. Seasonal profile of (**a**) $\kappa_{overall}$ (left) and (**b**) κ_{org} (right) in 2011.

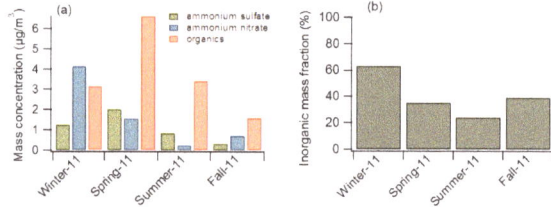

Figure 4. Seasonal average of (**a**) organic and inorganic mass concentration and (**b**) the inorganic mass fraction (%) in 2011.

2.4. Description of Models Used for Estimating N_{CCN}

To simplify both the comparison of the varied approaches used to estimate N_{CCN} and the comparison of the results, a common framework will be used to describe all of the approaches even though more straightforward descriptions would suffice for many of them. For all approaches a CCN spectrum, $N_{CCN}(S)$, was calculated for each size distribution measured by the SMPS. For the simplest approaches, the HTDMA data were not considered and a fixed hygroscopicity parameter was assumed. For all others the GF distributions were interpolated and extrapolated to each of the 90 size bins in the size distribution measurements and then converted to κ distributions using Equation (1). The differences among the approaches arise from the use and (any) averaging of the κ distributions, as summarized in Figure 5.

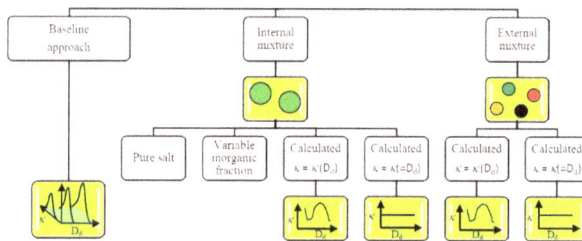

Figure 5. Overview of the N_{CCN} calculation approaches.

As a baseline for comparison with estimates from other approaches and with direct measurements, N_{CCN} was calculated using the full GF distributions without any averaging. Because all of the information in the GF distributions is retained, these estimates are expected to be more accurate than those from any of the other approaches, all of which rely on averaged or assumed hygroscopicity. For each measurement sequence the interpolated κ distributions were combined with the size distribution to create a matrix of number concentrations as a function of D_d and κ, such as that shown graphically in Figure 6a. A Köhler curve relating equilibrium S to droplet diameter, D, was calculated for each (D_d, κ) pair using Equation (3) below. The S_c for each pair was calculated as the maximum value of equilibrium S along the curve.

$$S(D) = \frac{D^3 - D_d^3}{D^3 - D_d^3(1 - \kappa)} \exp\left[\frac{4M_w \sigma_{sol}}{RT\rho_w D}\right] - 1 \tag{3}$$

As with calculations using Equation (1), σ_{sol} was assumed to be that of water, 0.072 J m^{-2}. The result can be viewed as a matrix with elements of S_c and the same D_d and κ arrays as used in the number concentration matrix described above. N_{CCN} was estimated for a prescribed S by integrating the number concentration $N(D_d, \kappa)$ for which $S > S_c$. This is presented graphically in Figure 6b, with N_{CCN} calculated by summing the N elements (whole or part) above and to the right of one of the four curves, each of which connects the elements having the same S_c. The resulting N_{CCN} estimates were first compared with direct measurements made with the CCNc and then with the results of the other estimate approaches outlined below.

For all other N_{CCN} estimates the aerosol was assumed to be either an internal mixture or an external mixture, as is generally required for regional and global scale climate models. The goals here were to assess the error introduced when making these simplifying assumptions and to identify the approach(es) most suitable for an aerosol similar to that found at SGP.

To treat the aerosol as an internal mixture the D_d-dependent κ distributions described above were replaced with a single κ value that is either dependent on D_d ($\kappa = \kappa(D_d)$) or the same for all D_d ($\kappa \neq \kappa(D_d)$). The former comes simply from the number concentration-weighted average of the κ distributions at each D_d, with the result depicted in Figure 6c in the same manner as for the baseline approach matrix. For approaches for which κ is assumed to be size independent, it was calculated either as the average of $\kappa(D_d)$ (Figure 6d) or, neglecting the hygroscopicity measurements, as that of particles composed of 20%, 50%, or pure soluble inorganics (~ammonium sulfate, AS) by volume ($\kappa = 0.12, 0.30$, and 0.60, respectively, with only $\kappa = 0.30$ shown in Figure 6g for clarity).

External mixtures were assumed to be comprised of two of the three particle types considered: insoluble, $\kappa = 0.0$, organic, $\kappa_{org} = 0.1$, and inorganic, $\kappa_{inorg} = 0.6$. As with the assumed internal mixture approaches, both size dependent and size independent scenarios were considered. For both, independent size distributions of the different particle types were calculated from the average $\kappa(= f(D_d)$ or $\neq f(D_d))$ using Equations (4) and (5) below.

Case 1: for $\kappa(D_d) > \kappa_{org}$

$$\left(\frac{dN}{d\log D_d}\right)_{inorg} = \frac{\kappa(D_d) - \kappa_{org}}{\kappa_{inorg} - \kappa_{org}}\left(\frac{dN}{d\log D_d}\right)_{SMPS}$$
$$\left(\frac{dN}{d\log D_d}\right)_{org} = \left(\frac{dN}{d\log D_d}\right)_{SMPS} - \left(\frac{dN}{d\log D_d}\right)_{inorg} \tag{4}$$
$$\left(\frac{dN}{d\log D_d}\right)_{insoluble} = 0$$

Case 2: for $\kappa(D_d) \leq \kappa_{org}$

$$\left(\frac{dN}{d\log D_d}\right)_{org} = \frac{\kappa(D_d)}{\kappa_{org}}\left(\frac{dN}{d\log D_d}\right)_{SMPS}$$
$$\left(\frac{dN}{d\log D_d}\right)_{insoluble} = \left(\frac{dN}{d\log D_d}\right)_{SMPS} - \left(\frac{dN}{d\log D_d}\right)_{org} \tag{5}$$
$$\left(\frac{dN}{d\log D_d}\right)_{inorg} = 0$$

The results for $\kappa(D_d)$ and $\kappa(\neq D_d)$ are depicted graphically in Figure 6e,f, respectively, where the lower and higher horizontal lines represent the organic and inorganic particle types, respectively. As with the other approaches, N_{CCN} was calculated by summing N elements above and to the right of the constant S_c curves. Contributions from the inorganic and organic particle types can also be calculated separately and then added to determine the total N_{CCN} (i.e., $N_{CCN} = N_{CCN.inorg} + N_{CCN.org}$).

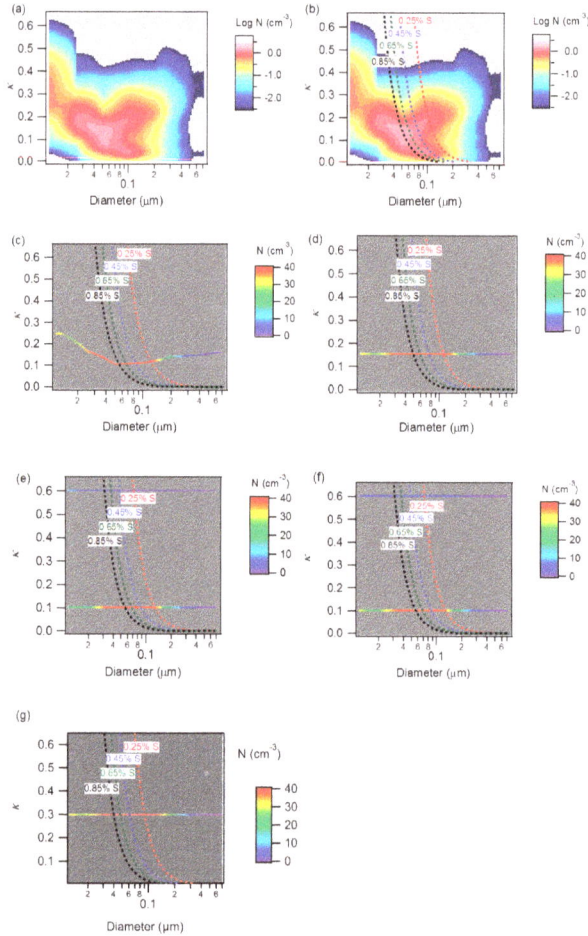

Figure 6. A graphical illustration of the approaches used to estimate N_{CCN}. (**a**) Number concentration matrix calculated from a set of measured size and GF distributions. The size distribution is partitioned among the κ bins according to the GF distributions interpolated to each D_p without any averaging or assumptions about mixing state or composition. The same matrix is shown in (**b**); with lines of constant S_c added. As with all of the estimation techniques, $N_{CCN}(S)$ is estimated by summing the concentration elements (whole or part) above and to the right of the constant S_c (= S) lines; The graphs in (**c**–**g**) are from the same set of size and GF distributions as those used for (**a**,**b**); but with various assumptions for mixing state and composition that collapse the κ distribution at each bin in the size distribution to one or two values; (**c**) internal mixture with $\kappa(D_d)$; (**d**) internal mixture with $\kappa(\neq D_d)$; (**e**) external mixture with $\kappa(D_d)$; (**f**) external mixture with $\kappa(\neq D_d)$; and (**g**) internal mixture with an assumed inorganic volume fraction of 0.5.

3. Results and Discussion

3.1. Comparison between Measured and Baseline N_{CCN} Estimate

The concentration measured by the CCNc was compared to that calculated using the baseline approach for all available data from 2009–2012. The results for May 2011 are shown in Figure 7 and for all of 2011 in Figure S3.

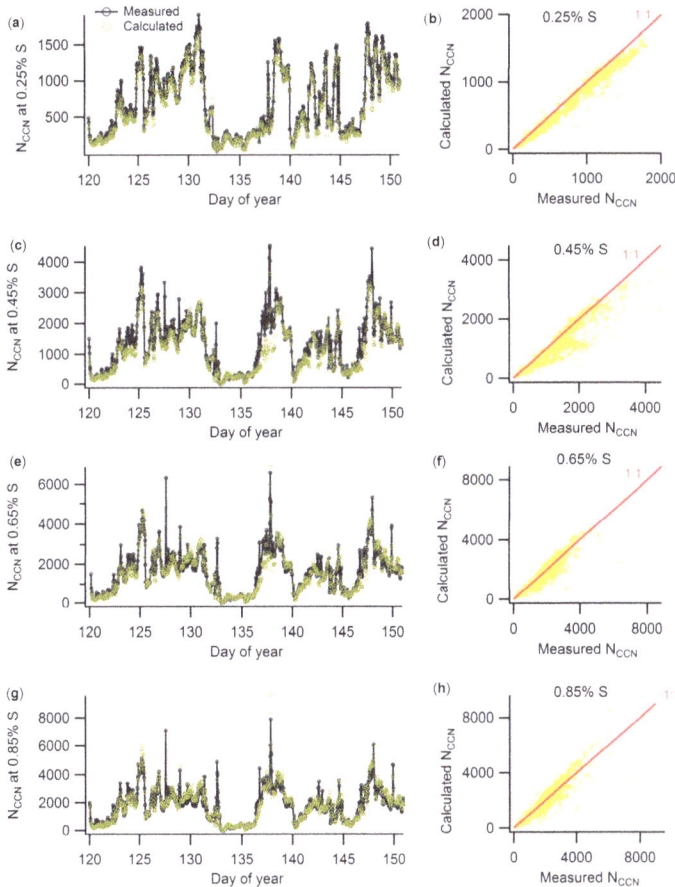

Figure 7. Comparison between measured and estimated N_{CCN} (cm^{-3}) for May 2011 at 0.25% (**a,b**); 0.45% (**c,d**); 0.65% (**e,f**); and 0.85% S (**g,h**). The dotted lines in the scatter plots represent +/− 20% relative to the 1:1 line.

The calculated N_{CCN} tracks that measured throughout the period considered. The deviation between the two is quantified as the Normalized Root Mean Square Error (NRMSE)

$$\text{NRMSE} = \left\{ \frac{1}{n} \sum_{i=0}^{n-1} \left(\frac{x_i - y_i}{x_i} \right)^2 \right\}^{0.5} \tag{6}$$

where x_i is the measured N_{CCN} at a given S, y_i the estimated N_{CCN} at the same S, and n the number of concentration pairs compared. The average NRMSE between the measured and calculated concentrations for May 2011 and for each of the four years analyzed are summarized in Table 3. The measured N_{CCN} at 0.25% S during 2012 was very noisy and was excluded from the analysis.

Table 3. NRMSE between measured and estimated N_{CCN} at 4 different S from 2009–2012 and for May 2011.

Year/Month	NRMSE @ % S			
	0.25	0.45	0.65	0.85
2009	0.53	0.36	0.35	0.35
2010	0.22	0.29	0.27	0.25
May 2011	**0.17**	**0.22**	**0.21**	**0.21**
2011	0.29	0.29	0.26	0.26
2012	-	0.33	0.38	0.36

In addition to measurement error, some possible reasons for deviations between the measured and calculated concentrations are (i) differences in assumed and actual properties of aerosol chemical species; (ii) interactions among components not captured by κ-Köhler Theory; (iii) the presence of low solubility organics that dissolve under the dilute conditions with S ~S_c but not in the more concentrated solution in the HTDMA at 90% RH; (iv) the presence of particles that contain slowly dissolving compounds or that are in an amorphous/glassy state for which hygroscopic growth and activation timescales may be comparable to or greater than the HTDMA and CCNc residence times; and (v) the presence of surface tension-reducing species, which influence S_c much more than GF. But as noted above, the primary goal here was not to assess closure but to confirm that the baseline spectra calculated from the size and GF distributions were reasonably accurate and were suitable to serve as the reference for comparison with the results from the simplified treatments of composition and mixing state.

3.2. Comparison of N_{CCN} Calculated from Different Approaches

Estimates of N_{CCN} assuming the aerosol is an internal or external mixture relative to those of the baseline approach for which no assumption about mixing state is made are presented in Figures 8 and 9 for 0.25% and 0.85%, respectively, and in Figures S4 and S5 for 0.45% and 0.65% S, respectively. Deviations from the 1:1 lines are interpreted as error introduced by the averaging or approximations required for all but the baseline calculations. Best fits through the data were assumed to be linear and were forced through the origin to facilitate interpretation of the results and simply because of the apparent linear correlations with minimal offset in the figures. Table 4 summarizes the slope ($m = \frac{N_{CCN \text{ from alternate approach}}}{N_{CCN \text{ from baseline approach}}}$) and goodness of fit ($r^2$) for each of the approaches with respect to the baseline estimate. Values of m above (below) 1 indicate that the simplified treatment results in N_{CCN} greater (less) than that from the baseline estimate for a given S. Treating the aerosol as an internal mixture with size dependent hygroscopicity resulted in the best agreement with the baseline estimates, with the average m closest to 1.0 and the highest r^2 (Table 4). Though there is considerably more scatter in the results for the internal and external mixtures with $\kappa(\neq D_d)$ than in those with $\kappa(D_d)$, the deviations are found to be tightly correlated. The relationships between the results for the assumed internal and external mixtures with $\kappa(\neq D_d)$ are shown in Figure 10e,f. The strong correlation in $N_{CCN}(S)$ obtained for those assumed internal and external mixtures could be interpreted as reflecting a general lack of sensitivity to the mixing state as has been reported in some closure studies. But, at least for these data, both estimates simply differ in the same direction from what is believed to be the most accurate estimate. For both mixing state assumptions, the estimated N_{CCN} with $\kappa(\neq D_d)$ is higher than that with $\kappa(D_d)$, as the κ averaged over the complete size range is typically higher than the $\kappa(D)$ over the ~0.05–0.20 µm size range that dominates N_{CCN}.

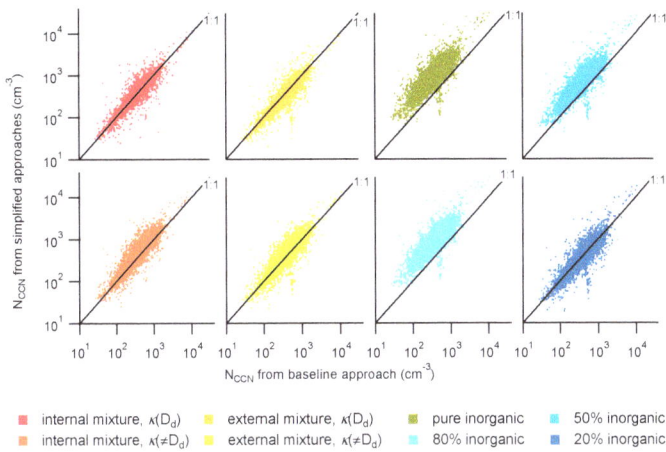

Figure 8. N_{CCN} estimated from simplified approaches vs. that from the baseline approach at 0.25% S for all 2011 data.

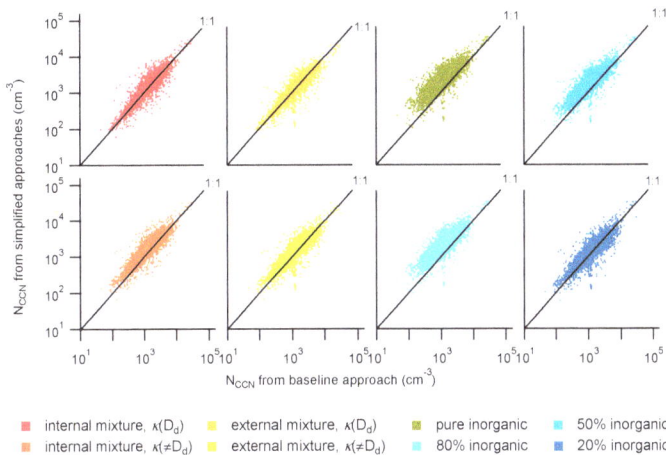

Figure 9. N_{CCN} estimated from simplified approaches vs. that from the baseline approach at 0.85% S for all 2011 data.

Table 4. Fit parameters of N_{CCN} estimate approaches for 2011 data

Model	Slope (*m*) @ % S				$\frac{m_{0.25\% \ S} - m_{0.85\% \ S}}{m_{0.25\% \ S}}$	Correlation Coefficient (r^2) @ % S			
	0.25	0.45	0.65	0.85		0.25	0.45	0.65	0.85
Baseline	1	1	1	1	0	1	1	1	1
Internal, $\kappa = \kappa(D_d)$	1.01	1.02	1.01	1.01	0.002	0.87	0.87	0.85	0.84
Internal, $\kappa = \kappa(\neq D_d)$	1.21	1.18	1.13	1.10	0.095	0.82	0.81	0.80	0.80
External, $\kappa = \kappa(D_d)$	0.90	0.92	0.90	0.90	0.009	0.81	0.81	0.78	0.77
External, $\kappa = \kappa(\neq D_d)$	1.06	1.07	1.02	1.00	0.061	0.80	0.80	0.79	0.77
Internal, pure AS	2.37	1.82	1.57	1.43	0.400	0.67	0.69	0.68	0.67
Internal, 50% AS	1.76	1.50	1.33	1.26	0.285	0.72	0.75	0.76	0.74
Internal, 20% AS	1.05	1.05	1.03	0.99	0.057	0.72	0.78	0.77	0.76

A correction of $1/m$ could be applied to results using any of the simplified treatments to increase accuracy. Of course, that sort of annual average correction would not reduce error that arises from variability in aerosol characteristics on seasonal or shorter time scales and is reflected in the r^2 values. For example, assuming the aerosol has a fixed κ of 0.12 (20% AS) results in average best fit slopes between 1.00 and 1.06, indicating that, on average, an error of only a few percent would be introduced if both the composition and hygroscopicity were not considered at all in the calculations. However, that represents the average of comparably large negative and positive errors for individual measurements or days, as is shown in Figure 11 that compares the 2011 daily averages of the results from the internal mixture with $\kappa(D_d)$ and the fixed 20% AS treatments relative to the baseline calculations. And as is shown in Figure 10, similar deviations from the 1.0 line are evident in the results for which the hygroscopicity was considered, but not its size dependence. Furthermore, the utility of any of the simplifying approaches is limited if not applicable over a wide range in S. The S dependence of the best fit slopes quantified as the fractional change over the full range in S considered is included in the 6th column in Table 4. That S dependence is much lower for the two cases for which the size dependence of κ is considered than for any of the others, which is expected because the change in N_{CCN} with S largely reflects inclusion or exclusion of particles within a rather narrow slice of the full size range.

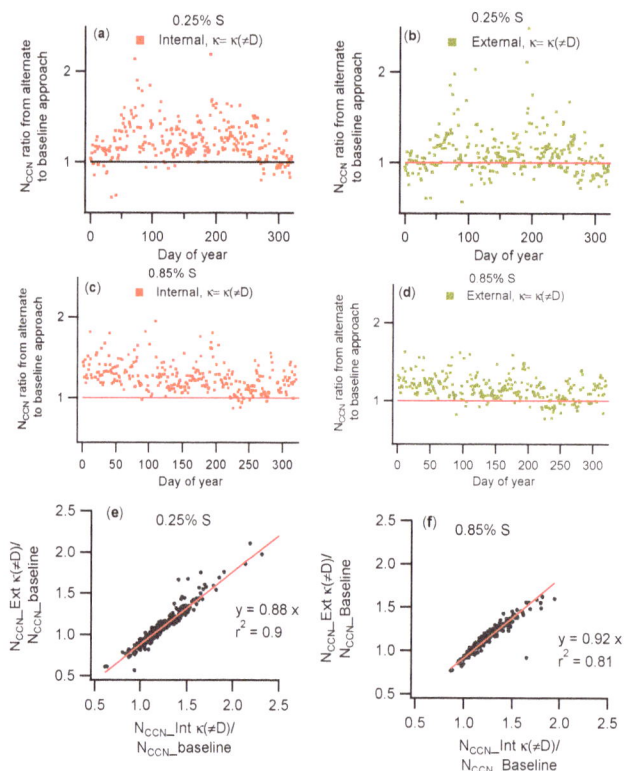

Figure 10. (**a–d**) Time series for 2011 of estimated N_{CCN} relative to that from the baseline approach: (**a**) internal mixture, $\kappa \neq \kappa(D)$ at 0.25% S; (**b**) external mixture, $\kappa \neq \kappa(D)$ at 0.25% S (**c**) internal mixture, $\kappa \neq \kappa(D)$ at 0.85% S; and (**d**) external mixture, $\kappa \neq \kappa(D)$ at 0.85% S; (**e**) Scatter plot of N_{CCN} ratio from (**a,b**); (**f**) scatter plot of N_{CCN} ratio from (**c,d**).

The results for the assumed external mixtures were found to be quite sensitive to the κ_{org} used. Specifically, substituting the seasonal minimum (0.06) for the annual average (0.10) of the estimated κ_{org} results in an average decrease in N_{CCN} of 8% when the size dependence of κ is considered and an increase of 10% when it is not. Similarly, use of the seasonal maximum of 0.16 results in an average decrease of 1% with $\kappa(D_d)$ and increase of 15% with $\kappa(\neq D_d)$. (The results for SGP are likely especially sensitive to the assumed value because it is similar to that average κ of 0.12 that was found to result in reasonable estimates averaged throughout the year). For any measurement a change in κ_{org} will cause a change in partitioning between the three particle types, with the net influence on N_{CCN} depending on the coupling of that change and the shape of the size distribution. Because the approach used allows for at most two particle types, an increase in κ_{org} that crosses the average κ of the aerosol will cause a shift in the assumed mixture from consisting of organic and inorganic ($\kappa = 0.6$) particles to organic and insoluble ($\kappa = 0.0$) particles, with a resulting increase in N_{CCN} due to the increased hygroscopicity of the organic particles, a decrease due to the loss of the inorganic particles, and a net change that could be positive or negative. To evaluate the sensitivity of the calculated N_{CCN} to the choice in κ_{inorg}, the analysis was repeated using values of 0.64 and 0.67 (instead of 0.6). The average change in N_{CCN} was about 0.6% when size-dependence is considered and about 1.0% when it is not.

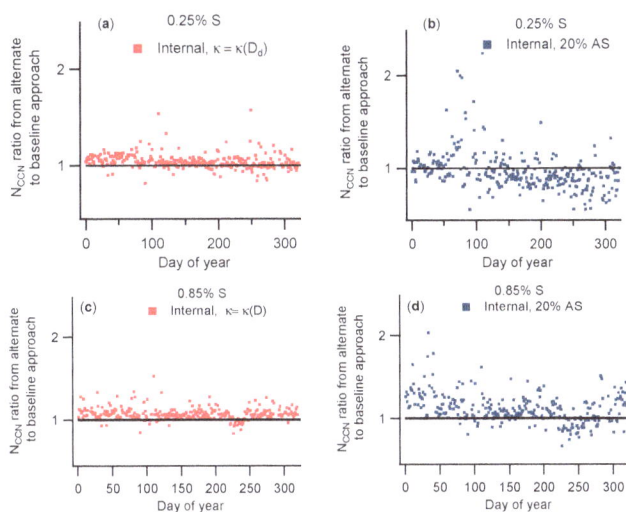

Figure 11. Time series for 2011 of estimated N_{CCN} relative to that from the baseline approach: (a) internal mixture, $\kappa = \kappa(D)$ at 0.25% S; (b) internal mixture, 20% AS at 0.25% S (c) internal mixture, $\kappa = \kappa(D)$ at 0.85% S; and (d) internal mixture, 20% AS at 0.85% S.

4. Summary and Conclusions

Size distributions and hygroscopic growth factor distributions measured from 2009 to 2012 at the SGP ARM site were used to estimate CCN concentrations over a range in supersaturation. An initial estimate of N_{CCN} that served as a basis for comparison used all of the information in the combined distributions without any averaging. For those estimates, matrices of $N(D_d, \kappa)$ and $S_c(D_d, \kappa)$ were calculated from the measured distributions and from κ-Köhler Theory, respectively, and $N_{CCN}(S)$ was calculated by integrating all of the N elements for which the corresponding S_c element < S. Comparisons of those estimates with direct measurements from a collocated CCN counter show that this baseline approach can reasonably predict N_{CCN} over a range of S.

The baseline spectra were then compared with those calculated using the same dataset but with aerosol treatments that are more commonly used for more efficient computation or simply because size-dependent composition or hygroscopicity distributions are not available. These included approximating the aerosol as an internal mixture with fixed inorganic volume fraction, as an internal mixture with size-dependent or size-independent hygroscopicity, and as an external mixture with size-dependent or size-independent hygroscopicity. Bias and variance relative to the baseline estimates were described with best fit slopes, m, and coefficients of determination, r^2, respectively, with both calculated from the thousands of N_{CCN} pairs over the period of analysis.

Though the results presented here are most directly applicable for the aerosol at SGP, that aerosol is quite diverse, with a range in sources, origins, and atmospheric processing prior to sampling. Thus, at least the overall ~10% uncertainty range accompanying assumption of either internal or external mixtures reported here is likely relevant for N_{CCN} prediction in other regions as well and can provide some guidance for interpretation of model results. And for prediction of N_{CCN} for an aerosol known to be similar to that at SGP the bias and scatter associated with each of the simplified treatments can be weighed together with computational cost for selection of the most appropriate approximation(s). Ongoing work is aimed at applying these results to improve the accuracy with which CCN spectra at SGP and other ARM sites can be predicted in the absence of direct measurements.

Supplementary Materials: The following are available online at www.mdpi.com/2073-4433/9/2/54/s1.

Acknowledgments: This research was supported by the office of biological and environmental research of the U.S. Department of Energy under grant DE-SC0016051 as part of the Atmospheric Radiation Measurement (ARM) climate research facility, an Office of Science scientific user facility. Data were obtained from the ARM climate research facility. All CCN concentration, chemical composition, size distribution, and hygroscopic growth factor distribution data used in our analysis were downloaded from the DOE ARM data archive at http://www.archive.arm.gov/. Spreadsheets containing many of the derived products are available at http://people.tamu.edu/~dcollins/Predicted_CCN_paper/. Other products can be obtained by contacting Don Collins (dcollins@tamu.edu).

Author Contributions: M.M., A.J., and D.R.C. conceived of the analysis and contributed to the data processing and use. M.M. wrote the paper with assistance from D.R.C. A.J. edited the paper and assisted with revision.

Conflicts of Interest: The authors declare no conflict of interest.

References

1. Köhler, H. The nucleus in and the growth of hygroscopic droplets. *Trans. Faraday Soc.* **1936**, *32*, 1152–1161. [CrossRef]

2. Hallquist, M.; Wenger, J.C.; Baltensperger, U.; Rudich, Y.; Simpson, D.; Claeys, M.; Dommen, J.; Donahue, N.M.; George, C.; Goldstein, A.H.; et al. The formation, properties and impact of secondary organic aerosol: Current and emerging issues. *Atmos. Chem. Phys.* **2009**, *9*, 5155–5236. [CrossRef]

3. Pennington, M.R.; Bzdek, B.R.; Depalma, J.W.; Smith, J.N.; Kortelainen, A.-M.; Ruiz, L.H.; Petäjä, T.; Kulmala, M.; Worsnop, D.R.; Johnston, M.V. Identification and quantification of particle growth channels during new particle formation. *Atmos. Chem. Phys.* **2013**, *13*, 10215–10225. [CrossRef]

4. Abbatt, J.; Broekhuizen, K.; Pradeepkumar, P. Cloud condensation nucleus activity of internally mixed ammonium sulfate/organic acid aerosol particles. *Atmos. Environ.* **2005**, *39*, 4767–4778. [CrossRef]

5. Prenni, A.J.; Petters, M.D.; Kreidenweis, S.M.; Demott, P.J.; Ziemann, P.J. Cloud droplet activation of secondary organic aerosol. *J. Geophys. Res.* **2007**, *112*. [CrossRef]

6. Bigg, E. Discrepancy between observation and prediction of concentrations of cloud condensation nuclei. *Atmos. Res.* **1986**, *20*, 81–86. [CrossRef]

7. Chan, M.N.; Kreidenweis, S.M.; Chan, C.K. Measurements of the Hygroscopic and Deliquescence Properties of Organic Compounds of Different Solubilities in Water and Their Relationship with Cloud Condensation Nuclei Activities. *Environ. Sci. Technol.* **2008**, *42*, 3602–3608. [CrossRef] [PubMed]

8. Raymond, T.M. Cloud activation of single-component organic aerosol particles. *J. Geophys. Res.* **2002**, *107*. [CrossRef]

9. Raymond, T.M. Formation of cloud droplets by multicomponent organic particles. *J. Geophys. Res.* **2003**, *108*. [CrossRef]

10. Roberts, G.C. Sensitivity of CCN spectra on chemical and physical properties of aerosol: A case study from the Amazon Basin. *J. Geophys. Res.* **2002**, *107*. [CrossRef]

11. Smith, J.N.; Dunn, M.J.; Vanreken, T.M.; Iida, K.; Stolzenburg, M.R.; Mcmurry, P.H.; Huey, L.G. Chemical composition of atmospheric nanoparticles formed from nucleation in Tecamac, Mexico: Evidence for an important role for organic species in nanoparticle growth. *Geophys. Res. Lett.* **2008**, *35*. [CrossRef]

12. Yli-Juuti, T.; Nieminen, T.; Hirsikko, A.; Aalto, P.P.; Asmi, E.; Hõrrak, U.; Manninen, H.E.; Patokoski, J.; Maso, M.D.; Petäjä, T.; et al. Growth rates of nucleation mode particles in Hyytiälä during 2003–2009: Variation with particle size, season, data analysis method and ambient conditions. *Atmos. Chem. Phys.* **2011**, *11*, 12865–12886. [CrossRef]

13. Ruehl, C.R.; Davis, J.F.; Wilson, K.R. An interfacial mechanism for cloud droplet formation on organic aerosols. *Science* **2016**, *351*, 1447–1450. [CrossRef] [PubMed]

14. Bilde, M.; Svenningsson, B. CCN activation of slightly soluble organics: The importance of small amounts of inorganic salt and particle phase. *Tellus* **2004**, *56*, 128–134. [CrossRef]

15. Hartz, K.E.H.; Tischuk, J.E.; Chan, M.N.; Chan, C.K.; Donahue, N.M.; Pandis, S.N. Cloud condensation nuclei activation of limited solubility organic aerosol. *Atmos. Environ.* **2006**, *40*, 605–617. [CrossRef]

16. Svenningsson, B.; Rissler, J.; Swietlicki, E.; Mircea, M.; Bilde, M.; Facchini, M.C.; Decesari, S.; Fuzzi, S.; Zhou, J.; Mønster, J.; Rosenørn, T. Hygroscopic growth and critical supersaturations for mixed aerosol particles of inorganic and organic compounds of atmospheric relevance. *Atmos. Chem. Phys.* **2006**, *6*, 1937–1952. [CrossRef]

17. Petters, M.D.; Kreidenweis, S.M. A single parameter representation of hygroscopic growth and cloud condensation nucleus activity. *Atmos. Chem. Phys.* **2007**, *7*, 1961–1971. [CrossRef]

18. Moore, R.H.; Bahreini, R.; Brock, C.A.; Froyd, K.D.; Cozic, J.; Holloway, J.S.; Middlebrook, A.M.; Murphy, D.M.; Nenes, A. Hygroscopicity and composition of Alaskan Arctic CCN during April 2008. *Atmos. Chem. Phys.* **2011**, *11*, 11807–11825. [CrossRef]

19. Bougiatioti, A.; Fountoukis, C.; Kalivitis, N.; Pandis, S.N.; Nenes, A.; Mihalopoulos, N. Cloud condensation nuclei measurements in the marine boundary layer of the Eastern Mediterranean: CCN closure and droplet growth kinetics. *Atmos. Chem. Phys.* **2009**, *9*, 7053–7066. [CrossRef]

20. Chang, R.Y.-W.; Slowik, J.G.; Shantz, N.C.; Vlasenko, A.; Liggio, J.; Sjostedt, S.J.; Leaitch, W.R.; Abbatt, J.P.D. The hygroscopicity parameter (κ) of ambient organic aerosol at a field site subject to biogenic and anthropogenic influences: Relationship to degree of aerosol oxidation. *Atmos. Chem. Phys.* **2010**, *10*, 5047–5064. [CrossRef]

21. Chuang, P.Y.; Collins, D.R.; Pawlowska, H.; Snider, J.R.; Jonsson, H.H.; Brenguier, J.L.; Flagan, R.C.; Seinfeld, J.H. CCN measurements during ACE-2 and their relationship to cloud microphysical properties. *Tellus* **2000**, *52*, 843–867. [CrossRef]

22. Covert, D.S.; Gras, J.L.; Wiedensohler, A.; Stratmann, F. Comparison of directly measured CCN with CCN modeled from the number-size distribution in the marine boundary layer during ACE 1 at Cape Grim, Tasmania. *J. Geophys. Res. Atmos.* **1998**, *103*, 16597–16608. [CrossRef]

23. Furutani, H.; Dallosto, M.; Roberts, G.; Prather, K. Assessment of the relative importance of atmospheric aging on CCN activity derived from field observations. *Atmos. Environ.* **2008**, *42*, 3130–3142. [CrossRef]

24. Gácita, M.S.; Longo, K.M.; Freire, J.L.M.; Freitas, S.R.; Martin, S.T. Impact of mixing state and hygroscopicity on CCN activity of biomass burning aerosol in Amazonia. *Atmos. Chem. Phys.* **2017**, *17*, 2373–2392. [CrossRef]

25. Kammermann, L.; Gysel, M.; Weingartner, E.; Herich, H.; Cziczo, D.J.; Holst, T.; Svenningsson, B.; Arneth, A.; Baltensperger, U. Subarctic atmospheric aerosol composition: 3. Measured and modeled properties of cloud condensation nuclei. *J. Geophys. Res.* **2010**, *115*. [CrossRef]

26. Kuwata, M.; Kondo, Y.; Miyazaki, Y.; Komazaki, Y.; Kim, J.H.; Yum, S.S.; Tanimoto, H.; Matsueda, H. Cloud condensation nuclei activity at Jeju Island, Korea in spring 2005. *Atmos. Chem. Phys.* **2008**, *8*, 2933–2948. [CrossRef]

27. Mircea, M.; Facchini, M.C.; Decesari, S.; Fuzzi, S.; Charlson, R.J. The influence of the organic aerosol component on CCN supersaturation spectra for different aerosol types. *Tellus* **2002**, *54*, 74–81. [CrossRef]

28. Moore, R.H.; Cerully, K.; Bahreini, R.; Brock, C.A.; Middlebrook, A.M.; Nenes, A. Hygroscopicity and composition of California CCN during summer 2010. *J. Geophys. Res. Atmos.* **2012**, *117*. [CrossRef]

29. Rissler, J.; Vestin, A.; Swietlicki, E.; Fisch, G.; Zhou, J.; Artaxo, P.; Andreae, M.O. Size distribution and hygroscopic properties of aerosol particles from dry-season biomass burning in Amazonia. *Atmos. Chem. Phys.* **2006**, *6*, 471–491. [CrossRef]

30. Roberts, G.C.; Day, D.A.; Russell, L.M.; Dunlea, E.J.; Jimenez, J.L.; Tomlinson, J.M.; Collins, D.R.; Shinozuka, Y.; Clarke, A.D. Characterization of particle cloud droplet activity and composition in the free troposphere and the boundary layer during INTEX-B. *Atmos. Chem. Phys.* **2010**, *10*, 6627–6644. [CrossRef]

31. Roberts, G.; Mauger, G.; Hadley, O.; Ramanathan, V. North American and Asian aerosols over the eastern Pacific Ocean and their role in regulating cloud condensation nuclei. *J. Geophys. Res.* **2006**, *111*. [CrossRef]

32. Rose, D.; Nowak, A.; Achtert, P.; Wiedensohler, A.; Hu, M.; Shao, M.; Zhang, Y.; Andreae, M.O.; Pöschl, U. Cloud condensation nuclei in polluted air and biomass burning smoke near the mega-city Guangzhou, China—Part 1: Size-resolved measurements and implications for the modeling of aerosol particle hygroscopicity and CCN activity. *Atmos. Chem. Phys.* **2010**, *10*, 3365–3383. [CrossRef]

33. Shantz, N.C.; Leaitch, W.R.; Phinney, L.; Mozurkewich, M.; Toom-Sauntry, D. The effect of organic compounds on the growth rate of cloud droplets in marine and forest settings. *Atmos. Chem. Phys.* **2008**, *8*, 5869–5887. [CrossRef]

34. Wang, J.; Cubison, M.J.; Aiken, A.C.; Jimenez, J.L.; Collins, D.R. The importance of aerosol mixing state and size-resolved composition on CCN concentration and the variation of the importance with atmospheric aging of aerosols. *Atmos. Chem. Phys.* **2010**, *10*, 7267–7283. [CrossRef]

35. Ervens, B.; Cubison, M.J.; Andrews, E.; Feingold, G.; Ogren, J.A.; Jimenez, J.L.; Quinn, P.K.; Bates, T.S.; Wang, J.; Zhang, Q.; et al. CCN predictions using simplified assumptions of organic aerosol composition and mixing state: A synthesis from six different locations. *Atmos. Chem. Phys.* **2010**, *10*, 4795–4807. [CrossRef]

36. Broekhuizen, K.; Chang, R.-W.; Leaitch, W.R.; Li, S.-M.; Abbatt, J.P.D. Closure between measured and modeled cloud condensation nuclei (CCN) using size-resolved aerosol compositions in downtown Toronto. *Atmos. Chem. Phys.* **2006**, *6*, 2513–2524. [CrossRef]

37. Cubison, M.J.; Ervens, B.; Feingold, G.; Docherty, K.S.; Ulbrich, I.M.; Shields, L.; Prather, K.; Hering, S.; Jimenez, J.L. The influence of chemical composition and mixing state of Los Angeles urban aerosol on CCN number and cloud properties. *Atmos. Chem. Phys.* **2008**, *8*, 5649–5667. [CrossRef]

38. Lance, S.; Nenes, A.; Mazzoleni, C.; Dubey, M.K.; Gates, H.; Varutbangkul, V.; Rissman, T.A.; Murphy, S.M.; Sorooshian, A.; Flagan, R.C.; et al. Cloud condensation nuclei activity, closure, and droplet growth kinetics of Houston aerosol during the Gulf of Mexico Atmospheric Composition and Climate Study (GoMACCS). *J. Geophys. Res.* **2009**, *114*. [CrossRef]

39. Padró, L.T.; Moore, R.H.; Zhang, X.; Rastogi, N.; Weber, R.J.; Nenes, A. Mixing state and compositional effects on CCN activity and droplet growth kinetics of size-resolved CCN in an urban environment. *Atmos. Chem. Phys.* **2012**, *12*, 10239–10255. [CrossRef]

40. Zaveri, R.A.; Barnard, J.C.; Easter, R.C.; Riemer, N.; West, M. Particle-resolved simulation of aerosol size, composition, mixing state, and the associated optical and cloud condensation nuclei activation properties in an evolving urban plume. *J. Geophys. Res.* **2010**, *115*. [CrossRef]

41. Bhattu, D.; Tripathi, S.N. CCN closure study: Effects of aerosol chemical composition and mixing state. *J. Geophys. Res. Atmos.* **2015**, *120*, 766–783. [CrossRef]

42. Gunthe, S.S.; King, S.M.; Rose, D.; Chen, Q.; Roldin, P.; Farmer, D.K.; Jimenez, J.L.; Artaxo, P.; Andreae, M.O.; Martin, S.T.; et al. Cloud condensation nuclei in pristine tropical rainforest air of Amazonia: Size-resolved measurements and modeling of atmospheric aerosol composition and CCN activity. *Atmos. Chem. Phys.* **2009**, *9*, 7551–7575. [CrossRef]

43. Medina, J.; Nenes, A.; Sotiropoulou, R.-E.P.; Cottrell, L.D.; Ziemba, L.D.; Beckman, P.J.; Griffin, R.J. Cloud condensation nuclei closure during the International Consortium for Atmospheric Research on Transport and Transformation 2004 campaign: Effects of size-resolved composition. *J. Geophys. Res.* **2007**, *112*. [CrossRef]

44. Stroud, C.A.; Nenes, A.; Jimenez, J.L.; Decarlo, P.F.; Huffman, J.A.; Bruintjes, R.; Nemitz, E.; Delia, A.E.; Toohey, D.W.; Guenther, A.B.; et al. Cloud Activating Properties of Aerosol Observed during CELTIC. *J. Atmos. Sci.* **2007**, *64*, 441–459. [CrossRef]

45. Conant, W.C.; Vanreken, T.M.; Rissman, T.A.; Varutbangkul, V.; Jonsson, H.H.; Nenes, A.; Jimenez, J.L.; Delia, A.E.; Bahreini, R.; Roberts, G.C.; et al. Aerosol-cloud drop concentration closure in warm cumulus. *J. Geophys. Res. Atmos.* **2004**, *109*. [CrossRef]

46. Dusek, U.; Frank, G.P.; Hildebrandt, L.; Curtius, J.; Schneider, J.; Walter, S.; Chand, D.; Drewnick, F.; Hings, S.; Jung, D.; et al. Size Matters More Than Chemistry for Cloud-Nucleating Ability of Aerosol Particles. *Science* **2006**, *312*, 1375–1378. [CrossRef] [PubMed]

47. Cantrell, W.; Shaw, G.; Cass, G.R.; Chowdhury, Z.; Hughes, L.S.; Prather, K.A.; Guazzotti, S.A.; Coffee, K.R. Closure between aerosol particles and cloud condensation nuclei at Kaashidhoo Climate Observatory. *J. Geophys. Res. Atmos.* **2001**, *106*, 28711–28718. [CrossRef]

48. Chang, R.-W.; Liu, P.; Leaitch, W.; Abbatt, J. Comparison between measured and predicted CCN concentrations at Egbert, Ontario: Focus on the organic aerosol fraction at a semi-rural site. *Atmos. Environ.* **2007**, *41*, 8172–8182. [CrossRef]

49. Liu, P.S.K.; Leaitch, W.R.; Banic, C.M.; Li, S.-M.; Ngo, D.; Megaw, W.J. Aerosol observations at Chebogue Point during the 1993 North Atlantic Regional Experiment: Relationships among cloud condensation nuclei, size distribution, and chemistry. *J. Geophys. Res. Atmos.* **1996**, *101*, 28971–28990. [CrossRef]

50. Rissler, J.; Swietlicki, E.; Zhou, J.; Roberts, G.; Andreae, M.O.; Gatti, L.V.; Artaxo, P. Physical properties of the sub-micrometer aerosol over the Amazon rain forest during the wet-to-dry season transition—Comparison of modeled and measured CCN concentrations. *Atmos. Chem. Phys.* **2004**, *4*, 2119–2143. [CrossRef]

51. Vanreken, T.M. Toward aerosol/cloud condensation nuclei (CCN) closure during CRYSTAL-FACE. *J. Geophys. Res.* **2003**, *108*. [CrossRef]

52. Wang, J.; Lee, Y.-N.; Daum, P.H.; Jayne, J.; Alexander, M.L. Effects of aerosol organics on cloud condensation nucleus (CCN) concentration and first indirect aerosol effect. *Atmos. Chem. Phys.* **2008**, *8*, 6325–6339. [CrossRef]

53. Ervens, B.; Cubison, M.; Andrews, E.; Feingold, G.; Ogren, J.A.; Jimenez, J.L.; Decarlo, P.; Nenes, A. Prediction of cloud condensation nucleus number concentration using measurements of aerosol size distributions and composition and light scattering enhancement due to humidity. *J. Geophys. Res.* **2007**, *112*. [CrossRef]

54. Almeida, G.P.; Brito, J.; Morales, C.A.; Andrade, M.F.; Artaxo, P. Measured and modelled cloud condensation nuclei (CCN) concentration in São Paulo, Brazil: The importance of aerosol size-resolved chemical composition on CCN concentration prediction. *Atmos. Chem. Phys.* **2014**, *14*, 7559–7572. [CrossRef]

55. Asa-Awuku, A.; Moore, R.H.; Nenes, A.; Bahreini, R.; Holloway, J.S.; Brock, C.A.; Middlebrook, A.M.; Ryerson, T.B.; Jimenez, J.L.; Decarlo, P.F.; et al. Airborne cloud condensation nuclei measurements during the 2006 Texas Air Quality Study. *J. Geophys. Res.* **2011**, *116*. [CrossRef]

56. Che, H.C.; Zhang, X.Y.; Wang, Y.Q.; Zhang, L.; Shen, X.J.; Zhang, Y.M.; Ma, Q.L.; Sun, J.Y.; Zhang, Y.W.; Wang, T.T. Characterization and parameterization of aerosol cloud condensation nuclei activation under different pollution conditions. *Sci. Rep.* **2016**, *6*. [CrossRef] [PubMed]

57. Mircea, M.; Facchini, M.C.; Decesari, S.; Cavalli, F.; Emblico, L.; Fuzzi, S.; Vestin, A.; Rissler, J.; Swietlicki, E.; Frank, G.; et al. Importance of the organic aerosol fraction for modeling aerosol hygroscopic growth and activation: A case study in the Amazon Basin. *Atmos. Chem. Phys.* **2005**, *5*, 3111–3126. [CrossRef]

58. Quinn, P.K.; Bates, T.S.; Coffman, D.J.; Covert, D.S. Influence of particle size and chemistry on the cloud nucleating properties of aerosols. *Atmos. Chem. Phys.* **2008**, *8*, 1029–1042. [CrossRef]

59. Mahish, M.; Collins, D. Analysis of a Multi-Year Record of Size-Resolved Hygroscopicity Measurements from a Rural Site in the U.S. *Aerosol Air Qual. Res.* **2017**, *17*, 1389–1400. [CrossRef]

60. Collins, D. *Tandem Differential Mobility Analyzer/Aerodynamic Particle Sizer (APS) Handbook*; No. DOE/ SC-ARM-TR-090; U.S. Department of Energy: Washington, DC, USA, 2010.

61. Gasparini, R.; Li, R.; Collins, D.R. Integration of size distributions and size-resolved hygroscopicity measured during the Houston Supersite for compositional categorization of the aerosol. *Atmos. Environ.* **2004**, *38*, 3285–3303. [CrossRef]

62. Roberts, G.; Nenes, A. A Continuous-Flow Streamwise Thermal-Gradient CCN Chamber for Atmospheric Measurements. *Aerosol Sci. Technol.* **2005**, *39*, 206–221. [CrossRef]

63. Jefferson, A. *Aerosol Observing System (AOS) Handbook*; Technical Report ARM-TR-014; U.S. Department of Energy: Washington, DC, USA, 2011.

64. Ng, N.L.; Herndon, S.C.; Trimborn, A.; Canagaratna, M.R.; Croteau, P.L.; Onasch, T.B. An Aerosol Chemical Speciation Monitor (ACSM) for Routine Monitoring of the Composition and Mass Concentrations of Ambient Aerosol. *Aerosol Sci. Technol.* **2011**, *45*, 780–794. [CrossRef]

65. Bond, T.C.; Anderson, T.L.; Campbell, D. Calibration and intercomparison of filter-based measurements of visible light absorption by aerosols. *Aerosol Sci. Technol.* **1999**, *30*, 582–600. [CrossRef]

66. Watson, T.B. *Aerosol Chemical Speciation Monitor (ACSM) Instrument Handbook*; No. DOE/SC-ARM-TR-196; U.S. Department of Energy: Washington, DC, USA, 2017.

67. Yang, M.; Howell, S.G.; Zhuang, J.; Huebert, B.J. Attribution of aerosol light absorption to black carbon, brown carbon, and dust in China—Interpretations of atmospheric measurements during EAST-AIRE. *Atmos. Chem. Phys.* **2009**, *9*, 2035–2050. [CrossRef]

68. Nakao, S.; Tang, P.; Tang, X.; Clark, C.H.; Qi, L.; Seo, E.; Asa-Awuku, A.; Cocker, D. Density and elemental ratios of secondary organic aerosol: Application of a density prediction method. *Atmos. Environ.* **2013**, *68*, 273–277. [CrossRef]

69. Fast, J.; Zhang, Q.; Tilp, A.; Shippert, T.; Parworth, C.; Mei, F. *Organic Aerosol Component (OACOMP) Value-Added Product*; No. DOE/SC-ARM-TR-131; U.S. Department of Energy: Washington, DC, USA, 2013.

atmosphere

MDPI

Article

Morphology, Composition, and Mixing State of Individual Aerosol Particles in Northeast China during Wintertime

Liang Xu [1], Lei Liu [1], Jian Zhang [1], Yinxiao Zhang [1], Yong Ren [1,2], Xin Wang [2] and Weijun Li [1,*]

[1] Environment Research Institute, Shandong University, Jinan 250100, China; xuliangsd@126.com (L.X.); liulei92459@163.com (L.L.); zjzhangjian666@163.com (J.Z.); zhangyinxiaosd@163.com (Y.Z.); reny2014@lzu.edu.cn (Y.R.)

[2] Key Laboratory for Semi-Arid Climate Change of the Ministry of Education, College of Atmospheric Sciences, Lanzhou University, Lanzhou 730000, China; wxin@lzu.edu.cn

* Correspondence: liweijun@sdu.edu.cn; Tel.: +86-531-8836-4675

Academic Editors: Swarup China and Claudio Mazzoleni
Received: 23 January 2017; Accepted: 20 February 2017; Published: 24 February 2017

Abstract: Northeast China is located in a high latitude area of the world and undergoes a cold season that lasts six months each year. Recently, regional haze episodes with high concentrations of fine particles ($PM_{2.5}$) have frequently been occurring in Northeast China during the heating period, but little information has been available. Aerosol particles were collected in winter at a site in a suburban county town (T1) and a site in a background rural area (T2). Morphology, size, elemental composition, and mixing state of individual aerosol particles were characterized by transmission electron microscopy (TEM). Aerosol particles were mainly composed of organic matter (OM) and S-rich and certain amounts of soot and K-rich. OM represented the most abundant particles, accounting for 60.7% and 53.5% at the T1 and T2 sites, respectively. Abundant spherical OM particles were likely emitted directly from coal-burning stoves. Soot decreased from 16.9% at the T1 site to 4.6% at the T2 site and sulfate particles decrease from 35.9% at the T2 site to 15.7% at the T1 site, suggesting that long-range transport air masses experienced more aging processes and produced more secondary particles. Based on our investigations, we proposed that emissions from coal-burning stoves in most rural areas of the west part of Northeast China can induce regional haze episodes.

Keywords: Northeast China; wintertime; individual aerosol particles; morphology; composition; mixing state

1. Introduction

Over the past 30 years, China has quickly undergone urbanization and industrialization. Rapid social and economic development led to heavy atmospheric pollution in East China. Air pollution has a great impact on human health, contributing to health conditions such as respiratory and cardiovascular disease [1]. Anthropogenic pollutants emitted from China may experience long-range transport to the Korean peninsula, Japan, and even into West America, which can cause a regional or global air pollution problem [2–4]. Previous studies on air pollution in China focused on three developed economic areas, consisting of North China Plain (around Beijing city), Yangtze River Delta (around Shanghai city), and Pearl River Delta (around Guangzhou city) [5–8]. However, only a few studies have investigated air pollution in some developing economic areas, even though their air quality has been worse than in the developed areas in China.

Although the economy in Northeast China area has undergone a slow increase in the past 15 years, air quality has deteriorated similar to the more developed economic areas. There is a sharp contrast in air quality in Northeast China between heating and non-heating periods. For example, the $PM_{2.5}$ concentration in urban areas climbed up to 1000 µg/m^3 during some serious haze-fog events during the heating period in 2015, in contrast to the common <75 µg/m^3 during non-heating periods. The first reason was that Northeast China is one of the coldest areas in China, with an average temperature around -15 °C in winter. Therefore, people in urban and rural areas need to consume much more fuel for heating (e.g., coal and biomass burning) in Northeast China than other places in North China. The second reason is that Northeast China has the longest heating period (about six months) in China. Large amounts of emissions from solid fuels for heating as a dominant source cause the formation of the regional severe hazes in Northeast China which could be different from the hazes in the North China Plain. Therefore, it is necessary to understand exactly what kinds of aerosol particles and what sources are existing in Northeast China.

To our knowledge, there is no study working on the regional haze episodes during the heating period in Northeast China, although a few studies have investigated aerosol optical properties and aerosol composition of $PM_{2.5}$ in some urban areas in non-heating periods [9–12]. In this study, we conducted a field campaign in regional haze episodes and chose a background site in the central area of Northeast China. Aerosol properties at the background site can reflect anthropogenic sources due to their low background around the sampling site. Also, aerosol properties at the background site can indicate the particle aging process during their long-range transports, which can help us understand the formation mechanism of regional haze episodes in Northeast China. A high-resolution transmission electron microscopy (TEM) was used to investigate morphology, composition, and mixing state of individual aerosol particles.

2. Experiments

2.1. Aerosol Sampling

The Northeast China Plain was divided into a western and an eastern part based on the distribution of its population and cities. Major cities, including the capital cities of three provinces (Heilongjiang, Jilin, and Liaoning), are located in the eastern part and only a few small cities and some villages are located in the western part. Therefore, we generally considered the eastern part to be an urban influence area and the western part to be a rural influence area. In this study, aerosol samples were collected from 28 January 2015 to 9 February 2015 in the western part of Northeast China. Two sampling sites were chosen for aerosol collection in Tongyu of Jilin Province (Figure 1): the suburban county town site (T1, 44.81° N, 123.09° E) and the background rural site (T2, 44.42° N, 122.87° E). T1 was located in a county town area which was mainly influenced by automobile exhaust and solid fuels (e.g., coal and biomass burning) for heating and cooking. T2 was situated in the rural area, 50 km south from the T1 site. Some small villages, but no industries, are situated in the vast rural area. Based on our local investigation, coals were the main fuels for cooking and heating in central Tongyu county, and a mixture of coals, wheat straw, and cornstalks were the main fuels for cooking and heating in rural areas of Tongyu county or in the villages.

The NOAA (National Oceanic and Atmospheric Administration) HYSPLIT (Hybrid Single Particle Lagrangian Integrated Trajectory) trajectory model was used to investigate transport pathways of air masses and their origins. Forty-eight hours backward trajectories arriving at 1000 m above ground level at 16 UTC were calculated for the T1 and T2 sites. During the sampling period, the air masses mainly came from northwestern areas including the northern part of Inner Mongolia, and Greater Khingan forests where some villages and towns are distributed (Figure 1 and Figure S1). During our sampling period, a regional haze occurred from 3 to 4 February 2015 (Figures S2 and S3).

Copper TEM grids coated with carbon film (carbon type-B, 300-mesh copper, TianId Co., Beijing, China) were used to collect aerosol particles by a single-stage cascade impactor with a 0.5 or 0.3 mm diameter jet nozzle and an air flow of 1.0 L/min. This sampler has a collection efficiency of 100% and a 0.5 µm aerodynamic diameter if the density of the particles is 2 g/cm^3. To avoid particles overlapping on the substrate, the sampling duration was controlled from 3 min to 20 min according to the visibility and PM concentration. In other words, the sampling duration on clear days due to the low $PM_{2.5}$ concentration was longer than on haze days. All samples were placed in sealed, dry plastic capsules and preserved in a desiccator at 25 °C and 20% ± 3% relative humidity (RH) for subsequent analysis. The interval between the sample collections at the T1 and T2 sites could vary across a few hours in a day.

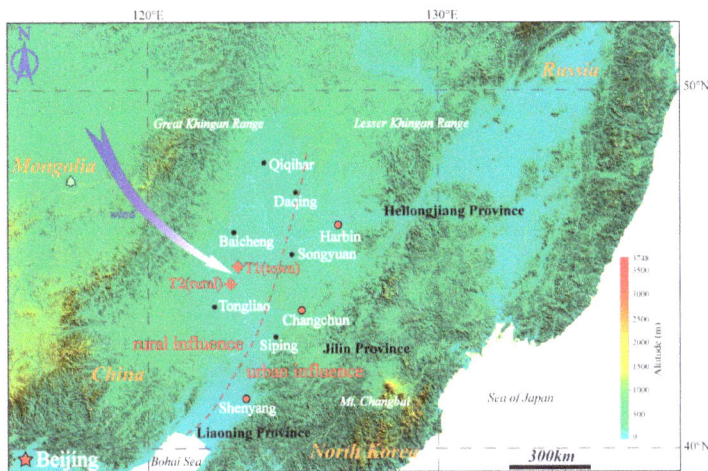

Figure 1. Locations of the two sampling sites and topography of the Northeast China area. The red line from the south to the north indicates rural and urban influence of air quality in Northeast China based on the distribution of population and cites. The blue arrow shows the main wind from northwest China and Mongolia during the sampling period. The T1 and T2 sites are located in a rural influence area.

2.2. Electron Microscopic Analyses

Aerosol particle samples were analyzed by JEOL JEM–2100 transmission electron microscopy (TEM, JEOL Ltd., Tokyo, Japan) operated at 200 kV. Elemental composition was determined semi-quantitatively by using an energy-dispersive X-ray spectrometer (EDS) (Oxford Instruments, Oxfordshire, UK) that can detect elements heavier than C ($Z \geq 6$). The relative percentages of the elements were estimated based on the EDS spectra acquired through the INCA software (Oxford Instruments, Oxfordshire, UK). The distribution of aerosol particles on TEM grids was not homogeneous: coarser particles occur near the center and finer particles are on the periphery. Therefore, to be more representative, three areas were chosen from the center to periphery of the sampling spot on each grid. iTEM software (Olympus Soft Imaging Solutions GmbH, Münster, Germany) was used to analyze the TEM images and to obtain the projected area, circularity, perimeter, and equivalent circle diameter of particles. A total of 2589 particles (1653 for T1 and 936 for T2) were analyzed at the two sites.

3. Results

3.1. Aerosol Particle Types

A regional light haze was simultaneously observed at the T1 and T2 sites from 3 to 4 February 2015 (Figures S2 and S3). Individual particle samples were collected from 28 January 2015 to 9 February 2015. Individual particles were divided into four major particle types depending on their different composition and morphology: S-rich, organic matter (OM), soot, and K-rich.

S-rich particles contained sulfur along with nitrogen, oxygen, and minor potassium. S-rich particles normally exhibited a rounded shape on the substrate (Figure 2). OM particles were composed of abundant carbon and minor amounts of oxygen. S-rich particles were easily sublimated under strong beam exposure, while OM particles were stable in the samples (Figure 2). The similar phenomenon has been described by Adachi and Buseck [13] and Fu et al. [8]. In our study, most of the OM particles were internally mixed with S-rich and K-rich particles (Figure 2). Based on the classifications of mixing structures of internally mixed particles proposed by Li et al. [14], we found two types of mixing structures between S-rich and OM particles, such as dumbbell-like particles (Figure 2b) and core-shell particles (e.g., OM particles coated by S-rich (Figure 2c) and S-rich particles coated by OM (Figure 2d)).

K-rich particles displayed irregular shapes and contained potassium (K), oxygen, sulfur, and/or nitrogen. Once K-rich particles are exposed longer times under the strong electron beam, they can be sublimed. Although the morphology of K-rich particles is similar to S-rich particles, their EDS were different. The EDS of K-rich particles usually displayed high O, K, and/or S (Figure 3a) and S-rich particles displayed high O and S with minor K (Figure 2). The chemical composition of these K-rich particles suggests that they could be KNO_3 and/or K_2SO_4 with minor amounts of KCl. TEM observations showed that most of the K-rich particles were internally mixed with organic, S-rich, and soot particles (Figure 3). In general, K-rich particles can be considered as an important tracer of biomass burning [13,15].

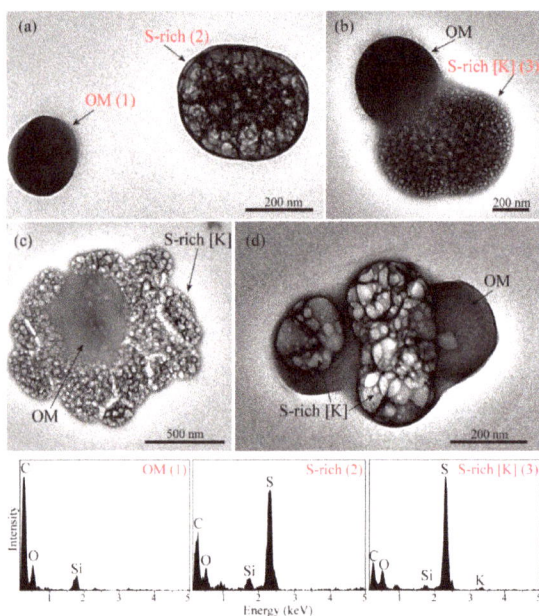

Figure 2. Transmission electron microscopy (TEM) images of organic matter (OM) and S-rich particles. (**a**) Externally mixed S-rich and OM particle; (**b**) dumbbell mixing structure between S-rich and OM particles; (**c**) OM particles surrounded by S-rich; (**d**) S-rich particle coated by OM. Energy-dispersive X-ray spectrometer (EDS) spectra show elemental composition of individual particles in red font in TEM images.

Soot particles, also known as black carbon (BC) or element carbon (EC), are a type of carbonaceous material with graphitic structures, which mainly come from incomplete combustion of fossil fuels and biomass [16,17]. Soot particles exhibit a chain-like aggregation morphology with a diameter of 10 to 100 nm [18]. Most soot particles observed in our study were internally mixed with S-rich and/or OM particles (Figure 4).

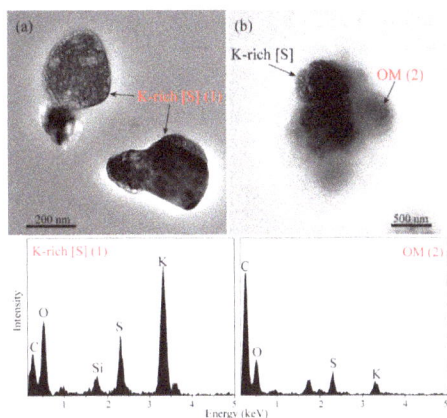

Figure 3. TEM images of K-rich particles. (**a**) K-rich particles; (**b**) K-rich particle internally mixed with OM particle. EDS spectra shows the elemental composition of individual particles in red font in TEM images.

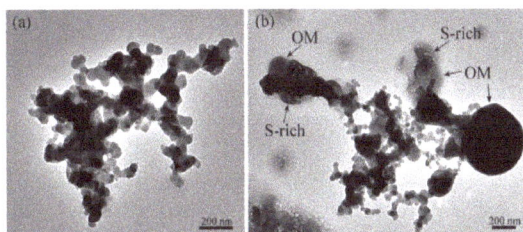

Figure 4. TEM images of soot particles. (**a**) Bare soot particle; (**b**) soot internally mixed with S-rich and OM particle.

3.2. Relative Abundance of Different Types of Aerosol Particles

In the samples, most of the individual particles were internal mixture which contains more than two different particle types (Figures 2–4). Here we named each particle according to their major aerosol component , as shown in Figure 5. In this study, 1653 and 936 particles were analyzed at the T1 (town) and T2 (rural) sites, respectively.

Figure 5 shows that the main aerosols are OM-dominated particles, accounting for 60.7% and 53.5% at the T1 and T2 sites, respectively. The percentage of soot particles was four times higher at the T1 site (16.9%) than at the T2 site (4.6%), and 35.9% of S-rich particles occurred at the T2 site, higher than 15.7% at the T1 site. At the two sampling sites, K-rich particles only accounted for about 6.7% and 6.0% of particles, respectively. It should be noted that most of the OM and S-rich particles (Figure 2) contained minor amounts of K.

Both spherical and irregular OM particles were considered as primary OM particles, they formed in cooling polluted plumes from coal combustion and biomass burning [19]. In contrast, the formation mechanism of OM coating was different from spherical OM and irregular OM, and they were normally considered secondary OM particles as the oxidized productions of volatile organic compounds (VOCs) [20,21]. In OM-dominated particles, we observed a high fraction of spherical OM particles and irregular OM particles (62%) at the T1 site and T2 site, suggesting that OM particles mainly came from primary emissions (e.g., coal combustion and biomass burning).

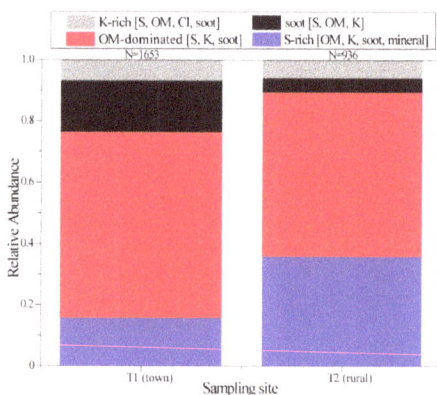

Figure 5. Relative abundance of aerosol particles at the T1 site and T2 site. Number (N) of the analyzed aerosol particles is shown above each column. The bracket indicates that individual particle possibly contained other particle components.

3.3. Size Distribution of S-rich Particles

S-rich particles represent secondary inorganic particles which mainly contain ammoniated sulfates and ammoniated nitrates [14]. The relative abundance of secondary particles has been used to suggest transports of air masses and their ageing properties [22]. We noticed that the relative abundance of S-rich particles was higher at the T2 site than at the T1 site. Therefore, it is reasonable that more S-rich particles occurred at the T2 site compared to the T1 site. Figure 6 shows that the size distribution of S-rich particles ranged from 0.04 to 1.5 μm with a median diameter of 0.24 μm at the T1 site, and from 0.04 μm to 1.6 μm with a median size of 0.31 μm at the T2 site. The size distribution of aerosol particles displays a broader peak at the T2 site than at the T1 site (Figure 6). In addition, TEM images show that most of the S-rich particles were internally mixed with OM particles (Figure 2). As we expected, long-range transport of air masses induced larger median size and broader peak in size distribution of individual particles at the T2 site than at the T1 site because additional secondary inorganic aerosols forming during long-range transports occurred on primary OM particles.

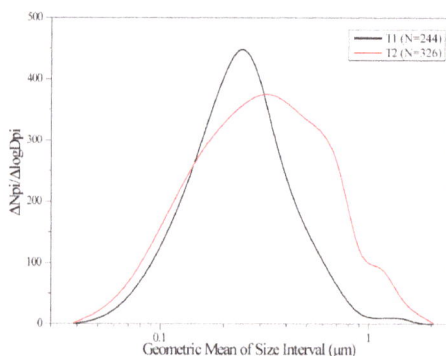

Figure 6. Size distributions of S-rich particles at the T1 site and T2 site.

4. Discussion

Based on the locations of the two sampling sites, we can know that air quality at the T1 site was influenced by anthropogenic air pollutants from long-range transport and local towns, and at the T2 site from long-range transport. The analyses were consistent with our results from TEM observations. We found more secondary particles (e.g., S-rich) and lesser primary soot particles at the T2 site than at the T1 site (Figures 5 and 7). In addition, these S-rich particles were internally mixed with primary OM particles at the T2 site (Figure 2b–d), thereby suggesting that OM particles become more aged during long-range transport.

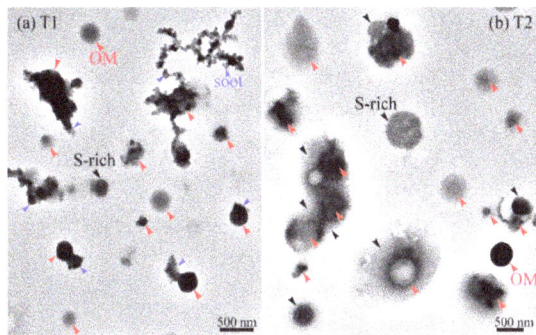

Figure 7. TEM images of aerosol particles at the T1 site (**a**) and the T2 site (**b**). The arrows in red, black, and blue indicate OM, S-rich, and soot particles, respectively.

Compared with haze particles collected in the North China Plain, our study did not observe fly ash and metal, but more abundance of OM-dominated particles [23,24]. As a result, we can exclude regional contribution from emissions of heavy industries and coal-fired power plants in upwind areas. TEM observations suggested that most of the OM-dominated particles were directly emitted from sources. In this study, we considered OM-dominated particles sourced from coal-burning for household heating and cooking at the T1 site [25]. Abundant OM-dominated particles at the T2 site likely were transported from villages and towns where people mainly used coals from household heating and cooking in the western part of Northeast China in winter [12]. In addition, the central heating from filter boilers are only supplied in the central part of the town and people in most areas of the town still used stoves for household heating. In rural areas, most of the people living in the villages used coal-burning stoves for household heating and cooking. Coal combustion in low-efficiency stoves for residential heating and cooking was considered to be an important source of air pollution in China [26]. Based on our investigation, similar phenomena appear to occur in all the small towns and rural villages in western part of Northeast China (Figure 1). As discussed in the previous report, the direct emissions from coal-burning stoves have several times higher emissions than filter boilers [27].

We noticed that 6%–7% of K-rich particles which are a tracer of biomass burning occurred at the T1 site and T2 site (Figure 5). In rural areas of Northeast China, the biomass materials (e.g., maize stalks and wheat straws) are often used for cooking in all seasons and for household heating from winter to spring [11]. Our results indicate that the regional contribution of biomass burning emissions should be far smaller than the contribution from coal burning emissions. However, the similar abundance of K-rich particles from the T1 and T2 sites indicated that continuous contributions from biomass burning emissions in villages should not be ignored in the whole Northeast China region.

5. Conclusions

Individual fine particles were collected at a suburban county town site (T1) and a background rural site (T2) in Northeast China during wintertime. Four types of individual fine particles such as S-rich, OM-dominated, soot, and K-rich particles were identified based on their morphology and composition. In the ambient air, most particles were internally mixed particles, which refers to particles that are composed of more than two types of particles. In particular, large amounts of visible primary spherical or irregular OM particles occurred at sites and were internally mixed with S-rich, soot, or K-rich particles. The phenomenon was quite different from the previous studies which suggested that secondary particles were dominant in the air in North China.

Our results revealed that OM particles were the most abundant aerosol particles, accounting for 60.7% and 53.5% of the particles at the T1 site and the T2 site, respectively. Abundant OM particles at the two sites suggested that they have similar emission sources at the background and town areas. Our investigations suggested that coal-burning for household heating and cooking in winter is likely the main source. We noticed a similar small fraction of K-rich particles (6%–7%) at the two sites, suggesting that regional contribution of biomass burning should be considered in the western part of Northeast China. Similar to coal-burning, these biomass burnings were used for household heating and cooking in rural areas. In comparing the aerosol particles at the two sampling sites, soot particles decrease from 16.9% at the T1 site to 4.6% at the T2 site and sulfate particles decrease from 35.9% at the T2 site to 15.7% at the T1 site. These results suggested that long-range transport particles experienced more aging processes and produced more secondary particles. We believe that the emissions from coal-burning for household heating and cooking can induce regional haze formation in the western part of Northeast China. In the future, central household heating with filtered boilers should be widely installed in towns in the western part of Northeast China.

In this study, we focused on sources and the formation of haze particles in a rural influence area of Northeast China (Figure 1). We further integrated our findings and regional air transports in winter in Northeast China. A conceptual model was proposed that can reflect emission and transport of air pollutants from western to eastern parts in Northeast China following the winter monsoon (Figure 8). It is interesting that the conceptual model shows that the anthropogenic pollutants emitted from a rural influence area can be transported and further impact the air quality of an urban influence area in Northeast China. The result is different from the traditional perspective that anthropogenic pollutants in urban areas influence the air quality of rural areas. Therefore, further study is needed to work on the extent to which the air quality of the urban influence areas are impacted through the transportation of pollutants from rural influence areas in winter in Northeast China.

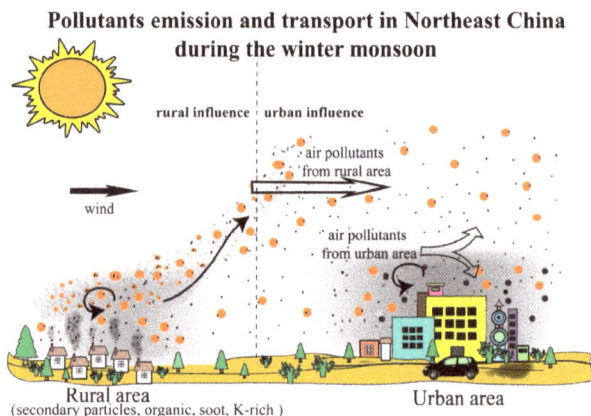

Figure 8. The conceptual graph of the emission and transport of pollutants in Northeast China during the winter monsoon. The direct emissions from rural area can influence the air quality of downwind urban areas.

Supplementary Materials: The following are available online at www.mdpi.com/2073-4433/8/3/47/s1. Figure S1: Forty-eight hour air mass backward trajectory for the T1 site and T2 site during our sampling period; Figure S2: China Meteorological Administration haze forecast map at 3 February 2015; Figure S3: Particulate matter (PM) concentration for Baicheng (a) and Tongliao (b), two cities which were close to our sampling sites.

Acknowledgments: This work was funded by grants from the National Natural Science Foundation of China (41575116 and 41622504), Shandong Provincial Science Fund for Distinguished Young Scholars, China (JQ201413), and Programs of Shandong University (2015WLJH37).

Author Contributions: Weijun Li and Liang Xu conceived and designed the experiments; Yong Ren, Xin Wang did the samples collection; Jian Zhang carried out TEM measurements; Lei Liu, Jian Zhang and Yinxiao Zhang carried out data analysis; Liang Xu and Weijun Li wrote the paper.

Conflicts of Interest: The authors declare no conflict of interest.

References

1. West, J.J.; Cohen, A.; Dentener, F.; Brunekreef, B.; Zhu, T.; Armstrong, B.; Bell, M.L.; Brauer, M.; Carmichael, G.; Costa, D.L.; et al. What We Breathe Impacts Our Health: Improving Understanding of the Link between Air Pollution and Health. *Environ. Sci. Technol.* **2016**, *50*, 4895–4904. [CrossRef] [PubMed]
2. Geng, H.; Kang, S.; Jung, H.J.; Choël, M.; Kim, H.; Ro, C.U. Characterization of individual submicrometer aerosol particles collected in Incheon, Korea, by quantitative transmission electron microscopy energy-dispersive X-ray spectrometry. *J. Geophys. Res.* **2010**, *115*. [CrossRef]
3. Heald, C.L.; Jacob, D.J.; Park, R.J.; Alexander, B.; Fairlie, T.D.; Yantosca, R.M.; Chu, D.A. Transpacific transport of Asian anthropogenic aerosols and its impact on surface air quality in the United States. *J. Geophys. Res.* **2006**, *111*. [CrossRef]
4. Lin, J.; Pan, D.; Davis, S.J.; Zhang, Q.; He, K.; Wang, C.; Streets, D.G.; Wuebbles, D.J.; Guan, D. China's international trade and air pollution in the United States. *Proc. Natl. Acad. Sci. USA* **2014**, *111*, 1736–1741. [CrossRef] [PubMed]
5. Huang, R.J.; Zhang, Y.; Bozzetti, C.; Ho, K.F.; Cao, J.J.; Han, Y.; Daellenbach, K.R.; Slowik, J.G.; Platt, S.M.; Canonaco, F.; et al. High secondary aerosol contribution to particulate pollution during haze events in China. *Nature* **2014**, *514*, 218–222. [CrossRef] [PubMed]
6. Guo, S.; Hu, M.; Zamora, M.L.; Peng, J.; Shang, D.; Zheng, J.; Du, Z.; Wu, Z.; Shao, M.; Zeng, L.; et al. Elucidating severe urban haze formation in China. *Proc. Natl. Acad. Sci. USA* **2014**, *111*, 17373–17378. [CrossRef] [PubMed]
7. Zhang, Y.; Hu, M.; Zhong, L.; Wiedensohler, A.; Liu, S.; Andreae, M.; Wang, W.; Fan, S. Regional integrated experiments on air quality over Pearl River Delta 2004 (PRIDE-PRD2004): Overview. *Atmos. Environ.* **2008**, *42*, 6157–6173. [CrossRef]
8. Fu, H.; Zhang, M.; Li, W.; Chen, J.; Wang, L.; Quan, X.; Wang, W. Morphology, composition and mixing state of individual carbonaceous aerosol in urban Shanghai. *Atmos. Chem. Phys.* **2012**, *12*, 693–707. [CrossRef]
9. Zhao, H.; Che, H.; Zhang, X.; Ma, Y.; Wang, Y.; Wang, X.; Liu, C.; Hou, B.; Che, H. Aerosol optical properties over urban and industrial region of Northeast China by using ground-based sun-photometer measurement. *Atmos. Environ.* **2013**, *75*, 270–278. [CrossRef]
10. Wu, Y.; Zhu, J.; Che, H.; Xia, X.; Zhang, R. Column-integrated aerosol optical properties and direct radiative forcing based on sun photometer measurements at a semi-arid rural site in Northeast China. *Atmos. Res.* **2015**, *157*, 56–65. [CrossRef]
11. Shen, Z.; Wang, X.; Zhang, R.; Ho, K.; Cao, J.; Zhang, M. Chemical composition of water-soluble ions and carbonate estimation in spring aerosol at a semi-arid site of Tongyu, China. *Aerosol Air Qual. Res.* **2011**, *11*, 360–368. [CrossRef]
12. Zhang, R.; Tao, J.; Ho, K.; Shen, Z.; Wang, G.; Cao, J.; Liu, S.; Zhang, L.; Lee, S. Characterization of atmospheric organic and elemental carbon of $PM_{2.5}$ in a typical semi-arid area of Northeastern China. *Aerosol Air Qual. Res.* **2012**, *12*. [CrossRef]
13. Adachi, K.; Buseck, P.R. Internally mixed soot, sulfates, and organic matter in aerosol particles from Mexico City. *Atmos. Chem. Phys.* **2008**, *8*, 6469–6481. [CrossRef]
14. Li, W.; Sun, J.; Xu, L.; Shi, Z.; Riemer, N.; Sun, Y.; Fu, P.; Zhang, J.; Lin, Y.; Wang, X.; et al. A conceptual framework for mixing structures in individual aerosol particles. *J. Geophys. Res.* **2016**, *121*, 13784–13798. [CrossRef]
15. Bi, X.; Zhang, G.; Li, L.; Wang, X.; Li, M.; Sheng, G.; Fu, J.; Zhou, Z. Mixing state of biomass burning particles by single particle aerosol mass spectrometer in the urban area of PRD, China. *Atmos. Environ.* **2011**, *45*, 3447–3453. [CrossRef]

16. Niu, H.; Shao, L.; Zhang, D. Soot particles at an elevated site in eastern China during the passage of a strong cyclone. *Sci. Total Environ.* **2012**, *430*, 217–222. [CrossRef] [PubMed]

17. China, S.; Scarnato, B.; Owen, R.C.; Zhang, B.; Ampadu, M.T.; Kumar, S.; Dzepina, K.; Dziobak, M.P.; Fialho, P.; Perlinger, J.A. Morphology and mixing state of aged soot particles at a remote marine free troposphere site: Implications for optical properties. *Geophys. Res. Lett.* **2015**, *42*, 1243–1250. [CrossRef]

18. Buseck, P.R.; Adachi, K.; Gelencsér, A.; Tompa, É.; Pósfai, M. Ns-Soot: A Material-Based Term for Strongly Light-Absorbing Carbonaceous Particles. *Aerosol Sci. Technol.* **2014**, *48*, 777–788. [CrossRef]

19. Chen, S.; Xu, L.; Zhang, Y.; Chen, B.; Wang, X.; Zhang, X.; Zheng, M.; Chen, J.; Wang, W.; Sun, Y.; et al. Direct observations of organic aerosols in common wintertime hazes in North China: Insights into direct emissions from Chinese residential stoves. *Atmos. Chem. Phys.* **2017**, *17*, 1259–1270. [CrossRef]

20. Adachi, K.; Zaizen, Y.; Kajino, M.; Igarashi, Y. Mixing state of regionally transported soot particles and the coating effect on their size and shape at a mountain site in Japan. *J. Geophys. Res.* **2014**, *119*, 5386–5396. [CrossRef]

21. Moffet, R.C.; Rödel, T.; Kelly, S.T.; Yu, X.-Y.; Carroll, G.; Fast, J.; Zaveri, R.A.; Laskin, A.; Gilles, M.K. Spectro-microscopic measurements of carbonaceous aerosol aging in Central California. *Atmos. Chem. Phys.* **2013**, *13*, 10445–10459. [CrossRef]

22. Niu, H.; Hu, W.; Zhang, D.; Wu, Z.; Guo, S.; Pian, W.; Cheng, W.; Hu, M. Variations of fine particle physiochemical properties during a heavy haze episode in the winter of Beijing. *Sci. Total Environ.* **2016**, *571*, 103–109. [CrossRef] [PubMed]

23. Li, W.; Zhou, S.; Wang, X.; Xu, Z.; Yuan, C.; Yu, Y.; Zhang, Q.; Wang, W. Integrated evaluation of aerosols from regional brown hazes over northern China in winter: Concentrations, sources, transformation, and mixing states. *J. Geophys. Res.* **2011**. [CrossRef]

24. Li, W.J.; Zhang, D.Z.; Shao, L.Y.; Zhou, S.Z.; Wang, W.X. Individual particle analysis of aerosols collected under haze and non-haze conditions at a high-elevation mountain site in the North China plain. *Atmos. Chem. Phys.* **2011**, *11*, 11733–11744. [CrossRef]

25. Shao, L.; Hou, C.; Geng, C.; Liu, J.; Hu, Y.; Wang, J.; Jones, T.; Zhao, C.; BéruBé, K. The oxidative potential of PM_{10} from coal, briquettes and wood charcoal burnt in an experimental domestic stove. *Atmos. Environ.* **2016**, *127*, 372–381. [CrossRef]

26. Liu, J.; Mauzerall, D.L.; Chen, Q.; Zhang, Q.; Song, Y.; Peng, W.; Klimont, Z.; Qiu, X.; Zhang, S.; Hu, M.; et al. Air pollutant emissions from Chinese households: A major and underappreciated ambient pollution source. *Proc. Natl. Acad. Sci. USA* **2016**, *113*, 7756–7761. [CrossRef] [PubMed]

27. Zhang, Y.; Schauer, J.J.; Zhang, Y.; Zeng, L.; Wei, Y.; Liu, Y.; Shao, M. Characteristics of Particulate Carbon Emissions from Real-World Chinese Coal Combustion. *Environ. Sci. Technol.* **2008**, *42*, 5068–5073. [CrossRef] [PubMed]

atmosphere

MDPI

Article

Physicochemical Characteristics of Individual Aerosol Particles during the 2015 China Victory Day Parade in Beijing

Wenhua Wang, Longyi Shao *, Jiaoping Xing, Jie Li, Lingli Chang and Wenjun Li

State Key Laboratory of Coal Resources and Safe Mining, College of Geoscience and Surveying Engineering, China University of Mining and Technology (Beijing), Beijing 100083, China; whwang91@126.com (W.W.); xingjiaoping@126.com (J.X.); jieli_lj@163.com (J.L.); linglilp@126.com (L.C.); liwenjun_620@126.com (W.L.)
* Correspondence: ShaoL@cumtb.edu.cn

Received: 17 December 2017; Accepted: 23 January 2018; Published: 25 January 2018

Abstract: During the 2015 China Victory Day parade control periods, the air quality in Beijing hit the best record, leading to 15 continuous good days with an average $PM_{2.5}$ mass concentration 18 $\mu g/m^3$, which provided a unique opportunity to study the ambient aerosols in megacity Beijing. The morphology and elemental composition of aerosol particles were investigated by transmission electron microscopy coupled with energy dispersive X-ray spectrometry (TEM-EDX). Five types of individual particles were identified, including homogeneous mixed S-rich particles (HS; 44.9%), organic coated S-rich particles (CS; 34.3%), mineral particles (10.5%), soot aggregates (7.21%) and organic particles (3.2%). The number percentage of secondary particles (including HS and CS) accounted for a large proportion with 79.2% during the control periods. The average diameter of secondary particles increased with relative humidity (RH), being 323 nm, 358 nm and 397 nm at the RH 34%, 43% and 53%, respectively, suggesting that the high RH might favor the growth of secondary particles. The higher proportion of CS particles may show great atmospheric implications and the CS particles may be formed by the condensation of secondary organic aerosols on pre-existing S-rich particles.

Keywords: morphology; aerosol particle; mixing state; S-rich particles

---

1. Introduction

Atmospheric aerosols, both derived from anthropogenic and natural sources, are composed of various kinds of organic and inorganic species. They have received much attention in megacities in China because they are involved in a multitude of climate and environmental issues [1]. Air pollution can cause a great impact on human health, such as respiratory and cardiovascular problems due to their association with toxic matters [2,3]. Aerosol particles can influence atmospheric chemistry because they can act as a medium for heterogeneous reactions. They can also change the climate indirectly by cloud condensation nuclei (CCN) activity and ice nuclei activity [4,5] and directly through light absorption and scattering [6,7]. Various anthropogenic aerosols can undergo long-rang transport, which can cause a regional and global environmental problem [8].

Recently, China has experienced serious atmospheric pollution due to rapid social and economic development, similar to that in developed nations [1]. The air pollution is characterized by high $PM_{2.5}$ mass concentration, frequent occurrences of haze days, expanded haze areas, and increased duration of a single haze event [7,9–12]. For example, extremely severe and persistent air pollution occurred in China during January 2013, with more than 1.3 million km^2 and 800 million people being affected [1]. Air pollution has been a concern for decades.

The tropospheric aerosol is a heterogeneous mixture of various particle types. Detailed knowledge of the chemical composition, physical state, morphology and size of individual aerosol particles should be known to fully understand the formation mechanism of regional haze episodes and to accurately predict the climate effects [13,14]. A high-resolution transmission electron microscopy (TEM) can be used to investigate the composition, morphology and mixing state of individual aerosol particles.

Understanding the relationship between source emissions and aerosol chemistry is important for mitigating air pollution in megacities [15]. In order to guarantee good air quality during some specific events, e.g., the Beijing Olympic Games in 2008 [16] and the Asia-pacific Economic Cooperation summit (APEC) in 2014 [17], the municipal government launched strong environmental control measures in Beijing and its surrounding areas, which provided some special experimental opportunities to study the impacts of source emission controls on air quality and the physicochemical properties of aerosol particles during clean-up control period. Similarly, a series of temporary control measures, including stopping construction activities, shutting down power plants and factories, limiting the number of vehicles on road and prohibiting outdoor barbecues, were implemented in Beijing and surrounding regions from 20 August to 3 September to ensure good air quality during the 2015 China Victory Day parade [15]. As a result, the air quality hit the best record in Beijing, leading to 15 continuous good days; the average $PM_{2.5}$ mass concentration was 18 μg/m^3 during the control period, with a 73.1% reduction compared with the same period in 2014 (http://www.gov.cn/xinwen/2015-09/07/content_2926447.htm). In this study, the characteristics of individual aerosol particles during the 2015 China Victory Day parade were analyzed by a high resolution TEM.

2. Experimental

2.1. Aerosol Sampling

The sampling site was located at China University of Mining & Technology, northwestern part of urban Beijing which is surrounded by commercial buildings and residential apartments. The sampler was placed on the top floor of a five-story teaching building. Cooper TEM grids coated with carbon film (300-mesh copper, Tianld Co., Beijing, China) were used to collect the aerosol samples. A single-stage cascade impactor with a 0.5 mm diameter jet nozzle and an air flow of 1.0 L/min was applied. The collection efficiency of this kind of sampler is close to 100% on a 0.5 μm aerodynamic diameter if the particle density is 2 g/cm^3 [18]. To avoid the particle overlapping on subtracts, the sampling time ranged from 60 to 90 s. A kestral 4000 pocket weather meter (Nielsen-kellermann Inc., Boothwyn, PA, USA) was simultaneously applied to record the RH, temperature (T) and pressure (P). All collected samples were placed in dry and sealed plastic tubes and stored in a desiccator of 25 °C and 20 ± 3% RH to avoid exposure to ambient air before laboratory analysis [19]. The sample information is shown in Table 1.

Table 1. Information on analyzed samples from Beijing.

ID	Collection Date	T (°C)	RH (%)	P (hPa)
A	2015/8/28	34.8	34.3	1000.1
B	2015/9/3	29.3	43.5	1005.1
C	2015/9/2	27.2	53.3	1006.5

2.2. TEM Analysis

Individual aerosol samples were analyzed by TEM (Hitachi H-8100, Hitachi, Ltd., Tokyo, Japan) with an accelerate voltage of 200 kV. Elemental composition was semi-quantitatively determined by energy-dispersive X-ray spectrometer (EDX) with a spectral acquisition time of ~30 s and elements heavier than C (Z ≥ 6) can be detected [20]. Copper (Cu) was not included from our analysis because the TEM grids were made of Cu. The distribution of aerosol particles on TEM grids was not even with coarser particles in the center and finer particles in the periphery. Therefore, three areas of each sample were selected from the center to the periphery to ensure representative data. An image analysis software (Leica Microsystems Image Solutions Ltd, Cambridge, UK) was used to obtain the equivalent circle diameter of particles [21]. A total of 810 individual particles were analyzed.

3. Results

3.1. Nature of Individual Aerosol Particles

TEM-EDX can be used to adequately characterize the individual aerosol particles. Based on their possible sources, we classified the aerosol particles into two groups: primary and secondary particles, which were further classified into five types as shown in Table 2. The primary particle group included soot, organic and mineral particles. The secondary particles included homogeneous mixed S-rich particles (HS) and organic coated S-rich particles (CS).

Table 2. Classification of individual aerosol particles.

Types	Sub-Types	Physical and Chemical Characteristics
Primary particles	Mineral	Irregular shaped; mainly composed O, Si, Al, Fe, Ca, Na, K, Mg; soil, road dust and construction dust sources
	Soot	Aggregate shaped spheres; composed of C and minor O; sourced from incomplete combustion.
	Organic	Spherical or near spherical shaped; composed of C and O; combustion sources
Secondary particles	Homogeneous mixed S-rich particles (HS)	Spherical or near spherical shaped; composed of C, S, O, N and sometimes with minor K; beam-sensitive; secondary formation in atmosphere
	Organic coated S-rich particles (CS)	Core-shell structure; composed of C, S, O, N and sometimes with minor K; beam-sensitive; secondary formation in atmosphere

Most of the mineral particles are irregularly shaped and tend to have a larger diameter than other particle types as shown in Figure 1a,b. Their elemental compositions are complex, consisting of O, Si, Al, Fe, Ca, Na, K, Mg and S. Mineral particles are mainly from suspension of soil, road dust, and construction dust. They can be also sourced from desert dust and undergo long rang transport [22]. Mineral particles are stable under strong electron beam irradiation.

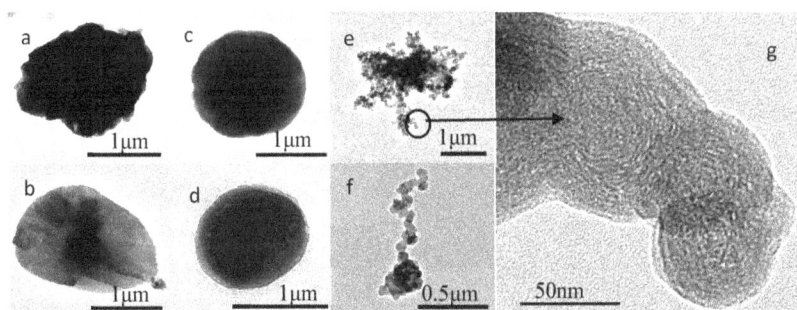

Figure 1. TEM images of primary individual particles: (**a,b**) mineral particles, (**c**) spherical organic particle, (**d**) near-spherical organic particle, (**e**) aged soot particle and (**f**) chain-like soot particle; (**g**) amplified soot spherical.

Soot aggregates contain numbers of C-dominated spheres with diameter less than 100 nm (Figure 1e,f). High magnification TEM images demonstrate that typical soot spheres display the onion-like structures with disordered graphitic layers as shown in Figure 1g. They are from incomplete combustion of fossil fuels and other materials [23].

Organic particles show higher O proportion than soot aggregates, though they are both dominated by C and O. Organic particles lack onion-like structure under high magnification TEM images, compared with soot particles. Most of the organic particles in this study are spherical or near spherical shaped. Li and Shao suggest that the primary organic particles normally have a spherical or irregular shape [24]. They are from combustion sources.

Different from primary particle types, the secondary particles are beam-sensitive when exposed to several seconds of beam irradiation. These secondary particles are mainly composed of element C, O, S and minor N and are defined as S-rich particles [18,20]. We infer that they indeed are mixtures of sulfate, nitrate, and/or organic matter (OM) in the atmosphere [18]. Although the secondary particles have similar elemental compositions, they show different mixing states and thus are further classified into HS and CS particles.

HS particles are foam-like as shown in Figure 2a,b. Although they are not strictly evenly distributed, we term these particles as homogeneous mixed S-rich particles (HS). Although HS particles look similar in the TEM images, they may display different optical and/or hygroscopic properties because of their mass fractions of sulfate, OM, and/or nitrate [22]. Apart from HS particles, we also find some secondary particles with organic coating. These core-shell particles clearly show an uneven surface, as shown in Figure 2c,d. When exposed to electron beam irradiation, the core disappeared, leaving some residues and an obvious shell structure as shown in Figure 2e,f. The organic shell and inorganic core have been certified by using a nanometer-scale secondary ion mass spectrometer (Nano-SIMS) according to Li et al. [25].

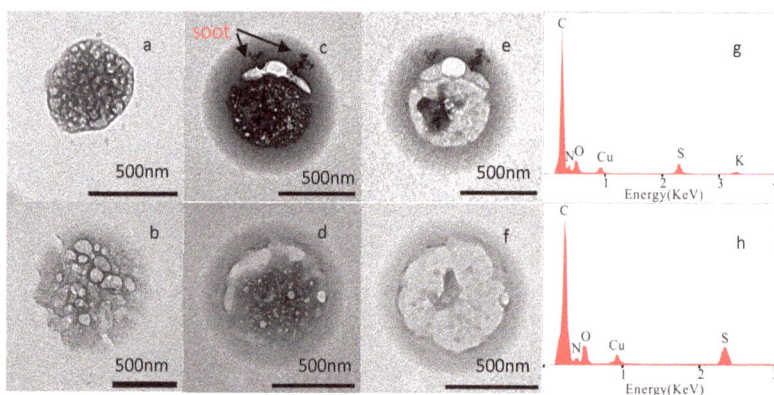

Figure 2. TEM images and elemental compositions of secondary particles: (**a,b**) HS particles, (**c,d**) CS particles, (**e,f**) images of (**c,d**) after electron beam exposure; (**g,h**) elemental compositions of (**c,d**).

3.2. Relative Number Percentage and Size Distribution

A total of 810 individual aerosol particles (336 for A, 246 for B and 228 for C) were analyzed in this study. Figure 3 shows the relative number percentage of individual aerosol particles. The results showed that there was a large proportion of secondary particles because of the strict emission control measures. During the 2015 China Victory Day parade control periods, HS particles and CS particles represent the major fraction with 44.9% and 34.3%, respectively, followed by mineral particles (10.5%), soot aggregates (7.21%) and organic particles (3.2%; Figure 3). Soot aggregates and organic particles show relative lower number percentage. It is consistent with Wang et al. [26], who reported that both the primary organic carbon and elemental carbon had the relative lower proportion during the control period.

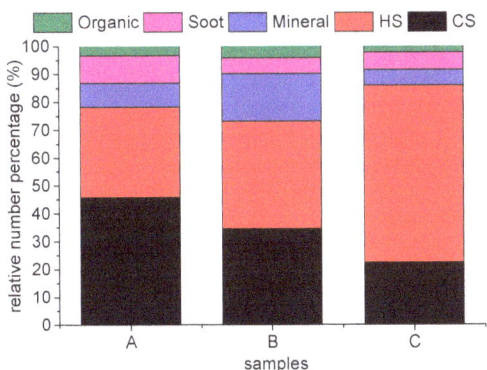

Figure 3. Relative number percentage of individual particles (Homogeneous mixed S-rich particles, HS; Core-shell structured S-rich particles, CS).

Equivalent circle diameter of secondary particles were obtained according to the image analysis software (Leica, UK). Number-size distributions of 639 secondary particles (including HS and CS particles) ranged from 0.1 to 1.0 μm are shown in Figure 4. The average diameter of secondary particles increases with RH, being 323 nm, 358 nm and 397 nm at the RH 34%, 43% and 53%, respectively. In addition, with the increase of RH, the peak shifts to the right side of the coordinate, suggesting that the high RH can favor the growth of secondary particles.

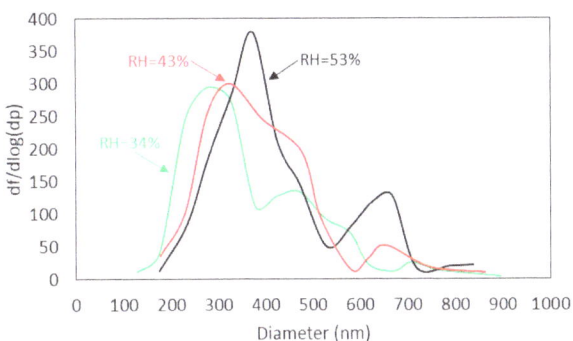

Figure 4. Size distributions of secondary particles at RH 34%, 43% and 53%.

4. Discussion and Atmospheric Implication

4.1. Influence of RH on Secondary Particle Formation

The secondary particles showed a higher proportion in number, accounting for 79.2%. This is consistent with Zhao et al., who reported that the primary species showed relatively more reductions (55–67%) than secondary aerosol species (33–44%) during the control periods [15]. The HS particles are the major fraction of aerosol particles and they increase with the increasing of RH, being 32.4%, 38.6% and 64.6% at RH 34%, 43% and 53%, respectively. The formation of S-rich particles might be affected by a number of factors. The wind direction is mainly from the southern part and the wind speed is comparable during our sampling time in Beijing [27]. The average hourly ozone concentration is 77, 51 and 74 µg/m^3, for samples A, B and C, respectively (https://www.aqistudy.cn/). Both the wind and ozone concentration showed no obvious correlation with the relative number percentage of HS particles. So, we suggest that high RH can favor the formation of HS particles. This is consistent with a previous study by Wang et al. in Beijing during the pollution periods that the S-rich particles were correlated with RH [28].

Water can be a major component of aerosol particles, and serves as a medium for aqueous-phase reactions [29]. Uptake of water by atmospheric particles modifies the mass of the aerosol particle and increases the particle size [30,31]. Previous studies have shown that particle diameter increases with increasing RH [32]. For example, experimental study demonstrated that when exposed to SO_2, NO_2, and NH_3, the growth factor of the seed particles increased with RH, with the values of near unity at RH < 20% and 2.3 at 70% RH [33]. We admit that the particles collected by impaction may potentially be flattened and the secondary particles at different RH might show different viscosity [34], having different flattened ability. However, the larger secondary particles in the atmosphere will generally have a larger flatten size on the filter. Therefore, the absolute diameter of the flattened particles may be different from the diameter of the particles in the atmosphere, but the increasing sizes for the impacted particles should represent the increasing trend of the particle diameters in atmosphere. Figure 4 showed that the average diameter of secondary particles increased with the increasing of RH, suggesting that higher RH can favor the growth of secondary particles. Our result is consistent with previous "bulk" sample analysis, which reported the higher RH promotes the secondary particle formation. For example, Wang et al. found the conversion of SO_2 to SO_4^{2-} not only contributes to the high production rate, but also enhances formations of NO_3^- and SOA on aqueous particles because of increasing particle hygroscopicity [33]. Liu et al. found that sulfates, together with nitrates, significantly contribute to wintertime air pollution in Beijing and suggested that the sulfate is due to heterogeneous reaction with abundant aerosol water under wet conditions [35].

4.2. Atmospheric Implication of CS Particles

The CS particles accounted for a large proportion with 34.3% in our samples. Core-shell morphology is prevalent configuration of tropospheric particles. They were also found in other sampling sites, but showed proportional difference; there are proportionally higher CS particles in mountain Tai, Greater Khingan and rural background sampling site, but less in urban and suburban-polluted air in Beijing and no CS particles in biomass and coal burning sourced particles [18].

When the atmospheric S-rich particles are coated with OM, they can lead to important implications for a wide range of climate-relevant endpoints as well as air quality predictions. The OM coating can change a host particle's early hygroscopic properties [36,37]. They can influence the RH at which crystallization and dissolution of sulfate solids occur in atmospheric particles [38], change the partitioning of semi volatile organic compounds between the gas and particle phase [39,40], and alter the scattering and absorption of solar radiation. In atmospheric chemistry, the heterogeneous reactions of gas-phase species (such as ozone or N_2O_5) on the surface of aerosol particles play an important role in the removal of species that affect the oxidative balance of the atmosphere [41]. If the ammonium sulfate is surrounded by the organic-rich phase, the N_2O_5 uptake may be reduced effectively influenced by the hygroscopicity and viscosity of the organic shell [42], leading to the increased concentrations of gas-phase NO_3 and N_2O_5 and changing the mixing ratios of N_2O_5, NO_3, particle phase nitrate, and volatile organic compounds [43].

4.3. Possible Phase Transformation of CS Particles

Secondary aerosols, formed by gas-to-particle phase conversion, account for a large proportion of total particles in our samples (Figure 3). The formation mechanism of the organic coated S-rich particles remains with some uncertainties. They may be formed by liquid-liquid phase separation under the various RH from high to low values [44] or by condensation of secondary organic compounds on the pre-existing inorganic aerosols [18,45].

Some researchers suggest that the well mixed secondary particles may undergo liquid-liquid phase separation under a lower RH, and the process would change the particles into the core-shell structures, typified by an organic core and an inorganic shell. For example, various "bulk" sample analyses conducted in the laboratory have confirmed the liquid-liquid phase separation do exist for the secondary particles [41,46,47]. The presence of the organic core and inorganic shell in individual particles have been confirmed in the laboratory by using optical microscopy and Raman spectroscopy [38,41,42], atomic force microscopy with infrared spectroscopy [48], and environmental scanning electron microscopy and scanning transmission X-ray microscopy [49].

However, previous studies have demonstrated that the liquid-liquid phase separation is not observed for the organic species having O/C ratio greater than 0.8 [44]. The average O/C ratios of secondary organic aerosols ranged from 0.84 to 1.0, higher than 0.8, during the control periods [15]. The higher O/C ratios of secondary organic aerosols during our sampling periods will not favor the Liquid-liquid phase separation process.

Hou et al. noticed that the secondary S-rich particles can be coated with the secondary organic aerosols, and that the thickness of the organic coatings will increase with the particle's aging [15]. During the control period, the organic aerosols accounted for a larger proportion in the total particles, and they were mostly secondary organic aerosols [26]. Therefore, we believe that the CS particles collected during the control periods were formed by the condensation of secondary organic compounds on pre-existing S-rich particles, instead of liquid-liquid phase separation. When the RH increases, the liquid-liquid mixing may occur on the CS particles again, forming a HS particle.

5. Conclusions

During the 2015 China Victory Day parade control periods, individual aerosol particles were analyzed. Based on the morphologies and elemental compositions, five types of individual particles were identified, which were further classified into two groups, including primary particles (mineral particles, soot aggregates, organic particles) and secondary particles (HS particles and CS particles). Among the 810 individual particles, secondary particles accounted for a large proportion at 79.2%. HS and CS particles were predominant with 44.9% and 34.3%, respectively, followed by mineral particles (10.5%), soot aggregates (7.2%) and organic particles (3.2%).

The average diameter of secondary particles increases with RH, being 323 nm, 358 nm and 397 nm at the RH 34%, 43% and 53%, respectively, suggesting that the high RH favor the growth of secondary particles. The CS particles may be formed by the condensation of secondary organic aerosols on pre-existing S-rich particles under the lower RH in this study.

Acknowledgments: This work was supported by the Projects of International Cooperation and Exchanges NSFC (Grant No. 41571130031) and National Basic Research Program of China (Grant No. 2013CB228503).

Author Contributions: Longyi Shao and Wenhua Wang conceived and designed the experiments; Wenhua Wang, Jie Li and Lingli Chang performed the experiments; Jiaoping Xing and Wenjun Li analyzed the data; Wenhua Wang and Longyi Shao wrote the paper.

Conflicts of Interest: The authors declare no conflict of interest.

References

1. Huang, R.J.; Zhang, Y.; Bozzetti, C.; Ho, K.F.; Cao, J.J.; Han, Y.; Daellenbach, K.R.; Slowik, J.G.; Platt, S.M.; Canonaco, F.; et al. High secondary aerosol contribution to particulate pollution during haze events in China. *Nature* **2014**, *514*, 218–222. [CrossRef] [PubMed]
2. West, J.J.; Cohen, A.; Dentener, F.; Brunekreef, B.; Zhu, T.; Armstrong, B.; Bell, M.L.; Brauer, M.; Carmichael, G.; Costa, D.L.; et al. What We Breathe Impacts Our Health: Improving Understanding of the Link between Air Pollution and Health. *Environ. Sci. Technol.* **2016**, *50*, 4895–4904. [CrossRef] [PubMed]

3. Shao, L.Y.; Hu, Y.; Shen, R.; Schäfer, K.; Wang, J.; Wang, J.; Schnelle-Kreis, J.; Zimmermann, R.; BéruBé, K.; Suppan, P. Seasonal variation of particle-induced oxidative potential of airborne particulate matter in Beijing. *Sci. Total Environ.* **2017**, *579*, 1152–1160. [CrossRef] [PubMed]
4. Planche, C.; Mann, G.W.; Carslaw, K.S.; Dalvi, M.; Marsham, J.H.; Field, P.R. Spatial and temporal CCN variations in convection-permitting aerosol microphysics simulations in an idealised marine tropical domain. *Atmos. Chem. Phys.* **2017**, *17*, 3371–3384. [CrossRef]
5. Phillips, V.T.J.; DeMott, P.J.; Andronache, C. An empirical parameterization of heterogeneous ice nucleation for multiple chemical species of aerosol. *J. Atmos. Sci.* **2008**, *65*, 2757–2783. [CrossRef]
6. Chen, B.; Zhu, Z.; Wang, X.; Andersson, A.; Chen, J.; Zhang, Q.; Gustafsson, Ö. Reconciling modeling with observations of radiative absorption of black carbon aerosols. *J. Geophys. Res. Atmos.* **2017**, *122*, 5932–5942. [CrossRef]
7. Tao, J.; Zhang, L.; Ho, K.; Zhang, R.; Lin, Z.; Zhang, Z.; Lin, M.; Cao, J.; Liu, S.; Wang, G. Impact of $PM_{2.5}$ chemical compositions on aerosol light scattering in Guangzhou—The largest megacity in South China. *Atmos. Res.* **2014**, *135*, 48–58. [CrossRef]
8. Heald, C.L.; Jacob, D.J.; Park, R.J.; Alexander, B.; Fairlie, T.D.; Yantosca, R.M.; Chu, D.A. Transpacific transport of Asian anthropogenic aerosols and its impact on surface air quality in the United States. *J. Geophys. Res. Atmos.* **2006**, *111*, D14. [CrossRef]
9. Ma, Q.X.; Wu, Y.; Zhang, D.; Wang, X.; Xia, Y.; Liu, X.; Tian, P.; Han, Z.; Xia, X.; Wang, Y. Roles of regional transport and heterogeneous reactions in the $PM_{2.5}$ increase during winter haze episodes in Beijing. *Sci. Total Environ.* **2017**, *599*, 246–253. [CrossRef] [PubMed]
10. Zhang, X.Y.; Wang, Y.Q.; Niu, T.; Zhang, X.C.; Gong, S.L.; Zhang, Y.M.; Sun, J.Y. Atmospheric aerosol compositions in China: Spatial/temporal variability, chemical signature, regional haze distribution and comparisons with global aerosols. *Atmos. Chem. Phys.* **2012**, *12*, 779–799. [CrossRef]
11. Zhang, F.; Wang, L.; Yang, J.; Chen, M.; Wei, Z.; Su, J. The characteristics of air pollution episodes in autumn over the southern Hebei, China. *World J. Eng.* **2015**, *12*, 221–236. [CrossRef]
12. Niu, H.; Cheng, W.; Pian, W.; Hu, W. The physicochemical properties of submicron particles from emissions of industrial furnace. *World J. Eng.* **2016**, *13*, 218–224. [CrossRef]
13. Posfai, M.; Buseck, P.R. Nature and Climate Effects of Individual Tropospheric Aerosol Particles. *Rev. Earth Planet. Sci.* **2010**, *38*, 17–43. [CrossRef]
14. Liu, S.; Aiken, A.C.; Gorkowski, K.; Dubey, M.K.; Cappa, C.D.; Williams, L.R.; Herndon, S.C.; Massoli, P.; Fortner, E.C.; Chhabra, P.S.; et al. Enhanced light absorption by mixed source black and brown carbon particles in UK winter. *Nat. Commun.* **2015**, *6*, 10. [CrossRef] [PubMed]
15. Zhao, J.; Du, W.; Zhang, Y.; Wang, Q.; Chen, C.; Xu, W.; Han, T.; Wang, Y.; Fu, P.; Wang, Z.; et al. Insights into aerosol chemistry during the 2015 China Victory Day parade: Results from simultaneous measurements at ground level and 260 m in Beijing. *Atmos. Chem. Phys.* **2017**, *17*, 3215–3232. [CrossRef]
16. Wang, T.; Nie, W.; Gao, J.; Xue, L.K.; Gao, X.M.; Wang, X.; Qiu, J.; Poon, C.N.; Meinardi, S.; Blake, D.; Wang, S.L. Air quality during the 2008 Beijing Olympics: Secondary pollutants and regional impact. *Atmos. Chem. Phys.* **2010**, *10*, 7603–7615. [CrossRef]
17. Chen, C.; Sun, Y.L.; Xu, W.Q.; Du, W.; Zhou, L.B.; Han, T.T.; Wang, Q.Q.; Fu, P.Q.; Wang, Z.F.; Gao, Z.Q.; et al. Characteristics and sources of submicron aerosols above the urban canopy (260 m) in Beijing, China, during the 2014 APEC summit. *Atmos. Chem. Phys.* **2015**, *15*, 12879–12895. [CrossRef]
18. Li, W.J.; Sun, J.; Xu, L.; Shi, Z.; Riemer, N.; Sun, Y.; Fu, P.; Zhang, J.; Lin, Y.; Wang, X.; et al. A conceptual framework for mixing structures in individual aerosol particles. *J. Geophys. Res. Atmos.* **2016**, *121*, 13784–13798. [CrossRef]
19. Li, W.J.; Shao, L.Y. Characterization of mineral particles in winter fog of Beijing analyzed by TEM and SEM. *Environ. Monit. Assess.* **2010**, *161*, 565–573. [CrossRef] [PubMed]
20. Xu, L.; Liu, L.; Zhang, J.; Zhang, Y.; Ren, Y.; Wang, X.; Li, W. Morphology, composition, and mixing state of individual aerosol particles in Northeast China during wintertime. *Atmosphere* **2017**, *8*, 10. [CrossRef]
21. Xing, J.P.; Shao, L.; Zheng, R.; Peng, J.; Wang, W.; Guo, Q.; Wang, Y.; Qin, Y.; Shuai, S.; Hu, M. Individual particles emitted from gasoline engines: Impact of engine types, engine loads and fuel components. *J. Clean. Prod.* **2017**, *149*, 461–471. [CrossRef]

22. Li, W.J.; Shao, L.; Zhang, D.; Ro, C.U.; Hu, M.; Bi, X.; Geng, H.; Matsuki, A.; Niu, H.; Chen, J. A review of single aerosol particle studies in the atmosphere of East Asia: Morphology, mixing state, source, and heterogeneous reactions. *J. Clean. Prod.* **2016**, *112*, 1330–1349. [CrossRef]

23. China, S.; Scarnato, B.; Owen, R.C.; Zhang, B.; Ampadu, M.T.; Kumar, S.; Dzepina, K.; Dziobak, M.P.; Fialho, P.; Perlinger, J.A.; et al. Morphology and mixing state of aged soot particles at a remote marine free troposphere site: Implications for optical properties. *Geophys. Res. Lett.* **2015**, *42*, 1243–1250. [CrossRef]

24. Li, W.J.; Shao, L.Y. Mixing and water-soluble characteristics of particulate organic compounds in individual urban aerosol particles. *J. Geophys. Res. Atmos.* **2010**, *115*, 9. [CrossRef]

25. Ghosal, S.; Weber, P.K.; Laskin, A. Spatially resolved chemical imaging of individual atmospheric particles using nanoscale imaging mass spectrometry: Insight into particle origin and chemistry. *Anal. Methods* **2014**, *6*, 2444–2451. [CrossRef]

26. Wang, Y.; Xue, Y.; Tian, H.; Gao, J.; Chen, Y.; Zhu, C.; Liu, H.; Wang, K.; Hua, S.; Liu, S.; et al. Effectiveness of temporary control measures for lowering $PM_{2.5}$ pollution in Beijing and the implications. *Atmos. Environ.* **2017**, *157*, 75–83. [CrossRef]

27. Wang, Z.S.; Li, Y.T.; Zhang, D.W.; Chen, T.; Wei, Q.; Sun, T.H.; Wang, B.Y.; Pan, J.X.; Cui, J.X.; Pi, S. Analysis on air quality in Beijing during the military parade period in 2015. *China Environ. Sci.* **2017**, *37*, 1628–1636. (In Chinese)

28. Wang, W.; Shao, L.; Guo, M.; Hou, C.; Xing, J.; Wu, F. Physicochemical properties of individual airborne particles in Beijing during pollution periods. *Aerosol Air Qual. Res.* **2017**, *17*, 3209–3219. [CrossRef]

29. Bian, Y.X.; Zhao, C.S.; Ma, N.; Chen, J.; Xu, W.Y. A study of aerosol liquid water content based on hygroscopicity measurements at high relative humidity in the North China Plain. *Atmos. Chem. Phys.* **2014**, *14*, 6417–6426. [CrossRef]

30. Villani, P.; Sellegri, K.; Monier, M.; Laj, P. Influence of semi-volatile species on particle hygroscopic growth. *Atmos. Environ.* **2013**, *79*, 129–137. [CrossRef]

31. Fajardo, O.A.; Jiang, J.K.; Hao, J.M. Continuous Measurement of Ambient Aerosol Liquid Water Content in Beijing. *Aerosol Air Qual. Res.* **2016**, *16*, 1152–1164. [CrossRef]

32. Liu, Y.C.; Wu, Z.; Tan, T.; Wang, Y.; Qin, Y.; Zheng, J.; Li, M.; Hu, M. Estimation of the $PM_{2.5}$ effective hygroscopic parameter and water content based on particle chemical composition: Methodology and case study. *Sci. China-Earth Sci.* **2016**, *59*, 1683–1691. [CrossRef]

33. Wang, G.H.; Zhang, R.; Gomez, M.E.; Yang, L.; Zamora, M.L.; Hu, M.; Lin, Y.; Peng, J.; Guo, S.; Meng, J.; et al. Persistent sulfate formation from London Fog to Chinese haze. *Proc. Natl. Acad. Sci. USA* **2016**, *113*, 13630–13635. [CrossRef] [PubMed]

34. Song, M.; Liu, P.F.; Hanna, S.J.; Li, Y.J.; Martin, S.T.; Bertram, A.K. Relative humidity-dependent viscosities of isoprene-derived secondary organic material and atmospheric implications for isoprene-dominant forests. *Atmos. Chem. Phys.* **2015**, *15*, 5145–5159. [CrossRef]

35. Liu, Z.R.; Hu, B.; Zhang, J.; Yu, Y.; Wang, Y. Characteristics of aerosol size distributions and chemical compositions during wintertime pollution episodes in Beijing. *Atmos. Res.* **2016**, *168*, 1–12. [CrossRef]

36. Semeniuk, T.A.; Wise, M.E.; Martin, S.T.; Russell, L.M.; Buseck, P.R. Water uptake characteristics of individual atmospheric particles having coatings. *Atmos. Environ.* **2007**, *41*, 6225–6235. [CrossRef]

37. Davies, J.F.; Miles, R.E.; Haddrell, A.E.; Reid, J.P. Influence of organic films on the evaporation and condensation of water in aerosol. *Proc. Natl. Acad. Sci. USA* **2013**, *110*, 8807–8812. [CrossRef] [PubMed]

38. Song, M.J.; Marcolli, C.; Krieger, U.K.; Lienhard, D.M.; Peter, T. Morphologies of mixed organic/inorganic/aqueous aerosol droplets. *Faraday Discuss.* **2013**, *165*, 289–316. [CrossRef] [PubMed]

39. Zuend, A.; Marcolli, C.; Peter, T.; Seinfeld, J.H. Computation of liquid-liquid equilibria and phase stabilities: Implications for RH-dependent gas/particle partitioning of organic-inorganic aerosols. *Atmos. Chem. Phys.* **2010**, *10*, 7795–7820. [CrossRef]

40. Schill, G.P.; Tolbert, M.A. Heterogeneous ice nucleation on phase-separated organic-sulfate particles: Effect of liquid vs. glassy coatings. *Atmos. Chem. Phys.* **2013**, *13*, 4681–4695. [CrossRef]

41. Stewart, D.J.; Cai, C.; Nayler, J.; Preston, T.C.; Reid, J.P.; Krieger, U.K.; Marcolli, C.; Zhang, Y.H. Liquid-Liquid Phase Separation in Mixed Organic/Inorganic Single Aqueous Aerosol Droplets. *J. Phys. Chem. A* **2015**, *119*, 4177–4190. [CrossRef] [PubMed]

42. Song, M.; Marcolli, C.; Krieger, U.K.; Zuend, A.; Peter, T. Liquid-liquid phase separation and morphology of internally mixed dicarboxylic acids/ammonium sulfate/water particles. *Atmos. Chem. Phys.* **2012**, *12*, 2691–2712. [CrossRef]

43. You, Y.; Renbaum-Wolff, L.; Carreras-Sospedra, M.; Hanna, S.J.; Hiranuma, N.; Kamal, S.; Smith, M.L.; Zhang, X.; Weber, R.J.; Shilling, J.E.; et al. Images reveal that atmospheric particles can undergo liquid-liquid phase separations. *Proc. Natl. Acad. Sci. USA* **2012**, *109*, 13188–13193. [CrossRef] [PubMed]

44. You, Y.; Smith, M.L.; Song, M.; Martin, S.T.; Bertram, A.K. Liquid-liquid phase separation in atmospherically relevant particles consisting of organic species and inorganic salts. *Int. Rev. Phys. Chem.* **2014**, *33*, 43–77. [CrossRef]

45. Hou, C.; Shao, L.; Hu, W.; Zhang, D.; Zhao, C.; Xing, J.; Huang, X.; Hu, M. Characteristics and aging of traffic-derived particles in a highway tunnel at a coastal city in southern China. *Sci. Total Environ.* **2018**, *619*, 1385–1393. [CrossRef]

46. Marcolli, C.; Krieger, U.K. Phase changes during hygroscopic cycles of mixed organic/inorganic model systems of tropospheric aerosols. *J. Phys. Chem. A* **2006**, *110*, 1881–1893. [CrossRef] [PubMed]

47. Hodas, N.; Zuend, A.; Mui, W.; Flagan, R.C.; Seinfeld, J.H. Influence of particle-phase state on the hygroscopic behavior of mixed organic-inorganic aerosols. *Atmos. Chem. Phys.* **2015**, *15*, 5027–5045. [CrossRef]

48. Bondy, A.L.; Kirpes, R.M.; Merzel, R.L.; Pratt, K.A.; Banaszak Holl, M.M.; Ault, A.P. Atomic Force Microscopy-Infrared Spectroscopy of Individual Atmospheric Aerosol Particles: Subdiffraction Limit Vibrational Spectroscopy and Morphological Analysis. *Anal. Chem.* **2017**, *89*, 8594–8598. [CrossRef] [PubMed]

49. O'Brien, R.E.; Wang, B.; Kelly, S.T.; Lundt, N.; You, Y.; Bertram, A.K.; Leone, S.R.; Laskin, A.; Gilles, M.K. Liquid-Liquid Phase Separation in Aerosol Particles: Imaging at the Nanometer Scale. *Environ. Sci. Technol.* **2015**, *49*, 4995–5002. [CrossRef] [PubMed]

![atmosphere logo] *atmosphere*

MDPI

Article

Elemental Mixing State of Aerosol Particles Collected in Central Amazonia during GoAmazon2014/15

Matthew Fraund [1], Don Q. Pham [1], Daniel Bonanno [1], Tristan H. Harder [2,3,†], Bingbing Wang [4,‡], Joel Brito [5,||], Suzane S. de Sá [6], Samara Carbone [5], Swarup China [4], Paulo Artaxo [5], Scot T. Martin [6], Christopher Pöhlker [7], Meinrat O. Andreae [7,8], Alexander Laskin [4,§], Mary K. Gilles [2] and Ryan C. Moffet [1,*]

[1] Department of Chemistry, University of the Pacific (UoP), Stockton, CA 95211, USA; m_fraund@u.pacific.edu (M.F.); D_pham5@u.pacific.edu (D.Q.P.); d_bonanno@u.pacific.edu (D.B.)
[2] Chemical Sciences Division, Lawrence Berkeley National Laboratory (LBNL), Berkeley, CA 94720, USA; Harder.tristan@gmail.com (T.H.H.); mkgilles@gmail.com (M.K.G.)
[3] Department of Chemistry, University of California, Berkeley, CA 94720, USA
[4] Environmental Molecular Sciences Laboratory, Pacific Northwest National Laboratory (PNNL), Richland, WA 99352, USA; Bingbing.wang@xmu.edu.cn (B.W.); Swarup.china@pnnl.gov (S.C.); alaskin@purdue.edu (A.L.)
[5] Institute of Physics, University of Sao Paulo (USP), São Paulo 05508-020, Brazil; J.brito@opgc.univ-bpclermont.fr (J.B.); samara.carbone@ufu.br (S.C.); artaxo@if.usp.br (P.A.)
[6] School of Engineering and Applied Sciences, Harvard University, Cambridge, MA 02138, USA; Suzane.s.desa@gmail.com (S.S.d.S); scot_martin@harvard.edu (S.T.M.)
[7] Biogeochemistry Department, Max Planck Institute for Chemistry (MPIC), Mainz 55020, Germany; C.Pohlker@mpic.de (C.P.); m.andreae@mpic.de (M.O.A.)
[8] Scripps Institution of Oceanography, University of California, La Jolla, CA 92093, USA
* Correspondence: rmoffet@pacific.edu; Tel.: +1-209-946-2006
† Current address: Physikalisches Institut, Universität Würzburg, Am Hubland, 97074 Würzburg, Germany.
‡ Current address: State Key Laboratory of Marine Environmental Science, College of Ocean and Earth Sciences, Xiamen University, Xiamen 361005, China.
§ Current address: Department of Chemistry, Purdue University (PU), West Lafayette, IN 47907, USA.
|| Current address: Laboratory for Meteorological Physics (LaMP), Université Clermont Auvergne, F-6300 Clermont-Ferrand 63000, France.

Received: 31 July 2017; Accepted: 12 September 2017; Published: 15 September 2017

Abstract: Two complementary techniques, Scanning Transmission X-ray Microscopy/Near Edge Fine Structure spectroscopy (STXM/NEXAFS) and Scanning Electron Microscopy/Energy Dispersive X-ray spectroscopy (SEM/EDX), have been quantitatively combined to characterize individual atmospheric particles. This pair of techniques was applied to particle samples at three sampling sites (ATTO, ZF2, and T3) in the Amazon basin as part of the Observations and Modeling of the Green Ocean Amazon (GoAmazon2014/5) field campaign during the dry season of 2014. The combined data was subjected to k-means clustering using mass fractions of the following elements: C, N, O, Na, Mg, P, S, Cl, K, Ca, Mn, Fe, Ni, and Zn. Cluster analysis identified 12 particle types across different sampling sites and particle sizes. Samples from the remote Amazon Tall Tower Observatory (ATTO, also T0a) exhibited less cluster variety and fewer anthropogenic clusters than samples collected at the sites nearer to the Manaus metropolitan region, ZF2 (also T0t) or T3. Samples from the ZF2 site contained aged/anthropogenic clusters not readily explained by transport from ATTO or Manaus, possibly suggesting the effects of long range atmospheric transport or other local aerosol sources present during sampling. In addition, this data set allowed for recently established diversity parameters to be calculated. All sample periods had high mixing state indices (χ) that were >0.8. Two individual particle diversity (D_i) populations were observed, with particles <0.5 μm having a D_i of ~2.4 and >0.5 μm particles having a D_i of ~3.6, which likely correspond to fresh and aged aerosols, respectively. The diversity parameters determined by the quantitative method presented here will serve to aid in the accurate representation of aerosol mixing state, source apportionment, and aging in both less polluted and more developed environments in the Amazon Basin.

Keywords: mixing state; Amazon; elemental composition; aerosol; STXM; SEM; EDX; diversity; aging

1. Introduction

Atmospheric aerosols are solid or liquid particles suspended in air and are comprised of mixtures of organic and/or inorganic species: organic molecules, salts, soot, minerals, and metals [1]. Aerosols have highly uncertain effects on radiative forcing [2]. Aerosol forcing occurs via two mechanisms: light can be scattered or absorbed directly by the aerosol particles (the "direct effect", also aerosol-radiation interactions) or indirectly through aerosol effects on cloud properties (the "indirect effect", also aerosol-cloud interactions) [3]. The latest Intergovernmental Panel on Climate Change (IPCC) report, released in 2013, shows that the extent of anthropogenic effects on cloud formation is currently the largest source of uncertainty for predictive understanding of global anthropogenic radiative forcing [2]. Both direct and indirect effects are heavily influenced by the composition of aerosols on a per-particle level [4–6]. To better understand and predict the influence of industrialization, one aspect of particular interest is the effect that anthropogenic emissions have on the per-particle composition of aerosols and their impacts on local and global climate [2,7].

One underlying reason for this uncertainty is the complex manner in which aerosol composition changes over time and distance through coagulation, condensation, and chemical reaction [8]. Because aerosol radiative forcing and cloud formation depend on the individual particle composition, it is important to know how atmospheric components are mixed within a population of aerosols. How these components are mixed plays a large role in determining the manner and extent to which radiative forcing is affected. For example, the coating of soot by organics can change the direct radiative forcing of those aerosols by as much as a factor of 2.4 over pure soot [5,9–11]. Hence, in this case, it is important to know whether soot and organics coexist in the same aerosol particle. How components are mixed in an aerosol sample is referred to as its mixing state. This mixing state can range anywhere from an internal mixture where each component is evenly distributed throughout all particles, to an external mixture where each component occupies its own population of particles. Many atmospheric models assume one of these extremes throughout their simulation [12–14]. Some models include a specific aspect of aerosol mixing such as the mixing state of black carbon [4,15], while other, nascent, models will account for a more complete mixing state [16]. Mixing state values for coated black carbon (BC) have been determined using a single-particle soot photometer (SP2) based on the time delay between light scattering and soot incandescence but thermodynamic properties of organic coatings must be assumed to infer coating thicknesses, making the technique qualitative [17,18]. This approach also becomes less applicable if inorganic dominant or non-soot containing particles are of interest. A real-time method for determining aerosol mixing state index has been achieved by using single particle mass spectrometry [19], although this technique is blind to detailed aerosol morphology.

Recently, more nuanced metrics were developed to quantify the mixing state of a population of aerosols [20]. Here, Riemer and West utilize an information theoretic approach to determine specific mixing states in populations of aerosols. Particle-specific mass fractions are used to calculate both bulk and individual particle diversity parameters. The mixing state of a population is then calculated from the ratio of individual and bulk diversities. This method of mixing state determination necessitates a mass quantitative method of determining per-particle composition. Spectromicroscopy techniques are uniquely suited to analyze both the morphology and the comprehensive mixing state of a population of aerosols. Here, these quantitative mixing state metrics are applied to microscopy images of particle samples collected in the central Amazon basin.

In this study, we determine the mass fractions of 14 elements on the exact same set of particles using the complementary techniques of Scanning Transmission X-ray Microscopy with Near-Edge X-ray Absorption Fine Structure spectroscopy (STXM/NEXAFS) and Scanning Electron Microscopy coupled with Computer Controlled Energy Dispersive X-ray spectroscopy (SEM/EDX). Each technique is limited in which elements it can investigate. STXM/NEXAFS is limited by the energy range of the synchrotron insertion device as well as the beamtime available for sample analysis. STXM/NEXAFS has the advantage of providing quantitative measurements of light, low Z (atomic number) elements (C, N, and O), as well as some heavier elements with L-shell absorption edges in the same energy range (e.g., K and Ca) [21,22]. Although SEM/EDX provides a faster method of per-particle spectromicroscopy, it is only considered quantitative for higher Z elements (Z > 11, Na) [7,23,24]. These two techniques are inherently complementary, with each technique providing mass information on elements that the other cannot adequately probe and both providing this information on an individual particle level. Both techniques have been used in tandem on microscopy samples previously [25]. In that study, O'Brien et al. [25] used STXM/NEXAFS and SEM/EDX to characterize individual particles from northern California. From this, mixing state parameters were calculated; however, because STXM and SEM were conducted on different particles within a given sample, separate mixing states were calculated from each technique. The current work combines STXM and SEM data at the single particle level in a similar way to Piens et al., 2016 [26], where both techniques were used together to determine hygroscopicity of individual particles. The per-particle elemental mass fractions determined herein are used to calculate an elemental mixing state for particles collected at three sampling sites.

Aerosol production in the Amazon basin plays an important role in global climate due to the large scale of biogenic emissions from the tropical forest often mixed with pollutants from vegetation fires (mostly related to deforestation and pasture burning) [27–30]. South America contributes significantly to the global aerosol carbon budget; ~17% of global soot emissions are produced in Central and South America combined [31]. Aerosols are also subject to long range transport and thus are of importance to global models [32]. This environmentally important region of Central Amazonia contains Manaus, a city with over two million people. Manaus is a large industrial manufacturing city as a consequence of its free trade status since the 1960s. The juxtaposition of pristine rainforest with a large anthropogenic center presents a unique circumstance for studying how native biogenic aerosols are affected by emissions from an industrial city [33]. To take advantage of this unique location, the Observations and Modeling of the Green Ocean Amazon (GoAmazon2014/5) field campaign was conducted from January 2014 through December 2015 [34,35]. The GoAmazon campaign was developed with multiple scientific objectives, two of which involve the biogenic and anthropogenic interactions studied here.

2. Experiments

2.1. Sampling Site Description

As part of the GoAmazon field campaign, two Intensive Operating Periods (IOPs) were conducted during 2014, with IOP2 taking place during the dry season from the 15 August through the 15 October 2014 [34]. This campaign was conducted over central Amazonia with multiple sampling sites around the city of Manaus (Figure 1). Northeasterly trade winds in this region dictate the general wind direction over the area and so the sampling sites were located with this in mind. These trade winds carry marine aerosols from the ocean inland and, during the wet season, can also carry supermicron mineral dust from the Sahara [36]. For the wet season, secondary organic aerosols (pure liquid or with a soot/inorganic core) dominate the submicrometer size range [37–39]. During the dry season, however, a large fraction of the aerosol population can be attributed to large scale biomass burning [30].

For this study, particle samples from three sampling sites were studied: the Amazon Tall Tower Observatory (ATTO; T0a), the Terrestrial Ecosystem Science site (ZF2; T0t), and the Atmospheric Radiation

Measurement (ARM) site located near Manacapuru (T3). The ATTO site is located approximately 150 km upwind of Manaus and serves as a background site. During the wet season, near-pristine conditions can be observed here; but, because the dry season is dominated by biomass burning particles, the ATTO site will serve as a regional background rather than a background of pure biogenic particles as might be expected [40]. The ZF2 site is located about 140 km directly downwind of the ATTO site. The final site, T3, is 70 km downwind of Manaus and often experiences the pollution plume from Manaus [34,35]. Site locations and characteristics are presented in Martin et al. 2016 [34]. For additional background information about the sampling sites, please see Andreae et al., 2015, Artaxo et al., 2013, and Martin et al., 2016 [28,34,40].

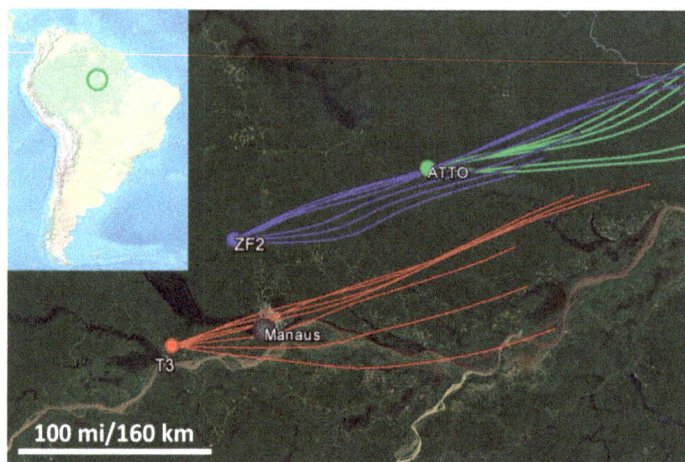

Figure 1. Positions of the three sampling sites located around the city of Manaus with representative National Oceanic and Atmospheric Administration (NOAA) Hybrid Single Particle Lagrangian Integrated Trajectory Model (HYSPLIT) back trajectories (14 September from 9:00 to 12:00 shown, 500 m starting elevation using the global data assimilation system data set) [41,42]. (Inset) Overview map of South America with the region of interest circled. Longer back trajectories as well as varied starting elevations for the three sites are shown in Figure S1 [28,34,40].

2.2. Sample Collection

At the three sampling sites, atmospheric particle samples were collected on silicon nitride (Si_3N_4) membranes overlaid on a 5 × 5 mm silicon chip frame with a central 0.5 × 0.5 mm window (100 nm thick membrane, Silson Inc., Southam, England). Samples were collected using a Micro-Orifice Uniform Deposit Impactor (MOUDI, MSP MOUDI-110, Shoreview, MN, USA) on the dates and times shown in Table 1. HYSPLIT back trajectories were examined for each sampling period to confirm the wind patterns seen in Figure 1. These samples were then analyzed sequentially with the two spectromicroscopy techniques discussed in the following sections.

2.3. STXM Data Collection and Image Processing

Samples were first imaged at the STXM beamline 5.3.2.2 (ALS, Berkeley, CA, USA) at the Advanced Light Source (ALS) [43]. The energy range of this STXM (200–600 eV) end station enables the quantitative study of carbon, nitrogen, and oxygen. Energy selected soft X-rays were focused down to a ~30 nm spot size and directed onto the sample surface. After a suitable 15 × 15 μm region was located, the sample stage was then raster scanned, with 40 nm steps, using piezo-electric stages to capture an image at a specific energy. This process was then repeated at multiple photon energies to produce a stack of images with an absorption spectrum associated with each 40 × 40 nm pixel.

For each element, photon energies were chosen before and after the k-shell absorption edge: 278 and 320 eV for carbon, 400 and 430 eV for nitrogen, and 525 and 550 eV for oxygen [44]. Additional images were also taken near the carbon edge at 285.4 and 288.5 eV, for the RC = CR and RCOOR C1s→π* transitions, respectively, in order to partly characterize the molecular speciation of carbon [45].

Table 1. Samples examined for this study. The nominal size range for Micro-Orifice Uniform Deposit Impactor (MOUDI) stage 7 is 0.56–0.32 µm and stage 8 is 0.32–0.18 µm.

Site	Date (2014)	Time Period (Local Time)	MOUDI Stage	# of Analyzed Particles
ATTO	14–15 Oct	19:00 (14 Oct)–19:00 (15 Oct)	7	501
T3	12–13 Sept	Night 18:00–6:00	7	334
	13 Sept	Day 8:00–12:00	7	279
	13 Sept	Day 8:00–12:00	8	59 [1]
	14 Sept	Day 9:00–12:00	7	182
	14 Sept	Day 9:00–12:00	8	50 [1]
ZF2	3–6 Oct	11:00 (3 Oct)–11:00 (6 Oct)	7	315
	6–8 Oct	14:00 (6 Oct)–12:00 (8 Oct)	7	309
	6–8 Oct	14:00 (6 Oct)–12:00 (8 Oct)	8	967

[1] Low particle counts are due to low particle loading of microscopy samples and time constraints. The # symbol is used to represent the word number here. Oct: October; Sept: September.

Any displacement between images within a stack is corrected by a routine based on Guizar-Sicarios' image registration algorithm [46]. Regions within a given stack were then identified as particles or substrate using Otsu's method on that stack's average intensity image over all 8 energies [47]. Background subtraction of a given element's pre-edge intensity image from its post-edge image is then performed to account for any absorbing species not attributed to that element.

The recorded intensity at each pixel determined to be a particle was converted to optical density using:

$$OD = -\ln\left(\frac{I}{I_0}\right) \tag{1}$$

where OD is optical density, I is intensity of the pixel, and I_0 is the background intensity. This is followed by a conversion to mass with the following formula:

$$m = \frac{OD * A}{\mu_{post} - \mu_{pre}} \tag{2}$$

where m is the mass of a specific element at that pixel, A is the area of that pixel, and μ_{pre} and μ_{post} refer to the mass absorption coefficients for that specific element before and after the absorption edge, respectively. Mass absorption coefficients have been both empirically and theoretically determined for a variety of elements as tabulated in Henke et al., 1993 [44].

Previously developed algorithms for determining the speciation of carbon using 278, 285.4, 288.5, and 320 eV were applied to each Field Of View (FOV) as well. This mapping technique uses a series of thresholds to identify inorganics, soot, and organic carbon. Total carbon is taken to be $OD_{320} - OD_{278}$, pixels with an OD_{278}/OD_{320} ratio 0.5 or greater are rich in inorganics, and pixels with an elevated (0.35) ratio of sp^2 bonding compared to total carbon $(OD_{288.5} - OD_{278})/(OD_{320} - OD_{278})$ are indicative of soot [45].

2.4. SEM/EDX Data Collection

The same sample windows previously imaged with STXM were imaged again with a computer controlled scanning electron microscope (FEI, Quanta 3D FEG, Hillsboro, AL, USA) coupled with energy dispersive X-ray spectroscopy (CCSEM/EDX, PNNL, Richland, CA, USA). The SEM utilized a field emission tip to produce an electron beam which was directed and focused onto a sample with an accelerating voltage of 20 kV which can cause core shell atomic electrons to be ejected from the sample. Higher shell electrons then relax into the newly created orbital hole, releasing an elementally characteristic photon recorded by an energy dispersive X-ray detector (EDAX PV7761/54 ME with Si(Li) detector, Mahwah, NJ, USA). As the electron beam was scanned over the sample, the transmitted electron image was used to identify the exact same FOVs from the previous STXM images. Once a FOV previously analyzed with STXM is located, a 10,000× image (30 nm/pixel resolution) was captured. This image combines both transmitted and backscattered electron images to improve particle detection [23]. A threshold contrast level was then set to identify which areas of the collected image counted as particles using the "Genesis" software from EDAX, Inc. A software filter was then applied which discounts particles that are too small (e.g., noise spikes) or too large (e.g., multiple nearby particles counted as a single large particle). The electron beam was then directed towards each identified particle in sequence and an EDX spectrum was collected. Afterwards, software was used to fit the peaks of eleven relevant elements selected for this study: Na, Mg, P, S, Cl, K, Ca, Mn, Fe, Ni, and Zn. Some elements of interest have been included in the spectral fit, but omitted from quantitation, including Al, Si, and Cu due to background sources of these elements: (1) the STXM sample holder where the Si_3N_4 windows sat was made of Al and was inserted into the SEM as well, (2) the mounting stage that holds samples inside the microscope was fabricated from beryllium-copper alloy, (3) the EDX data was collected using a Si(Li) detector with a 10 mm^2 active area. Each of these circumstances could contribute background signal for the elements in question.

After data has been collected from both SEM and STXM, individual particle mass information is contained in two sets of images: one from STXM and one from SEM. Due to differing contrast mechanisms, image resolution, and other factors, particles do not necessarily appear the same between images taken with the two techniques. The manual matching of particles was performed using pattern recognition to ensure proper alignment of the image sets from both techniques.

2.5. Quantifying Higher Z Elements

Using the aforementioned methods, STXM yields quantitative, absolute mass information on a sub-particle basis. SEM/EDX is more limited in this aspect, being quantitative for elements with $Z > 11$ (Na) but only semi-quantitative for C, N, and O [23]. Due to the EDAX software used for EDX data collection and analysis, there is an additional caveat to the quantitation of $Z > 11$ elements: the software reports only the relative mass percentages compared to the elements chosen during data processing. In order to properly quantify the mixing state, the absolute mass of each element in each particle is necessary. To determine these absolute masses, a system of equations was set up using the following equation types:

$$OD_i = \rho t \sum_{a=1}^{A} f_a \mu_{a,i} \tag{3}$$

$$\frac{f_x}{f_y} = \frac{rel.\%_x}{rel.\%_y} \tag{4}$$

$$\sum_{a=1}^{A} f_a = 1 \tag{5}$$

For each pixel, OD_i is the optical density taken at energy i, ρ is the density and t is the thickness of the sample (at that pixel), f_a is the mass fraction of element a, and $\mu_{a,i}$ is the mass absorption coefficient of element a at energy i. Equation (4) is a general relationship, which equates the ratio of two absolute mass fractions (f_x and f_y) with the ratio of relative mass percentages ($rel.\%_x$ and $rel.\%_y$) produced by the EDAX software. Equation (3) utilizes the quantitative nature of STXM whereas the relative mass percent of elements with $Z > 11$ were used in Equation (4) to combine the quantitative abilities of SEM/EDX. This system was then solved for the 14 absolute mass fractions (f_a) of each element chosen in this study.

Equation (5) is an assumption that is valid when the 14 elements analyzed comprise close to 100% of the particle's composition. Here, systematic error in the calculated mass fractions of specific particles can be introduced in particles where elements not considered represent a significant portion of that particle's mass (e.g., mineral dust and Si or Al). During the Amazonian dry season, Al and Si represent 0.3% and 0.4% of the average fine mode particle mass [48]. This mass fraction error becomes negligible, however, when the ensemble diversity values or mixing state index is considered due to the overwhelming mass of C, N, and O in each particle.

After both sets of images are matched and the corresponding light and heavy element information has been processed quantitatively, mass information for each FOV is contained in sets of maps, one for each element analyzed.

2.6. Mixing State Parameterization

The method of parameterizing mixing state used here is based on calculating mass fractions for different groupings of the individual components defined and is reproduced from Riemer and West [20]. The absolute mass of a given component a, within a given particle i, is labeled as m_i^a where $a = 1, ..., A$ (and A is the total number of components) and $i = 1, ..., N$ (the total number of particles). From this, the following relationships are established:

$$\sum_{a=1}^{A} m_i^a = m_i \left(Mass\ of\ i^{th}\ particle \right) \tag{6}$$

$$\sum_{i=1}^{N} m_i^a = m^a \left(Mass\ of\ a^{th}\ component \right) \tag{7}$$

$$\sum_{a=1}^{A} \sum_{i=1}^{N} m_i^a = m\ (Total\ mass\ of\ sample) \tag{8}$$

Mass fractions are then established from these relationships with:

$$f_i = \frac{m_i}{m},\ f^a = \frac{m^a}{m},\ f_i^a = \frac{m_i^a}{m_i} \tag{9}$$

where f_i is the mass fraction of a particle within a sample, f^a is the mass fraction of component a within a sample, and f_i^a is the mass fraction of component a within particle i.

These mass fractions are used to calculate the Shannon entropy (also called information entropy) for each particle, each component, and for the bulk using Equations (10)–(12), respectively.

$$H_i = \sum_{a=1}^{A} -f_i^a \ln f_i^a \tag{10}$$

$$H_\alpha = \sum_{i=1}^{N} f_i H_i \tag{11}$$

$$H_\gamma = \sum_{a=1}^{A} -f^a \ln f^a \tag{12}$$

Each type of mass fraction can be thought of as a probability, and thus the collection of mass fractions defines a probability distribution. The Shannon entropy of a probability distribution quantifies how uniform the distribution is. Shannon entropy is maximized if every element in the distribution is equally probable, and the entropy decreases the more likely any individual element becomes [20]. With this information entropy, diversity values are defined with the following equations:

$$D_i = e^{H_i},\ D_\alpha = e^{H_\alpha},\ D_\gamma = e^{H_\gamma} \tag{13}$$

The diversity values contain the same type of information, but represent it in another way. Each diversity value represents the effective number of species (weighted by mass) within a given population (i.e., D_i represents the number of species within a specific particle, D_α is the average number of species within any given particle, and D_γ represents the number of species within the entire sample). From these diversity values, the mixing state index is defined as

$$\chi = \frac{D_\alpha - 1}{D_\gamma - 1} \tag{14}$$

This definition compares how many species exist, on average, within individual particles, with the total number of species identified in the sample. χ is at a minimum of 0 when D_α is 1, corresponding to each particle being comprised of exactly one species. A mixing state index of 1 occurs when D_α and D_γ are equal, meaning that each particle has the same composition as the bulk sample.

2.7. Error in Mixing State Index, χ

The measurement uncertainty of χ due to STXM, EDX, or the system of equations was found to be insignificant compared to the statistical uncertainty of χ within each cluster and thus only the statistical uncertainty is considered here. To determine this uncertainty, the statistical uncertainty in D_α, and D_γ were found separately.

Determining statistical uncertainty in D_γ starts with f^a from Equation (9). From Riemer and West [20], f^a is a ratio of the total mass of the ath component and the total mass of the sample, however this is equivalent to the ratio of the mean mass of the ath component and the mean mass of particles within the sample:

$$f^a = \frac{m^a}{m} = \frac{\sum_{i=1}^N m_i^a}{\sum_{i=1}^N \sum_{a=1}^A m_i^a} = \frac{\frac{1}{N}\sum_{i=1}^N m_i^a}{\frac{1}{N}\sum_{i=1}^N \sum_{a=1}^A m_i^a} = \frac{\overline{m^a}}{\overline{m}} \tag{15}$$

where $\overline{m^a}$ is the mean mass of the ath component and $\overline{m}$ is the mean mass of particles within the sample. From this, the standard error (for a 95% confidence level) can be determined for $\overline{m^a}$ and $\overline{m}$ which is then propagated through Equations (9)–(13).

The statistical uncertainty in D_α was found by first rearranging and combining Equations (11) and (13):

$$H_\alpha = \sum_{i=1}^N f_i \ln D_i \tag{16}$$

and, because this takes the form of an expected value $E(x) = \sum f_x x$, the error in H_α can be found with Equation (15) and then propagated with Equation (16) to determine the error in D_α.

2.8. k-Means Clustering

All analyzed particles were combined and a k-means clustering algorithm was then used to group particles into clusters [49]. A vector of 18 variables were used for k-means clustering: the quantitative elemental mass fractions composition of the 14 elements chosen, the Circular Equivalent Diameter (CED) [1], D_i, the mass fraction of carbon attributed to soot, and the area fraction of the particle dominated by inorganics. In this way, particles were clustered based on size, elemental composition, as well as on how carbon speciation was distributed. The square root of these parameters was used in the clustering algorithm to enhance trace elements in accordance with Rebotier et al. [50]. CED is used here as the descriptor of particle size due to it being readily calculable from STXM data. While aerodynamic equivalent diameter is the physical parameter determining MOUDI sampling, it is difficult to retrieve from microscopy data.

The correct number of clusters was initially chosen based on a combination of two common methods: the elbow method, and the silhouette method [51]. Using these two methods, 12 clusters were identified.

3. Results

As a general trend, during the dry season, the whole Amazon Basin experiences a significantly higher aerosol number concentration and $CO_{(g)}$ concentration compared to the wet season, largely due to in-Basin fires [30,40]. Furthermore, in addition to biomass burning, emissions from Manaus are often observed at the T3 site (downwind of Manaus), sporadically at the ZF2 site (upwind but near the city) and rarely at ATTO (upwind and ~150 km away). Table 2 outlines some supporting measurements made at the three sites.

Table 2. Supporting data during sampling times for each sampling site. Values listed under sample dates are averages over that sampling period. Values in the adjacent (Avg.) column are monthly averages only for the hours coinciding with the sampling times listed in Table 1 (e.g., the average particle concentration between the hours of 8:00 and 12:00 averaged over the entire month for the Avg. column next to 13 September). Ammonium, chloride, organics, sulfate, and Black Carbon (BC) pertain to aerosol measurements whereas $CO_{(g)}$ and $O_{3(g)}$ are gas phase measurements. Measurements with blank values were not available during the period of this study. Information regarding collection conditions can be found in Andreae et al., 2015, Artaxo et al., 2013, Martin et al., 2016 [28,34,40].

Data Product	T3						ZF2				ATTO	
	12 Sept (night)	Avg.	13 Sept (day)	Avg.	14 Sept (day)	Avg.	3–6 Oct	Avg.	6–8 Oct	Avg.	15 Oct	Avg.
Particle Conc. (cm^{-3})	2400 [10]	3400 [10]	2400 [10]	3400 [10]	5800 [10]	3400 [10]	-	-	-	-	1100 [11]	1400 [11]
Ammonium (μg m^{-3})	0.33 [1]	0.45 [1]	0.34 [1]	0.42 [1]	0.28 [1]	0.42 [1]	-	-	-	-	0.23 [2]	0.20 [2]
Chloride (ng m^{-3})	14.9 [1]	20 [1]	17.0 [1]	27 [1]	24 [1]	27 [1]	-	-	-	-	14.4 [2]	14.9 [2]
Nitrate (μg m^{-3})	0.11 [1]	0.16 [1]	0.20 [1]	0.19 [1]	0.28 [1]	0.19 [1]	-	-	-	-	0.16 [2]	0.15 [2]
Organics (μg m^{-3})	7.9 [1]	10.7 [1]	7.6 [1]	10.0 [1]	14.1 [1]	10.0 [1]	-	-	-	-	3.8 [2]	4.4 [2]
Sulfate (μg m^{-3})	1.0 [1]	1.4 [1]	0.86 [1]	1.1 [1]	0.71 [1]	1.1 [1]	-	-	-	-	0.53 [2]	0.61 [2]
$CO_{(g)}$ (ppb)	178 [3]	210 [3]	211 [3]	257 [3]	558 [3]	254 [3]	178 [4]	169 [4]	159 [4]	168 [4]	141 [4]	138 [4]
$O_{3(g)}$ (ppb)	8 [5]	9 [5]	36 [5]	29 [5]	43 [5]	32 [5]	17 [6]	13 [6]	12 [6]	13 [6]	-	-
BC (μg m^{-3})	0.8 [7]	0.9 [7]	1.0 [7]	1.0 [7]	1.2 [7]	1.0 [7]	0.5 [8]	0.4 [8]	0.6 [8]	0.4 [8]	0.5 [9]	0.4 [9]

[1] Aerosol Mass Spectrometer (AMS), [2] Aerosol Chemical Speciation Monitor (ACSM), [3] ARM/Mobile Aerosol Observatory System (MAOS) Los Gatos ICOS™ Analyzer, [4] Picarro Cavity Ringdown Spectrometer (CRDS), [5] ARM/MAOS Ozone Analyzer, [6] Thermo 49i, [7] ARM/AOS Aethalometer, [8] MultiAngle Absorption Photometer (MAAP)-5012, [9] MAAP-5012, [10] ARM/MAOS Scanning Mobility Particle Sizer (SMPS), [11] SMPS. Sept: September; Oct: October.

Most values listed for the 12 September and 13 September sample at T3 are consistent with their sample-period monthly averages, even considering time of day each sample was collected. The data from 14 September, however, shows a marked increase in particle concentration, nitrate, organic, and CO concentration, along with a small increase in BC. This is indicative of a heavy pollution plume which, in this case, had recently passed over the T3 sampling site (see Figure S2). AMS and particle concentration data for the T3 sites show a reasonable agreement with either background or polluted conditions previously reported, as do ozone measurements [52,53]. The monthly average values for 13 September and 14 September are often similar due to the similar (though not identical) sampling time from 8:00 to 12:00 and from 9:00 to 12:00, respectively. The similarity between monthly average particle concentrations for 12 September and 13 September are purely coincidental.

Particle concentration and AMS/ACSM data were not available for the ZF2 site during this study. Concentrations of CO, ozone, and BC values agreed well with their sample-period monthly averages with the lone exception of ozone levels for the 3–6 October sample period. This increase is also reflected, albeit to a much lesser degree, in an increase of CO and BC levels. From Figure S3, it appears that sample collection began in the middle of a period of higher than average pollution levels. Temporary enhancements in BC due to emissions from Manaus have been observed previously at ZF2 [54]. Overall the levels of CO, ozone, and BC at ZF2 are smaller with respect to T3 values as is expected. ACSM data collected for past studies at site ZF fits well with the trends seen in the limited data presented here [28]. Concentrations of aerosol components can fluctuate depending on the day but September (2012) averages for ammonium, chloride, nitrate, organics, and sulfate have been reported as 0.46, 0.01, 0.22, 13.9, 0.37 μg m^{-3}, respectively by Artaxo et al., 2013 [28].

Unsurprisingly, the ATTO site shows the lowest levels of almost all presented aerosol and gas components. The sample collected on 15 October also appears to be fairly average with respect to the sample-period monthly average for this collection time. The low particle concentration suggests a sample with low pollution levels which makes this sample ideal for our purposes. The values presented for the ATTO site are also in fair agreement with previously published data [40].

The supporting data tabulated here has been collected from different instruments at the three sites and so a direct comparison could be suspect, especially in the case of BC measurements [55]. However, considering the agreement with published literature and qualitative use of Table 2 in the current work, we believe any associated error is acceptable.

Figure 2 shows an example FOV and the type of data calculated for all three sites. Each particle has an OD map (which is proportional to mass, refer to Equation (2) for C, N, and O as well as a C speciation map. In Figure 2, the STXM grayscale image shown is the average intensity map over the four C edge images. There is a correlation between the brightest spots and the identification of soot in the C speciation map. This speciated image is possible due to the sub-particle spatial resolution achievable with STXM mapping, which is highlighted in the C, N, and O maps. Potential inter-site differences can be seen in this figure: the ATTO sample shows large inorganic inclusions coated by organics along with Na, Mg, and Cl representing the bulk of the higher Z elements. The particles present at ATTO also often look like either inorganic aerosols from biomass burning events or small biogenic K salt particles (due to the KCl or NaCl inorganic cores), or secondary organics, with a few particles appearing to be sea spray [56–58]. The ZF2 sample has a consistent circular morphology with appreciable mixing between the three carbon species. ZF2 particles often look amorphous with some particles appearing to be sulfate-based aerosols [59]. Lastly, the T3 sample is the most varied in terms of morphology and in elemental composition with S, P, and K all present in many of the particles sampled. Unsurprisingly, soot inclusions are much more common in the T3 sample. Particles from this site often look like biomass burning particles with a few fractal soot particles as well [60]. It is important to keep in mind that the particle morphologies presented have possibly changed from their original state when collected. This change could be due to the impaction of particles during sampling, or the changes in relative humidity experienced as these particles are collected, stored, transported, and placed in vacuum before STXM or SEM images can be obtained. Liquid particles can spread upon impact making them appear larger on microscopy substrates. Particles with high water content can effloresce at lower relative humidity leaving solid phase inorganic components. Loss of highly volatile organic carbon or volatile inorganics like ammonium nitrate is also possible, making concentrations detected here a lower limit for inorganic and organic species.

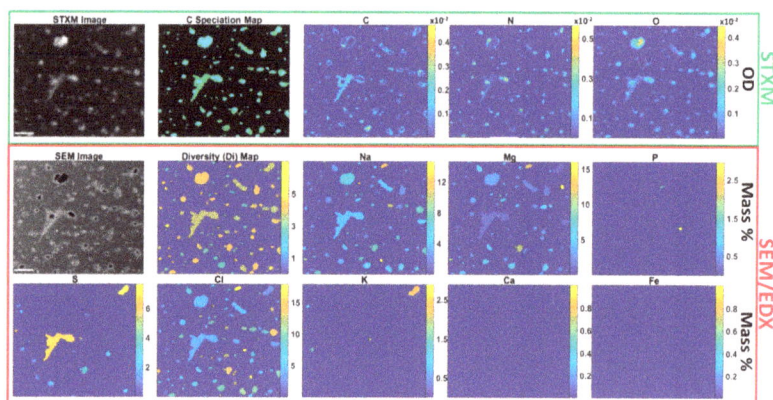

(a)

Figure 2. *Cont.*

(b)

(c)

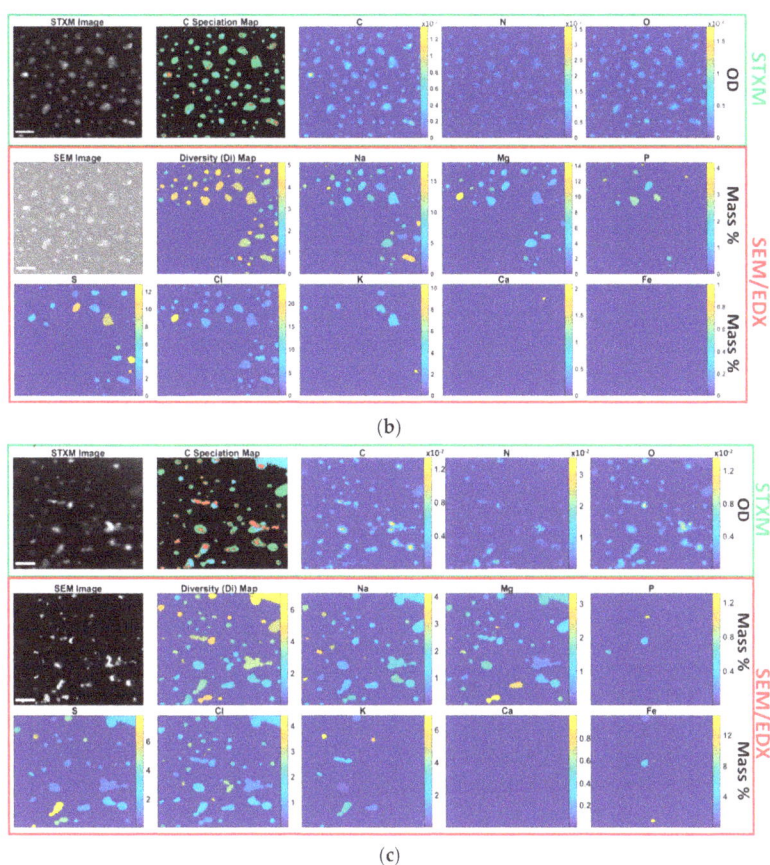

Figure 2. Raw and processed image maps for selected Field of Views (FOVs) from (**a**) the Amazon Tall Tower Observatory (ATTO) site collected on 15 October 2014, (**b**) the ZF2 site collected on 3–6 October 2014, and (**c**) the T3 site collected on 13 September 2014. Raw images for Scanning Transmission X-ray Microscopy (STXM) and Scanning Electron Microscopy (SEM) are shown (with 2 μm scale bar in bottom left) along with false color maps showing the sub-particle (for C, N, and O) or per-particle (for higher Z elements) mass distribution. Also shown is a color coded carbon speciation map showing soot (red), inorganic (teal), and organic (green) carbon. The calculated individual particle diversity (D_i) is also shown. Note the large spot in the upper right corner of the T3 sample, this was most likely the edge of the Si_3N_4 window and was removed from calculations. Also note the empty lower left corner in the ZF2 sample Energy Dispersive X-ray spectroscopy (EDX) data lacking for those particles; because of this they were removed. Zn, Mn, and Ni maps are omitted here as they were not detected in these FOVs.

The SEM grayscale image shows the slightly different views presented by the two techniques, with particle shapes appearing different between them along with a higher spatial resolution image (10 nm vs. 40 nm with STXM). Soot inclusions identified in the C speciation map are also seen as bright spots in the SEM grayscale image in addition to many of the inorganic inclusions [23]. From the EDX data collection, mass fraction maps for each element (on a per-particle basis) were used to calculate individual particle diversity (D_i) values for each particle. Another aspect of the maps is the varying background level between SEM images, seen especially in the high background of the ZF2 image.

This is a consequence of the brightness and contrast levels being set before EDX acquisition and was performed to ensure that the maximum number of particles were detected by the CCSEM particle detection software.

3.1. Clustering and Source Attribution

For each of the 12 clusters, determined by the k-means algorithm, a random representative sample of 40 particles (taken from any sample or sampling site) was selected for the images shown in Figure 3. The average elemental composition of each cluster is shown in Figure 4 along with the fraction of each cluster collected at the three sampling sites. Finally, Table 3 outlines the assigned colors and labels, as well as some relevant descriptive statistics for each cluster. As can be seen in the average particle diversity column in Table 3, most clusters have a D_α value near either 2.4 or 3.6 (with a single exception). These two values define the "low" or "high" diversity referred to in the cluster names and are discussed in more detail in Section 3.5. A similar source apportionment was discussed in a previous SEM based study; however, it was conducted during the wet season when biogenic aerosols dominate [37]. During the dry season, these biogenic particles are still present but are overwhelmed by aerosols derived from biomass burning.

One notable aspect of Figure 4 is the ratio of elemental Cl to S in each of the clusters shown. From the EDX spectroscopy data presented here, the mass fraction of Cl is often greater or at least similar to that of S. This is apparently contradicted by Table 2, where the concentrations of chloride are an order of magnitude less than the concentrations of sulfate. There are, however, a few extenuating circumstances for this comparison. Firstly, the chloride level in Table 2 is that of non-refractory material owing to the AMS's method of volatilizing particles at ~600 °C. This is well below the vaporization temperature for NaCl and KCl, two major sources of Cl (and inorganics in general). Hence, Table 2 AMS data underestimates Cl mass fractions. Another requirement to allow direct comparison is to change concentration of sulfate to S by multiplying by the ratio of molar masses (~32/96), reducing the concentrations seen in Table 2 to about one third. The third circumstance is the potential for beam damage using the two sequential microscopy techniques here. Some of the inorganic inclusions/cores detected using STXM/NEXAFS spectroscopy may be particularly sensitive to electron beam damage. These sensitive inorganics (particularly ammonium sulfate) could have been volatilized during the scanning/locating phase of SEM and therefore would not be well characterized with subsequent EDX spectroscopy. This carries two consequences: a possible underestimation in the mass fraction of S, and the identification of inorganic regions with STXM without the detection of many inorganic elements by EDX to explain the inclusions. This issue of S quantification is further highlighted when the S/Cl ratios in Figure 4 are compared with previous Particle Induced X-ray Emission (PIXE) measurements which report aerosol S concentrations an order of magnitude greater than Cl concentrations [61,62]. In addition to PIXE, Artaxo et al., 1994 used factor analysis to determine broad particle classes including soil dust, biogenic, marine, and biomass burning classes. One finding of relevance is the high degree of correlation between biogenic particles and S concentration [62]. For the current work, this could suggest an underestimation in the number of clusters hypothesized to contain biogenic particles.

3.1.1. Soot Clusters (LDS1, LDS2, HDS)

The HDS cluster is characterized by a thick organic coating around a soot core. The high levels of organics and K suggests that this cluster mostly originated from biomass burning, but may also contain urban emissions [63]. This, combined with the overwhelming majority of HDS particles being from site T3, suggests that they are predominantly anthropogenic in nature. This cluster's enhanced mass fraction of Na, Cl, Mg, and S indicates a contribution from marine aerosols. Together with the appreciable amount of P and K, this results in the higher particle diversity seen in this cluster. Further supporting this cluster's identity is that enhancement of these elements have been

observed previously during biomass burning events in the presence of a background rich in marine
aerosols [64,65].

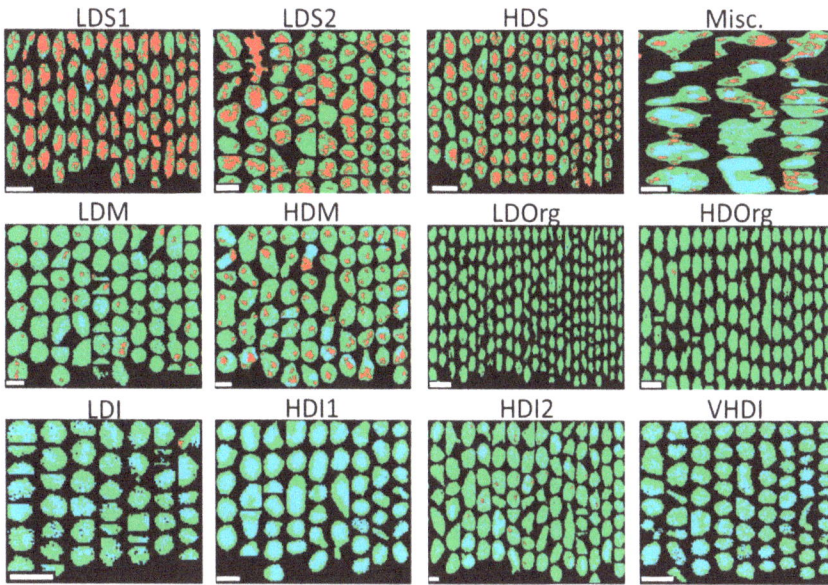

Figure 3. Random sample of ~40 particles from each cluster showing sub-particle carbon speciation as
either soot (red), inorganic (teal), or organic (green). 1 μm scale bars are shown in the bottom left of
each image. Cluster identification (image labels) is provided in Table 3.

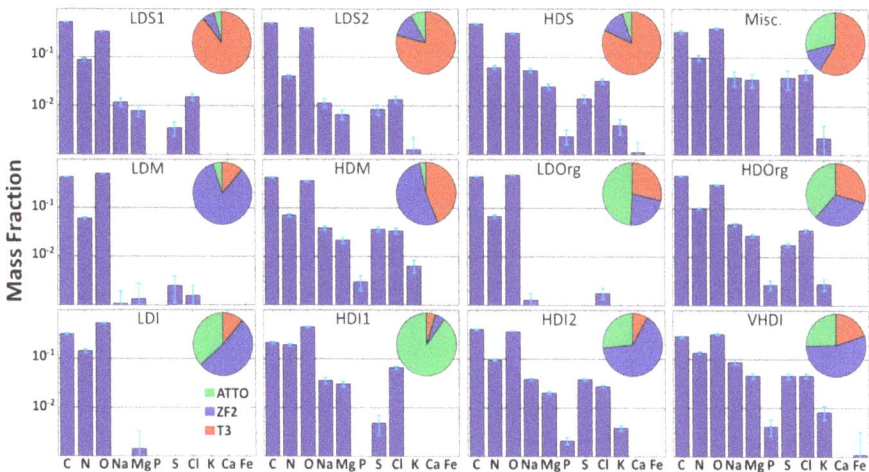

Figure 4. Average elemental composition of each cluster with inset pie chart showing each
cluster's representation at the three sampling sites: ATTO (green), ZF2 (blue), and T3 (red). Al and Si
were not included due to the background from the Al sample holder and the Si_3N_4 substrate. Mn, Ni,
and Zn were not detected and so are omitted here. Cluster identifications (image labels) are provided
in Table 3.

Table 3. Cluster identifying information.

Cluster Name	Label	Avg. Diversity, Dα (Std Err)	CED, μm (Std Err)	O/C Ratio [2]	N
Sub-μm Low Diversity Soot	LDS1	2.67 (0.24)	0.37 (0.01)	0.49	261
Super-μm Low Diversity Soot [1]	LDS2	2.74 (0.29)	1.04 (0.06)	0.60	180
High Diversity Soot	HDS	3.49 (0.51)	0.52 (0.02)	0.48	183
Low Diversity Organics	LDOrg	2.36 (0.08)	0.29 (0.01)	0.81	540
High Diversity Organics	HDOrg	3.49 (0.36)	0.34 (0.01)	0.50	647
Low Diversity Inorganics	LDI	2.57 (0.17)	0.39 (0.02)	1.26	160
High Diversity Coated Inorganics	HDI1	3.87 (0.60)	0.62 (0.03)	1.57	201
High Diversity Inorganics	HDI2	3.75 (0.26)	0.75 (0.02)	0.69	655
Very High Diversity Inorganics	VHDI	4.83 (1.92)	0.45 (0.03)	0.86	212
Low Diversity Mixed	LDM	2.43 (0.13)	0.91 (0.04)	0.88	221
High Diversity Mixed	HDM	3.73 (0.72)	0.94 (0.04)	0.63	209
Miscellaneous	Misc.	3.83 (2.10)	2.35 (0.22)	0.89	47

[1] While MOUDI stages 7 and 8 are nominally submicron stages, it is possible for larger particles to bounce from upper stages and be found in smaller stages. The sub/supermicron descriptor here is also based on circular equivalent diameter rather than aerodynamic diameter like the MOUDI stage cut-off values. [2] O/C ratio calculated for entire particles including organics and inorganics.

In addition to LDS1 lacking the P, K, and Ca that HDS has, LDS1 also has a smaller amount of Na, Mg, S, and Cl which results in the lower average particle diversity. The decreased abundance of K and Cl may indicate urban combustion sources such as diesel engines as opposed to biomass burning [56]. The coatings of organics around the soot cores seen in this cluster are much thinner compared to other soot containing clusters which suggests that particles in this cluster are less aged [66]. The vast majority of particles from this cluster type are from site T3; given that T3 is downwind of Manaus, this suggests fresh urban soot emissions as this cluster's source [35].

The large and multiple soot inclusions and the presence of fractal soot are indications that LDS2 is comprised of particles with a contribution from combustion [67,68]. Particles in LDS2 are found mostly at site T3 which points towards Manaus being the source of these aerosols. The sometimes substantial organic coating on many of these particles is most likely due to condensation during travel of fresh aerosols from Manaus travel to the T3 sampling site. With the exception of the one night time sample, all T3 samples were collected during the mid to late morning (~9:00 to 12:00) which has been seen in other urban areas to correspond to an increased in aged soot over fresh soot owing to the increase in photochemistry [5].

3.1.2. Organic Clusters (LDOrg, HDOrg)

HDOrg is comprised of small particles with their carbon being entirely organic dominant. This cluster has a substantial amount of the heavier elements (Z = 11 (Na) and above) driving the diversity up. The presence of P specifically is important as these elements, coupled with the carbon speciation, suggest that the particles from this cluster are biological in origin [61]. This cluster also contains the highest number of particles with collocated Na and S, which has been previously shown to suggest biogenic particles [69]. The HDOrg cluster also contains an appreciable amount of K and, given that one sample was taken at night, could include biological particles derived from fungal spores [58].

LDOrg is similar in carbon speciation, morphology, and size but lacks the heavier elements contained in HDOrg. The almost entirely C, N, and O composition, small size, and the organic carbon speciation suggest that particles in this cluster are secondary organic aerosols. This is supported by a slight majority of particles in this cluster coming from the general direction of the ATTO site, where a dominant appearance of biogenic secondary organic aerosols and a smaller influence from anthropogenic emissions is expected. The high O/C ratio, along with the dearth of inorganics, may make this cluster comparable to aged ambient oxygenated aerosols [70].

As discussed further on in Section 3.2, both organic clusters are unique in that they make up a sizeable fraction of particles at all sampling sites.

3.1.3. Inorganic Clusters (LDI, HDI1, HDI2, VHDI)

Other than C, N, O, and trace levels of Mg, no other elements are observed in the LDI cluster. Carbon speciation, however, shows a clear inorganic core with an organic coating. This leaves only a few options for the identity of the inorganic cores seen here. One possibility is that the inorganic core is composed of elements not analyzed here, such as Si or Al. Due to the Al mounting plate and the Si(Li) detector used, we are not able to quantitatively detect Al and Si. However, a more likely possibility is that, as mentioned above, the inorganic cores that were initially detected with STXM were particularly sensitive to electron beam damage leading to these sensitive inorganics (possibly ammonium sulfate) being poorly characterized by EDX.

HDI1 is characterized by large, vaguely cubic, inorganic inclusions coated with organics. This cluster has a fairly high fraction of O, Na, and Mg while containing the highest amount of Cl of any cluster. The HD1 cluster is also unique in that particles in it were collected almost exclusively at the ATTO site. Because of this, we suspect these particles represent marine aerosols [71–73]. The organic coating here is substantial and is likely due to aging as aerosols are transported inland to the ATTO site. The transport of particles inland over a large distance is also reflected in the O/C ratio, where the particles (specifically the organic coatings) may have oxidized more than other clusters.

HDI2 is characterized by many small inorganic inclusions speckled throughout the particles which are not as localized as with the HDI1 cluster. There are small soot inclusions and an increased presence of P, K, and S as compared to HDI1. These particles are mainly seen at the ZF2 site with a smaller portion present at ATTO. Thus, it is possible that this cluster is associated with spore rupturing but further investigation is needed to apportion this cluster [74].

The VHDI cluster is unique in that it possesses the highest D_α value of any of the clusters at 4.83, well above both the nominal "high diversity" value of 3.6 and the second highest D_α value of 3.83. This cluster also has a large statistical error of 1.92 (at a 95% confidence level), which could indicate multiple disparate groups are present in this cluster. This cluster is comprised mostly of particles from ZF2, but ATTO and T3 particles contribute substantially as well. The VHDI cluster's elemental composition is similar to that of HDI2, but with a decreased C and O mass fraction and an enhancement of the other elements, especially K (often seen in inorganic salt grains from biomass burning) [30,75]. Inorganics are seen both as large localized inclusions, and as many small inclusions speckled throughout the particle. This cluster's high diversity and larger statistical spread may also be indicative of the varied biomass burning fuels and burning conditions present.

3.1.4. Mixed Clusters (LDM, HDM)

The LDM cluster is characterized by all three carbon speciation types being present in many of the particles. The presence of inorganic inclusions along with the lack of heavier elements suggests ammonium nitrate (and possibly ammonium sulfate) as the identity of the inorganics. This cluster is seen almost entirely at the ZF2 site which, along with its low diversity and few elemental constituents, may indicate a local aerosol source near site ZF2. In which case, particles would have little time or distance to age and scavenge new elements. The presence of soot in the LDM cluster might suggest these aerosols come from the same source as the HDM cluster.

The species of carbon found in the HDM cluster's particles are well mixed with soot, non-carbonaceous inorganic, and organic carbon found in varying ratios. The large soot inclusions, high diversity, and substantial presence of higher Z elements may point to an industrial or automotive origin. Although the sizeable representation of the HDM cluster at T3 supports this, a slightly larger representation is seen at site ZF2. This raises the possibility that some emissions from service vehicles driving to or past the ZF2 site, or nearby generators may be collected at the ZF2 site. Emissions from Manaus are not uncommon either and could account for this cluster's presence at ZF2 [38].

3.1.5. Miscellaneous (Misc.) Clusters

Particles placed in this cluster were most likely grouped due to their supermicron size rather than their composition or diversity. This cluster is comprised of some large rectangular crystals, particles which did not fit well into the other clusters, as well as cases of particles with multiple large inclusions (inorganic or soot) each encased within individual lobes. Since we could not include Al and Si in our analysis, this cluster may also contain mineral dust particles (e.g., quartz and kaolinite) coated with organic material.

This last particle type may contain adjacent particles being erroneously deemed a single particle by our detection algorithm because of overlap of the organic coating upon impaction. This grouping of multiple individual particles into agglomerations much larger than expected for the given MOUDI stage could have caused them to be placed in the Misc. cluster.

One notable particle type seen in this cluster is a collection of particles with a rectangular inorganic core with a small patch of organic carbon in the center. Some of these inorganic cores wrap around the carbon center while some others have a side missing but they all retain the same basic shape. The elemental composition of the inorganic portion contains small amounts of Na and Mg, a relatively large amount of S along with most of the particle's N and O mass fraction. These particles are observed on the only night time sample that was collected. This, along with the particles being found mainly at site T3 could suggest an industrial process whose emissions become easier to identify at night when other sources of aerosols (automotive) experience a decrease. Fragments of ruptured biological particles also may be a possibility based on their elemental composition [74].

3.2. Cluster Type Dependence on Sampling Site

The cluster contributions at each sampling site are shown in Figure 5 separated by stage. Although particles from all cluster types were seen at each location, some particle types were predominantly associated with a particular sampling site. The clusters labeled LDOrg, HDOrg, HDI1, and HDI2 account for ~86% of the particles seen at the ATTO site. To account for a similar share of particles at site ZF2 one must consider the clusters labeled: HDI2, VHDI, LDOrg, HDOrg, LDM, and HDM. Site T3 requires LDS1, LDS2, HDS, LDOrg, HDOrg, and HDM to be accounted for to define a similar portion of particles.

As the ATTO sampling site is less polluted and representative of biogenic aerosols, the presence of both organic clusters as well as two inorganic clusters with possible biogenic origins is expected. Conversely, the relative absence of soot clusters or the mixed clusters further highlights the ATTO site's remoteness from regional anthropogenic (urban) influences. However, even this site is far from being pristine, as shown by the presence of significant amounts of BC, presumably from long-range transport.

While the ZF2 site contains many of the same clusters present at the ATTO site, there are some notable differences. The presence of the HDI1 cluster is diminished (~1% as compared to ATTO's 26%), and both mixed clusters are seen in substantial amounts. The largest difference between the two stages is the enhancement of the LDM cluster in stage 7 data and the minor increase in all three soot clusters in stage 8 particles.

Site T3 shows the presence of many clusters, with all three soot clusters present in substantial amounts. This is expected, as automotive exhaust or energy production through fuel oil burning will produce soot particles that travel to site T3 [76]. Both organic clusters are present with a slight enhancement in stage 8 particles. Because both organic clusters are seen in reasonable amounts at each sampling site, these particles may be part of the aerosol background inherent to sampling in a heavily forested region. Stage 8 particles are also devoid of LDI, HDI1, and Misc. clusters, but few of these were seen in stage 7 and so this absence may be due to insufficient sampling.

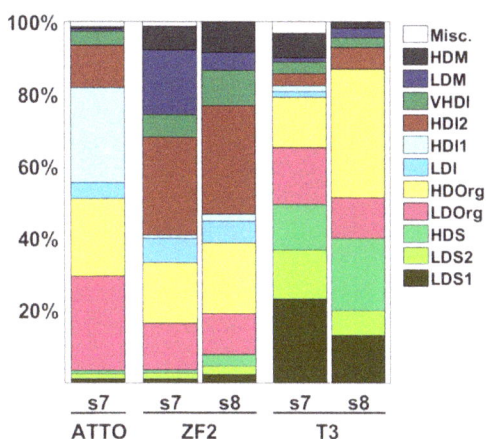

Figure 5. Contribution of the twelve particle-type clusters identified in the samples from stage 7 (nominal aerodynamic size range: 560–320 nm) and stage 8 (320–180 nm) at each sampling site.

Another aspect of Figure 5 is how many clusters make up most of each site's aerosol population, for which we use the following metric. Each site's cluster contribution is sorted in descending order and an effective number of clusters is found using $E(r) = \sum r f_r$, where r is the rank of each cluster's contribution to that site's population (with 1 assigned to the cluster with the largest contribution), f_r is the fraction of that site's population that cluster r accounts for, and $E(r)$ is the effective number of clusters. This metric will vary, in this case from 1 to 12, where the lower the effective number of clusters, the better a given site is characterized by fewer clusters. The values calculated from this metric are listed in Table 4. This metric highlights the increased diversity of sites T3 and ZF2 with respect to the ATTO site. Site ZF2's cluster composition is more varied. This is possibly due to specific events occurring during sampling, or by virtue of being closer to Manaus and therefore more susceptible to anthropogenic emissions. Site ZF2 samples were also collected over multiple days meaning some of the cluster variability may be due to the inclusion of both day and nighttime aerosols. This higher cluster variety could also be attributed to a local aerosol source as mentioned previously. Site T3 stage 7 shows the highest cluster variability due to T3's proximity to (and location downwind of) the anthropogenic center of Manaus. Stage 8 of site T3, in contrast, shows the lowest cluster variability. This may be the influence of fresh emissions coming from Manaus, which tend to be smaller in size and similar in composition (placing many of them in the same cluster).

Table 4. Effective number of clusters for the available sampling site and stage data. The lower the value, the fewer clusters needed to characterize a majority of the sample.

Stage Number	ATTO	ZF2	T3
Stage 7	2.90	3.46	3.80
Stage 8	-	3.37	2.86

3.3. Cluster Size Dependence

Although relatively few supermicron particles were collected, most clusters included some fraction of both sub- and supermicron particles. Only one cluster (Misc) was exclusively supermicron in size, whereas three clusters (LDS1, LDI, LDOrg) included exclusively submicron particles. Referencing Figure 6, the only clusters observed in the supermicron size range were those labeled: Misc (located around 2 μm with a very small percentage), LDM, HDM, HDI2, LDS2, and HDI1.

Supermicron particles in the clusters HDS, HDOrg, and VHDI particles were also observed, but in very small numbers. Many clusters that make up the supermicron range represent more aged species.

The submicron range is composed of many more clusters relative to the supermicron range. Many clusters in the submicron range were often labeled as less aged than the ones found in the supermicron range. This qualitative observation is supported by Figure 7 where there is an increasing trend in individual particle diversity (D_α) with increasing particle size and the notion that D_α is correlated with the extent of particle aging.

3.4. Composition and Diversity Size Dependence

Submicron particles, as seen in Figure 7, have a high fraction of C, N, and O. With D_α values, calculated for both sub- and supermicron particles being 3.3 and 3.4, respectively, submicron particles appear to be the least diverse. However, the error analysis described below, renders this merely suggestive rather than conclusive. As particle size increases, two things are observed: (1) average particle diversity increases slightly and (2) the fraction of inorganics increases. Because of the ubiquity of C, N, and O in aerosol particles, the average particle diversity will almost always be slightly above 3. Given the relatively constant ratios of C, N, and O (with O/C ≈ 0.91 and N/C ≈ 0.22), individual particle diversity is not dependent on these elements; with the exception of soot. Rather, it is mainly the presence of heavier elements which are responsible for any increase in diversity. These larger particles, often represented by more aged clusters like LDS2, HDM, or HDI1, have had sufficient time and travel distance to acquire additional elements during the aging process. A similar conclusion was observed during the Carbonaceous Aerosol and Radiative Effects Study (CARES) conducted in 2010 in California, where heavier elements appreciably affected the mixing state of particles and increased with size, while C, N, and O remained constant [25,37].

After clustering, most clusters were assigned so that their average particle diversity (D_α) was close to one of the two modes present in Figure 8. This clear distinction between the two diversity modes is what the high and low diversity cluster names are referring to.

The bimodality seen here may represent the separation between fresh and more aged aerosol particles. The diversity values of the lower and upper mode of the combined data set were 2.4 and 3.6, respectively. Considering that the three elements C, N, and O dominate the mass fractions of most particles, it is fitting for one mode to be below and one mode to be above 3. Particles in the lower diversity group are mostly C, N, and O with very little presence of other elements. The differing mass fraction between each of these elements causes the diversity to drop below 3.

This bimodality is absent in the T3 samples, having only the less diverse mode. The production of soot from transportation or fuel combustion is most likely the cause of this enhancement of lower diversity particles because of soot's relative elemental purity.

The two dimensional histograms between D_i and CED in Figure 8 serve to reinforce the idea that smaller particles tend to be less diverse. These smaller, less diverse particles are also less spread out whereas the more diverse particles show a wider spread in both diversity and size.

The increased spread seen in the aged aerosol group may be due to the variety of ways that aerosols can age and differences in distances traveled from the aerosols origin. Because the same variability isn't seen in the smaller, less diverse, fresh aerosol group we suspect these particles have sources closer to where they were sampled. By sampling particles with nearby sources, the elemental composition and, by extension, particle diversity will be determined by the method of production and therefore be much less variable.

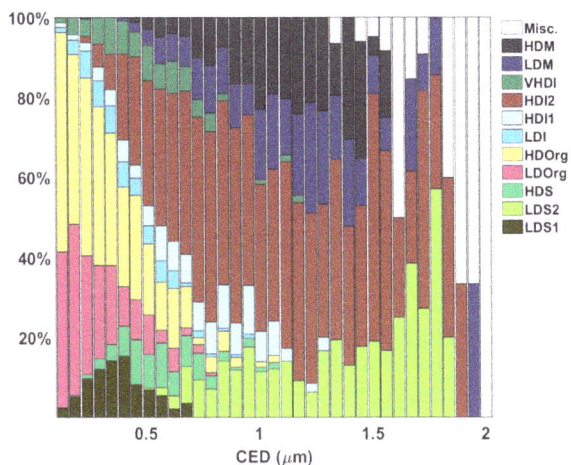

Figure 6. Cluster contribution for varying particle size. Particles >2 μm have been omitted due to their very small abundance and to highlight submicron cluster composition.

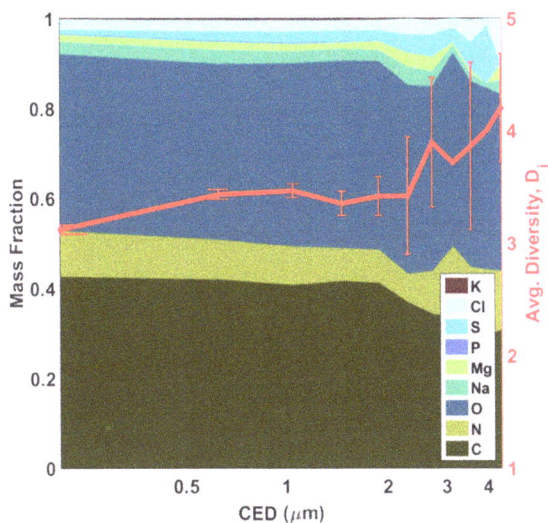

Figure 7. Elemental composition (mass fraction) and average individual particle diversity D_i (red trace) as a function of Circular Equivalent Diameter (CED) for all particles analyzed across all samples. Of note is that only 32 particles with diameters >2 μm were analyzed which is why this region is fairly noisy. Error bars are not shown when only a single particle of that size was measured. Only 9 elements are labeled (with P and K seen as small slivers) whereas the others are too small to be seen in the figure.

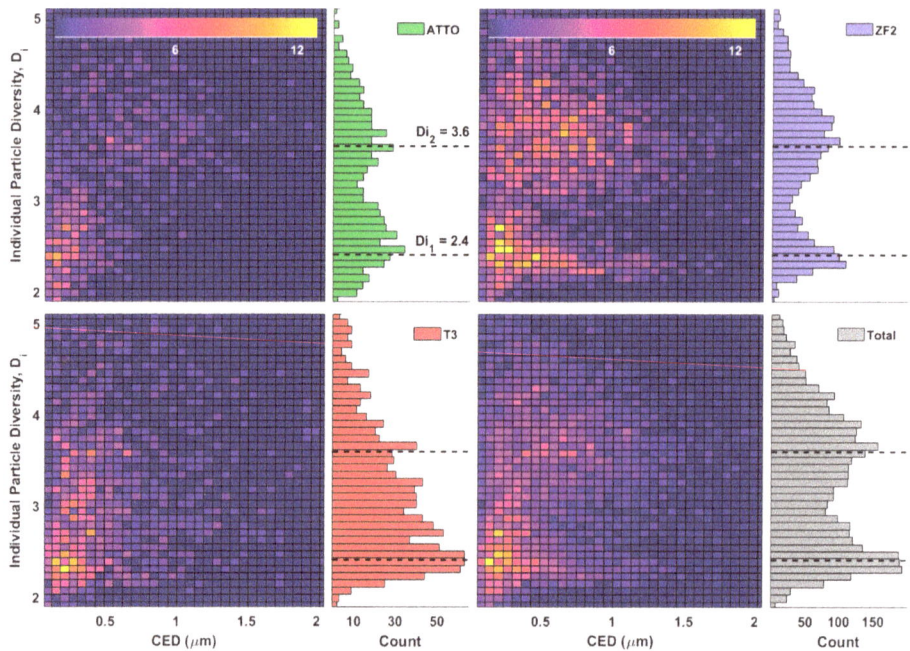

Figure 8. Histograms of individual particle diversity values for each sampling site and the combined data set of all three sites. The individual diversity values for the two modes are indicated with dashed lines and were calculated by fitting two Gaussian distributions to the total data set histogram. Bin counts for 2D histograms are represented by color values shown at the top.

3.5. Mixing State of Particles at Different Sampling Sites

Entropy metrics were used to quantify mixing state for each sample analyzed here. Figure 9 shows the mixing state index (χ) corresponding to particles in each sample. In this case, the variation in mixing state index is small, with all samples having a χ bounded between 0.8 and 0.9. This is a result of D_α and D_γ consistently being around 3.4 and 3.9, respectively.

In the previous study by O'Brien et al. [25], similar sets of STXM and SEM/EDX data were collected for the CARES field campaign. In that study, the same diversity and mixing state parameterization was used except that STXM data (elements C, N, and O) and SEM/EDX (elements Na, Mg, S, P, Cl, K, Fe, Zn, Al, Si, Mn, and Ca) data were analyzed independently. They found that mixing state index values for heavier elements usually ranged from 0.4–0.6 with some values as high as 0.9. The values of χ for C, N, and O generally ranged from 0.75 to 0.9. The mixing state indices retrieved exclusively from STXM data closely matches the values determined in this paper. This suggests that χ is almost entirely determined by C, N, and O due to these three elements dominating the mass of the individual particles and the sample as a whole.

A point of note is the small spread of D_γ values within a given sampling site. With D_γ representing bulk diversity, these values serve to compare the average elemental composition (for the 14 elements chosen) of all aerosols, condensed down into one number. For a given site, samples analyzed were taken during the same season of the same year with similar wind trajectories, sampling times and sampling duration. It is expected then, that there will be a consistency in how much of any given element is present in the aerosol population, based on how much of each element is emitted and included in the particulate material. For D_γ to vary wildly from one day to another, or from one sampling period to another, would require an event or aerosol source producing substantially more of one element than usual.

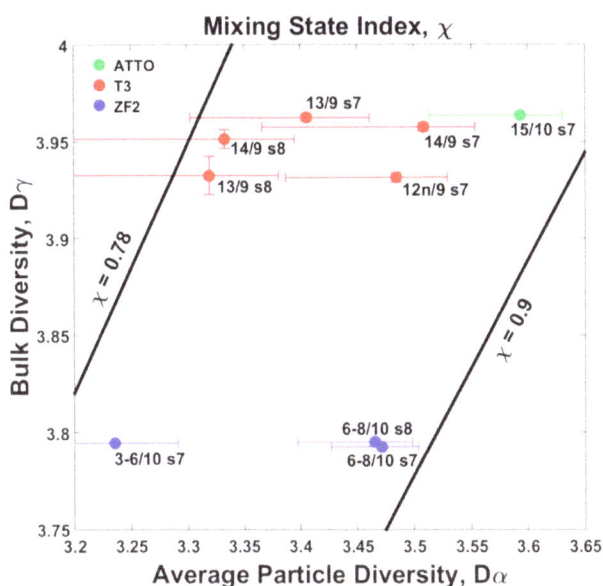

Figure 9. Mixing state index of each sample (color coded by site) with associated error bars, adjusted to one-tenth of their size for readability. All average particle diversities are not significantly different. The site ZF2 bulk diversities are significantly different from the T3 and ATTO bulk diversities. Samples are labeled with day/month and stage number. The horizontal and vertical axes are essentially the numerator and denominator of the definition of χ (refer to Equation (14)).

The spread in D_α values among samples within a sampling site is much wider than that of D_γ. A large spread in D_α is expected when a singular diversity value is calculated from samples containing the variety of distinct particle types seen in Figure 3. This value is more susceptible to change from one sampling date to another compared to D_γ and depends on how much of each aerosol type is collected during a given sampling time.

The increase in the average particle diversity (D_α) with respect to increasing particle size is hinted at here, albeit in less certain terms. Focusing on samples collected where both stage 7 and 8 data were analyzed, average D_α values appear to be larger for stage 7 particles.

Samples collected at site T3 were expected to have a lower mixing state than either the ATTO or the ZF2 site. This hypothesis was borne from the quantity of fresh emissions in Manaus, specifically soot production from combustion, which would serve to drive the mixing state downwards towards total external mixing. However, the end result of the error calculations in Section 2.7 is that the values of χ for each point in Figure 9a become statistically unresolvable, as can be seen from the large error bars. This is also not an issue that would be solved with any reasonable amount of extra data collection but is instead mainly the consequence of the intrinsic spread in D_α.

4. Conclusions

Presented here is a quantitative combination between two complementary per-particle spectromicroscopy techniques, STXM/NEXAFS and SEM/EDX, on the exact same data set. STXM/NEXAFS data was collected at the C, N, and O K-edges on a sub-particle level. This allowed not only the quantitative determination of C, N, and O absolute masses, but also carbon speciation and morphology. SEM/EDX allowed the approximate composition of the inorganic fraction to be determined and then quantified along with the STXM data. The combination of these two techniques enables almost all atmospherically relevant elements to be quantitatively probed on a per-particle basis. The potential issue with S detection discussed above could be mitigated entirely in future measurements by conducting STXM measurements at the S L-edge to obtain S mass fractions. This combined technique could be especially useful for identifying aerosol sources using elemental tracers or unique elemental compositions.

Using particle-specific elemental composition, size, carbon speciation, and individual particle diversity (D_i), k-means clustering was used to separate particles into 12 clusters. The cluster average of these same parameters allowed for potential sources to be assigned. It was found that the stage 7 of the T3 site had a more varied population of particles (as defined by the effective cluster number) and contained more soot-containing clusters than either the ATTO or ZF2 site. Clusters also exhibited size dependence, with a large portion of supermicron particles assigned to high diversity clusters which have been hypothesized to represent more aged particles. This approach could be used for even larger data sets, especially those located at long-standing measurement facilities. From this, diurnal, seasonal, or yearly changes in the aerosol population could be monitored directly. Application of this combined technique would be especially fruitful near large pollution sources, as these anthropogenic sources are difficult to model without the size-resolved composition presented here [77]. The clustering presented here offers an opportunity not only to classify particles but also to identify sources, which can be invaluable in determining the effects of trade or environmental protection policies. The largest detriments to the utility of this composite technique are the long analysis times needed and the requirement for two separate instruments as well as beam time at a synchrotron light source.

Utilizing the composite data set to determine a quantitative mixing state index revealed that particles at site T3 were more externally mixed than at the other two sites. Error analysis, however, shows a fairly large uncertainty in the elemental mixing state for all samples, with statistical errors in χ ranging from 0.3 to 0.8. These error estimates do show that when calculating mixing state by using the 14 elements listed here, mixing state values are close together and show most samples to be highly (between 80 and 90%) internally mixed. Size-dependent trends were also observed in individual particle diversity, with larger particles being slightly more diverse (3.3 and 3.4 for sub and supermicron particles, respectively). This size-dependent trend in diversity was seen even more drastically within the fine mode, with particles <0.5 μm having an average D_i value of around 2.4 and particles >0.5 μm having about a 3.6 diversity value, with a much larger spread of diversity values for larger particles. This difference may identify a separation between fresh and aged aerosols in terms of diversity. This result and the experimental method could be useful for climate models, allowing an experimental mixing state and size-resolved particle composition to be used rather than assumed to improve model performance [78–80]. Even though this type of individual particle microscopy study is time-consuming, regions that are important to global climate models (such as the Amazon) may benefit from the improved accuracy of an experimentally determined mixing state.

The quantitative mixing state index presented is a useful tool, but its utility can be readily expanded. Two of the advantages that this combined spectromicroscopy technique has are the ability to identify morphology both of the particles as a whole and of the constituents within the particle. Due to the general nature of the mixing state parameterization, the mixing state index and its interpretation is heavily dependent on what components were used. In this study, 14 elements were used, however the omission or addition of just a few elements (especially abundant elements such as carbon) could drastically alter the value of χ. Because of this, specifics about which parameters were used and how relevant they are to the samples being studied must be examined before interpreting the value of χ. How well mixed individual elements are also may have limited usefulness to modelers or in general. With the exception of elemental carbon, mass fractions of specific elements (like nitrogen) are less chemically relevant than the molecules or ions they may be found in (nitrate vs. ammonium for example). Future work will build upon this composite technique to instead determine masses

Atmosphere **2017**, *8*, 173

and a molecular mixing state for chemically and atmospherically relevant species such as nitrate, carbonate, sulfate, and etc. Modification in this way could allow our current combined technique to determine an aerosol population's radiative forcing contribution due to both direct and indirect effects. Specific aspects about the indirect effect like hygroscopicity or the number of effective Cloud Condensation Nuclei (CCN) within a population of aerosols could also be gleaned from this method [81]. This type of modification would bolster the usefulness of this technique as well as the usefulness of χ for climate modelers.

Supplementary Materials: The MatLab scripts used to analyze this data are available online at: github.com/MFraund/ElementalMixingStateofAerosolsDuringGoamazon_2017. The following are available online at www.mdpi.com/2073-4433/8/9/173/s1, Figure S1: 72 h HYSPLIT back trajectories starting at 500, 1000, and 1500 m Above Ground Level (AGL) for sites (a) ATTO, (b) ZF2, and (c) T3, Figure S2: (a) AMS and BC and (b) Ozone, CO, and particle concentration time series for the month of September 2014 for the T3 Site. Light red bands represent sample collection periods. From right to left: 12 September 18:00–6:00, 13 September 8:00–12:00, 14 September 9:00–12:00, Figure S3: (a) AMS and BC and (b) CO and particle concentration time series for the month of October 2014 for the ATTO Site. The vertical light red band represents the sample collection period from 14 October 19:00–15 October 19:00, Figure S4: (a) BC and (b) Ozone and CO time series for the month of October 2014 for the ZF2 Site. Vertical light red bands represent sample collection periods. From right to left: 3 October 11:00–6 October 11:00 and 6 October 14:00–8 October 12:00.

Acknowledgments: Funding for sample collection during GoAmazon2014/5 study was provided by the Atmospheric Radiation Measurement Program sponsored by the U.S. Department of Energy (DOE), Office of Science, Office of Biological and Environmental Research (OBER), Climate and Environmental Sciences Division (CESD). Funding for the data analysis was provided by the U.S. DOE's Atmospheric System Research Program, BER under grant DE-SC0013960. The STXM/NEXAFS particle analysis was performed at beamline 5.3.2.2 and at 11.0.2.2 at the Advanced Light Source (ALS) at Lawrence Berkeley National Laboratory. The work at the ALS was supported by the Director, Office of Science, Office of Basic Energy Sciences, of the U.S. DOE under contract DE-AC02-05CH11231. We thank A.L.D. Kilcoyne, Young-Sang Yu, and David Shapiro for their assistance with STXM experiments. Data were obtained from the Atmospheric Radiation Measurement (ARM) Program sponsored by the U.S. Department of Energy, Office of Science, Office of Biological and Environmental Research, Climate and Environmental Sciences Division. The CCSEM/EDX analysis was performed at Environmental Molecular Sciences Laboratory, a National Scientific User Facility sponsored by OBER at PNNL. PNNL is operated by the US Department of Energy by Battelle Memorial Institute under contract DE-AC06-76RL0. This work has been supported by the Max Planck Society (MPG). For the operation of the ATTO site, we acknowledge the support by the German Federal Ministry of Education and Research (BMBF contract 01LB1001A) and the Brazilian Ministério da Ciência, Tecnologia e Inovação (MCTI/FINEP contract 01.11.01248.00) as well as the Amazon State University (UEA), FAPEAM, LBA/INPA and SDS/CEUC/RDS-Uatumã. This paper contains results of research conducted under the Technical/Scientific Cooperation Agreement between the National Institute for Amazonian Research, the State University of Amazonas, and the Max-Planck-Gesellschaft e.V.; the opinions expressed are the entire responsibility of the authors and not of the participating institutions. We highly acknowledge the support by the Instituto Nacional de Pesquisas da Amazônia (INPA). We would like to especially thank all the people involved in the technical, logistical, and scientific support of the ATTO project. The authors gratefully acknowledge the NOAA Air Resources Laboratory (ARL) for the provision of the HYSPLIT transport and dispersion model and/or READY website (http://www.ready.noaa.gov) used in this publication. P. Artaxo acknowledges funding from FAPESP and CNPq. The work was conducted under 001030/2012-4 of the Brazilian National Council for Scientific and Technological Development (CNPq). The authors thank Glauber Cirino for help in collecting samples at ZF2. The authors also thank Rachel O'Brien for her participation in sample collection and Jost V. Lavric, Stefan Wolff, Jorge Saturno, and David Walter for providing ATTO peripheral data and Luciana Rizzo for providing ZF2 peripheral data.

Author Contributions: M.F. led data collection, analyses, and writing. D.Q.P. and D.B. collected STXM data. T.H.H. and B.W. collected STXM/SEM data and assisted with analysis. M.O.A. and C.P. provided ATTO microscopy samples and ATTO peripheral data. J.B. and S.C. (USP) provided ATTO peripheral data as well. A.L. provided ZF2 samples. S.C. (PNNL) assisted in collecting SEM data. M.K.G. provided T3 samples and collected STXM/SEM data on samples. S.S.d.S. and S.T.M. provided T3 supporting data. P.A. provided ZF2 supporting data. R.C.M. conceived the experiment, collected STXM data, and administered the project. All authors provided input on the project and edited the manuscript.

Conflicts of Interest: The authors declare no conflict of interest. The founding sponsors had no role in the design of the study; in the collection, analyses, or interpretation of data; in the writing of the manuscript, and in the decision to publish the results.

References

1. Seinfeld, J.H.; Pandis, S.N. *Atmospheric Chemistry and Physics: From Air Pollution to Climate Change*; John Wiley & Sons: Hoboken, NJ, USA, 2012.
2. Intergovernmental Panel on Climate Change (IPCC). Contribution of working group I to the fifth assessment report of the intergovernmental panel on climate change. In *Climate Change 2013*; Stocker, T., Qin, D., Plattner, G., Tignor, M., Allen, S., Boschung, J., Nauels, A., Xia, Y., Bex, V., Midgley, P., et al., Eds.; Cambridge University Press: Cambridge, UK, 2013.
3. Pöschl, U. Atmospheric aerosols: Composition, transformation, climate and health effects. *Angew. Chem. Int. Edit.* **2005**, *44*, 7520–7540. [CrossRef] [PubMed]
4. Jacobson, M.Z. Strong radiative heating due to the mixing state of black carbon in atmospheric aerosols. *Nature* **2001**, *409*, 695–697. [CrossRef] [PubMed]
5. Moffet, R.C.; Prather, K.A. In-situ measurements of the mixing state and optical properties of soot with implications for radiative forcing estimates. *Proc. Natl. Acad. Sci. USA* **2009**, *106*, 11872–11877. [CrossRef] [PubMed]
6. Fierce, L.; Riemer, N.; Bond, T.C. Explaining variance in black carbon's aging timescale. *Atmos. Chem. Phys.* **2015**, *15*, 3173–3191. [CrossRef]
7. Ault, A.P.; Axson, J.L. Atmospheric aerosol chemistry: Spectroscopic and microscopic advances. *Anal. Chem.* **2016**, *89*, 430–452. [CrossRef] [PubMed]
8. Warren, D.R.; Seinfeld, J.H. Simulation of aerosol size distribution evolution in systems with simultaneous nucleation, condensation, and coagulation. *Aerosol Sci. Technol.* **1985**, *4*, 31–43. [CrossRef]
9. Liu, D.; Whitehead, J.; Alfarra, M.R.; Reyes-Villegas, E.; Spracklen, D.V.; Reddington, C.L.; Kong, S.; Williams, P.I.; Ting, Y.-C.; Haslett, S. Black-carbon absorption enhancement in the atmosphere determined by particle mixing state. *Nat. Geosci.* **2017**, *10*, 184–188. [CrossRef]
10. Peng, J.; Hu, M.; Guo, S.; Du, Z.; Zheng, J.; Shang, D.; Zamora, M.L.; Zeng, L.; Shao, M.; Wu, Y.S. Markedly enhanced absorption and direct radiative forcing of black carbon under polluted urban environments. *Proc. Natl. Acad. Sci. USA* **2016**, *113*, 4266–4271. [CrossRef] [PubMed]
11. Raatikainen, T.; Brus, D.; Hyvärinen, A.-P.; Svensson, J.; Asmi, E.; Lihavainen, H. Black carbon concentrations and mixing state in the Finnish Arctic. *Atmos. Chem. Phys.* **2015**, *15*, 10057–10070. [CrossRef]
12. D'Almeida, G.A.; Koepke, P.; Shettle, E.P. *Atmospheric Aerosols: Global Climatology and Radiative Characteristics*; A Deepak Publishing: Hampton, VA, USA, 1991.
13. Cappa, C.D.; Onasch, T.B.; Massoli, P.; Worsnop, D.R.; Bates, T.S.; Cross, E.S.; Davidovits, P.; Hakala, J.; Hayden, K.L.; Jobson, B.T. Radiative absorption enhancements due to the mixing state of atmospheric black carbon. *Science* **2012**, *337*, 1078–1081. [CrossRef] [PubMed]
14. Schmidt, G.A.; Ruedy, R.; Hansen, J.E.; Aleinov, I.; Bell, N.; Bauer, M.; Bauer, S.; Cairns, B.; Canuto, V.; Cheng, Y. Present-day atmospheric simulations using GISS modelE: Comparison to in situ, satellite, and reanalysis data. *J. Clim.* **2006**, *19*, 153–192. [CrossRef]
15. Jacobson, M.Z. GATOR-GCMM: A global-through urban-scale air pollution and weather forecast model: 1. Model design and treatment of subgrid soil, vegetation, roads, rooftops, water, sea ice, and snow. *J. Geophys. Res. Atmos.* **2001**, *106*, 5385–5401. [CrossRef]
16. Zaveri, R.A.; Barnard, J.C.; Easter, R.C.; Riemer, N.; West, M. Particle-resolved simulation of aerosol size, composition, mixing state, and the associated optical and cloud condensation nuclei activation properties in an evolving urban plume. *J. Geophys. Res. Atmos.* **2010**, *115*, 1383–1392. [CrossRef]
17. Schwarz, J.; Gao, R.; Fahey, D.; Thomson, D.; Watts, L.; Wilson, J.; Reeves, J.; Darbeheshti, M.; Baumgardner, D.; Kok, G.; et al. Single-particle measurements of midlatitude black carbon and light-scattering aerosols from the boundary layer to the lower stratosphere. *J. Geophys. Res. Atmos.* **2006**, *111*, D16207. [CrossRef]
18. Sedlacek, A.J.; Lewis, E.R.; Kleinman, L.; Xu, J.; Zhang, Q. Determination of and evidence for non-core-shell structure of particles containing black carbon using the single-particle soot photometer (SP2). *Geophys. Res. Lett.* **2012**, *39*, L06802. [CrossRef]
19. Healy, R.M.; Riemer, N.; Wenger, J.C.; Murphy, M.; West, M.; Poulain, L.; Wiedensohler, A.; O'Connor, I.P.; McGillicuddy, E.; Sodeau, J.R.; et al. Single particle diversity and mixing state measurements. *Atmos. Chem. Phys.* **2014**, *14*, 6289–6299. [CrossRef]

20. Riemer, N.; West, M. Quantifying aerosol mixing state with entropy and diversity measures. *Atmos. Chem. Phys.* **2013**, *13*, 11423–11439. [CrossRef]
21. Hopkins, R.J.; Tivanski, A.V.; Marten, B.D.; Gilles, M.K. Chemical bonding and structure of black carbon reference materials and individual carbonaceous atmospheric aerosols. *J. Aerosol Sci.* **2007**, *38*, 573–591. [CrossRef]
22. Moffet, R.C.; Tivanski, A.V.; Gilles, M.K. Scanning transmission X-ray microscopy: Applications in atmospheric aerosol research. In *Fundamentals and Applications of Aerosol Spectroscopy*; Reid, J., Signorell, R., Eds.; Taylor and Francis Books: Didcot, UK, 2011.
23. Laskin, A.; Cowin, J.P.; Iedema, M.J. Analysis of individual environmental particles using modern methods of electron microscopy and X-ray microanalysis. *J. Electron. Spectros. Relat. Phenomena* **2006**, *150*, 260–274. [CrossRef]
24. Falcone, R.; Sommariva, G.; Verità, M. WDXRF, EPMA and SEM/EDX quantitative chemical analyses of small glass samples. *Microchim. Acta* **2006**, *155*, 137–140. [CrossRef]
25. O'Brien, R.E.; Wang, B.; Laskin, A.; Riemer, N.; West, M.; Zhang, Q.; Sun, Y.; Yu, X.Y.; Alpert, P.; Knopf, D.A.; et al. Chemical imaging of ambient aerosol particles: Observational constraints on mixing state parameterization. *J. Geophys. Res. Atmos.* **2015**, *120*, 9591–9605. [CrossRef]
26. Piens, D.S.; Kelly, S.T.; Harder, T.H.; Petters, M.D.; O'Brien, R.E.; Wang, B.; Teske, K.; Dowell, P.; Laskin, A.; Gilles, M.K. Measuring mass-based hygroscopicity of atmospheric particles through in situ imaging. *Environ. Sci. Technol.* **2016**, *50*, 5172–5180. [CrossRef] [PubMed]
27. Streets, D.; Bond, T.; Lee, T.; Jang, C. On the future of carbonaceous aerosol emissions. *J. Geophys. Res. Atmos.* **2004**, *109*, D24212. [CrossRef]
28. Artaxo, P.; Rizzo, L.V.; Brito, J.F.; Barbosa, H.M.; Arana, A.; Sena, E.T.; Cirino, G.G.; Bastos, W.; Martin, S.T.; Andreae, M.O. Atmospheric aerosols in Amazonia and land use change: From natural biogenic to biomass burning conditions. *Faraday Discuss.* **2013**, *165*, 203–235. [CrossRef] [PubMed]
29. Wang, J.; Krejci, R.; Giangrande, S.; Kuang, C.; Barbosa, H.M.; Brito, J.; Carbone, S.; Chi, X.; Comstock, J.; Ditas, F.; et al. Amazon boundary layer aerosol concentration sustained by vertical transport during rainfall. *Nature* **2016**, *539*, 416–419. [CrossRef] [PubMed]
30. Martin, S.T.; Andreae, M.O.; Artaxo, P.; Baumgardner, D.; Chen, Q.; Goldstein, A.H.; Guenther, A.; Heald, C.L.; Mayol-Bracero, O.L.; McMurry, P.H.; et al. Sources and properties of Amazonian aerosol particles. *Rev. Geophys.* **2010**, *48*, RG2002. [CrossRef]
31. Bond, T.; Venkataraman, C.; Masera, O. Global atmospheric impacts of residential fuels. *Energy Sustain. Dev.* **2004**, *8*, 20–32. [CrossRef]
32. Staudt, A.; Jacob, D.J.; Logan, J.A.; Bachiochi, D.; Krishnamurti, T.; Sachse, G. Continental sources, transoceanic transport, and interhemispheric exchange of carbon monoxide over the Pacific. *J. Geophys. Res. Atmos.* **2001**, *106*, 32571–32589. [CrossRef]
33. Kuhn, U.; Ganzeveld, L.; Thielmann, A.; Dindorf, T.; Schebeske, G.; Welling, M.; Sciare, J.; Roberts, G.; Meixner, F.; Kesselmeier, J.; et al. Impact of manaus city on the Amazon green ocean atmosphere: Ozone production, precursor sensitivity and aerosol load. *Atmos. Chem. Phys.* **2010**, *10*, 9251–9282. [CrossRef]
34. Martin, S.; Artaxo, P.; Machado, L.; Manzi, A.; Souza, R.; Schumacher, C.; Wang, J.; Andreae, M.; Barbosa, H.; Fan, J.; et al. Introduction: Observations and modeling of the green ocean Amazon (GoAmazon2014/5). *Atmos. Chem. Phys.* **2016**, *16*, 4785–4797. [CrossRef]
35. Martin, S.T.; Artaxo, P.; Machado, L.; Manzi, A.; Souza, R.; Schumacher, C.; Wang, J.; Biscaro, T.; Brito, J.; Calheiros, A.; et al. The green ocean Amazon experiment (GoAmazon2014/5) observes pollution affecting gases, aerosols, clouds, and rainfall over the rain forest. *Bull. Am. Meteorol. Soc.* **2017**, *98*, 981–997. [CrossRef]
36. Swap, R.; Garstang, M.; Greco, S.; Talbot, R.; Kållberg, P. Saharan dust in the Amazon Basin. *Tellus B* **1992**, *44*, 133–149. [CrossRef]
37. Pöschl, U.; Martin, S.; Sinha, B.; Chen, Q.; Gunthe, S.; Huffman, J.; Borrmann, S.; Farmer, D.; Garland, R.; Helas, G.; et al. Rainforest aerosols as biogenic nuclei of clouds and precipitation in the Amazon. *Science* **2010**, *329*, 1513–1516. [CrossRef] [PubMed]
38. Martin, S.; Andreae, M.; Althausen, D.; Artaxo, P.; Baars, H.; Borrmann, S.; Chen, Q.; Farmer, D.; Guenther, A.; Gunthe, S.; et al. An overview of the Amazonian aerosol characterization experiment 2008 (AMAZE-08). *Atmos. Chem. Phys.* **2010**, *10*, 11415–11438. [CrossRef]

39. Chen, Q.; Farmer, D.; Rizzo, L.; Pauliquevis, T.; Kuwata, M.; Karl, T.G.; Guenther, A.; Allan, J.D.; Coe, H.; Andreae, M.; et al. Submicron particle mass concentrations and sources in the Amazonian wet season (AMAZE-08). *Atmos. Chem. Phys.* **2015**, *15*, 3687–3701. [CrossRef]

40. Andreae, M.; Acevedo, O.; Araùjo, A.; Artaxo, P.; Barbosa, C.; Barbosa, H.; Brito, J.; Carbone, S.; Chi, X.; Cintra, B.; et al. The Amazon Tall Tower Observatory (ATTO): Overview of pilot measurements on ecosystem ecology, meteorology, trace gases, and aerosols. *Atmos. Chem. Phys.* **2015**, *15*, 10723–10776. [CrossRef]

41. Stein, A.; Draxler, R.R.; Rolph, G.D.; Stunder, B.J.; Cohen, M.; Ngan, F. NOAA's HYSPLIT atmospheric transport and dispersion modeling system. *Bull. Am. Meteorol. Soc.* **2015**, *96*, 2059–2077. [CrossRef]

42. Rolph, G.; Stein, A.; Stunder, B. Real-time environmental applications and display system: Ready. *Environ. Model. Softw.* **2017**, *95*, 210–228. [CrossRef]

43. Kilcoyne, A.; Tyliszczak, T.; Steele, W.; Fakra, S.; Hitchcock, P.; Franck, K.; Anderson, E.; Harteneck, B.; Rightor, E.; Mitchell, G.; et al. Interferometer-controlled scanning transmission X-ray microscopes at the Advanced Light Source. *J. Synchrotron Radiat.* **2003**, *10*, 125–136. [CrossRef] [PubMed]

44. Henke, B.L.; Gullikson, E.M.; Davis, J.C. X-ray interactions: Photoabsorption, scattering, transmission, and reflection at E = 50–30,000 ev, Z = 1–92. *Atom. Data Nucl. Data Tables* **1993**, *54*, 181–342. [CrossRef]

45. Moffet, R.C.; Henn, T.; Laskin, A.; Gilles, M.K. Automated chemical analysis of internally mixed aerosol particles using X-ray spectromicroscopy at the carbon K-Edge. *Anal. Chem.* **2010**, *82*, 7906–7914. [CrossRef] [PubMed]

46. Guizar-Sicairos, M.; Thurman, S.T.; Fienup, J.R. Efficient subpixel image registration algorithms. *Opt. lett.* **2008**, *33*, 156–158. [CrossRef] [PubMed]

47. Otsu, N. A threshold selection method from gray-level histograms. *Automatica* **1975**, *11*, 23–27. [CrossRef]

48. Artaxo, P.; Martins, J.V.; Yamasoe, M.A.; Procópio, A.S.; Pauliquevis, T.M.; Andreae, M.O.; Guyon, P.; Gatti, L.V.; Leal, A.M.C.; et al. Physical and chemical properties of aerosols in the wet and dry seasons in Rondônia, Amazonia. *J. Geophys. Res. Atmos.* **2002**, *107*. [CrossRef]

49. Hartigan, J.A.; Wong, M.A. Algorithm as 136: A k-means clustering algorithm. *J. R. Stat. Soc. Ser. C Appl. Stat.* **1979**, *28*, 100–108. [CrossRef]

50. Rebotier, T.P.; Prather, K.A. Aerosol time-of-flight mass spectrometry data analysis: A benchmark of clustering algorithms. *Anal. Chim. Acta* **2007**, *585*, 38–54. [CrossRef] [PubMed]

51. Kodinariya, T.M.; Makwana, P.R. Review on determining number of cluster in K-means clustering. *Int. J.* **2013**, *1*, 90–95.

52. De Sá, S.S.; Palm, B.B.; Campuzano-Jost, P.; Day, D.A.; Newburn, M.K.; Hu, W.; Isaacman-VanWertz, G.; Yee, L.D.; Thalman, R.; Brito, J.; et al. Influence of urban pollution on the production of organic particulate matter from isoprene epoxydiols in central Amazonia. *Atmos. Chem. Phys.* **2017**, *17*, 6611–6629. [CrossRef]

53. Kirkman, G.; Gut, A.; Ammann, C.; Gatti, L.; Cordova, A.; Moura, M.; Andreae, M.; Meixner, F. Surface exchange of nitric oxide, nitrogen dioxide, and ozone at a cattle pasture in Rondonia, Brazil. *J. Geophys. Res. Atmos.* **2002**, *107*. [CrossRef]

54. Ahlm, L.; Nilsson, E.; Krejci, R.; Mårtensson, E.; Vogt, M.; Artaxo, P. A comparison of dry and wet season aerosol number fluxes over the Amazon rain forest. *Atmos. Chem. Phys. Discuss.* **2009**, *10*, 3063–3079. [CrossRef]

55. Saturno, J.; Pöhlker, C.; Massabò, D.; Brito, J.; Carbone, S.; Cheng, Y.; Chi, X.; Ditas, F.; de Angelis, I.H.; Morán-Zuloaga, D.; et al. Comparison of different aethalometer correction schemes and a reference multi-wavelength absorption technique for ambient aerosol data. *Atmos. Meas. Tech.* **2017**, *10*, 2837. [CrossRef]

56. Li, J.; Pósfai, M.; Hobbs, P.V.; Buseck, P.R. Individual aerosol particles from biomass burning in southern Africa: 2, Compositions and aging of inorganic particles. *J. Geophys. Res. Atmos.* **2003**, *108*, D13. [CrossRef]

57. Ault, A.P.; Moffet, R.C.; Baltrusaitis, J.; Collins, D.B.; Ruppel, M.J.; Cuadra-Rodriguez, L.A.; Zhao, D.; Guasco, T.L.; Ebben, C.J.; Geiger, F.M. Size-dependent changes in sea spray aerosol composition and properties with different seawater conditions. *Environ. Sci. Technol.* **2013**, *47*, 5603–5612. [CrossRef] [PubMed]

58. Pöhlker, C.; Wiedemann, K.T.; Sinha, B.; Shiraiwa, M.; Gunthe, S.S.; Smith, M.; Su, H.; Artaxo, P.; Chen, Q.; Cheng, Y.; et al. Biogenic potassium salt particles as seeds for secondary organic aerosol in the Amazon. *Science* **2012**, *337*, 1075–1078. [CrossRef] [PubMed]

59. Brooks, S.D.; Garland, R.M.; Wise, M.E.; Prenni, A.J.; Cushing, M.; Hewitt, E.; Tolbert, M.A. Phase changes in internally mixed maleic acid/ammonium sulfate aerosols. *J. Geophys. Res. Atmos.* **2003**, *108*, D15. [CrossRef]

60. Pósfai, M.; Gelencsér, A.; Simonics, R.; Arató, K.; Li, J.; Hobbs, P.V.; Buseck, P.R. Atmospheric tar balls: Particles from biomass and biofuel burning. *J. Geophys. Res. Atmos.* **2004**, *109*, D6. [CrossRef]

61. Artaxo, P.; Gerab, F.; Rabello, M.L. Elemental composition of aerosol particles from two atmospheric monitoring stations in the Amazon Basin. *Nucl. Instrum. Methods Phys. Res. B* **1993**, *75*, 277–281. [CrossRef]

62. Artaxo, P.; Gerab, F.; Yamasoe, M.A.; Martins, J.V. Fine mode aerosol composition at three long-term atmospheric monitoring sites in the Amazon Basin. *J. Geophys. Res. Atmos.* **1994**, *99*, 22857–22868. [CrossRef]

63. Andreae, M.O. Soot carbon and excess fine potassium: Long-range transport of combustion-derived aerosols. *Science* **1983**, *220*, 1148–1151. [CrossRef] [PubMed]

64. Maudlin, L.; Wang, Z.; Jonsson, H.; Sorooshian, A. Impact of wildfires on size-resolved aerosol composition at a coastal California site. *Atmos. Environ.* **2015**, *119*, 59–68. [CrossRef]

65. Wonaschütz, A.; Hersey, S.; Sorooshian, A.; Craven, J.; Metcalf, A.; Flagan, R.; Seinfeld, J. Impact of a large wildfire on water-soluble organic aerosol in a major urban area: The 2009 station fire in Los Angeles county. *Atmos. Chem. Phys.* **2011**, *11*, 8257–8270. [CrossRef]

66. Rudich, Y.; Donahue, N.M.; Mentel, T.F. Aging of organic aerosol: Bridging the gap between laboratory and field studies. *Annu. Rev. Phys. Chem.* **2007**, *58*, 321–352. [CrossRef] [PubMed]

67. Zhang, R.; Khalizov, A.F.; Pagels, J.; Zhang, D.; Xue, H.; McMurry, P.H. Variability in morphology, hygroscopicity, and optical properties of soot aerosols during atmospheric processing. *Proc. Natl. Acad. Sci. USA* **2008**, *105*, 10291–10296. [CrossRef] [PubMed]

68. Soto-Garcia, L.L.; Andreae, M.O.; Andreae, T.W.; Artaxo, P.; Maenhaut, W.; Kirchstetter, T.; Novakov, T.; Chow, J.C.; Mayol-Bracero, O.L. Evaluation of the carbon content of aerosols from the burning of biomass in the Brazilian Amazon using thermal, optical and thermal-optical analysis methods. *Atmos. Chem. Phys.* **2011**, *11*, 4425–4444. [CrossRef]

69. Sorooshian, A.; Crosbie, E.; Maudlin, L.C.; Youn, J.S.; Wang, Z.; Shingler, T.; Ortega, A.M.; Hersey, S.; Woods, R.K. Surface and airborne measurements of organosulfur and methanesulfonate over the western United States and coastal areas. *J. Geophys. Res. Atmos.* **2015**, *120*, 8535–8548. [CrossRef] [PubMed]

70. Aiken, A.C.; Decarlo, P.F.; Kroll, J.H.; Worsnop, D.R.; Huffman, J.A.; Docherty, K.S.; Ulbrich, I.M.; Mohr, C.; Kimmel, J.R.; Sueper, D.; et al. O/C and OM/OC ratios of primary, secondary, and ambient organic aerosols with high-resolution time-of-flight aerosol mass spectrometry. *Environ. Sci. Technol.* **2008**, *42*, 4478–4485. [CrossRef] [PubMed]

71. Artaxo, P.; Storms, H.; Bruynseels, F.; Van Grieken, R.; Maenhaut, W. Composition and sources of aerosols from the Amazon Basin. *J. Geophys. Res. Atmos.* **1988**, *93*, 1605–1615. [CrossRef]

72. Garstang, M.; Greco, S.; Scala, J.; Swap, R.; Ulanski, S.; Fitzjarrald, D.; Martin, D.; Browell, E.; Shipman, M.; Connors, V. The Amazon boundary-layer experiment (ABLE 2B): A meteorological perspective. *Bull. Am. Meteorol. Soc.* **1990**, *71*, 19–32. [CrossRef]

73. Artaxo, P.; Fernandas, E.T.; Martins, J.V.; Yamasoe, M.A.; Hobbs, P.V.; Maenhaut, W.; Longo, K.M.; Castanho, A. Large-scale aerosol source apportionment in Amazonia. *J. Geophys. Res. Atmos.* **1998**, *103*, 31837–31847. [CrossRef]

74. China, S.; Wang, B.; Weis, J.; Rizzo, L.; Brito, J.; Cirino, G.G.; Kovarik, L.; Artaxo, P.; Gilles, M.K.; Laskin, A. Rupturing of biological spores as a source of secondary particles in Amazonia. *Environ. Sci. Technol.* **2016**, *50*, 12179–12186. [CrossRef] [PubMed]

75. Pósfai, M.; Simonics, R.; Li, J.; Hobbs, P.V.; Buseck, P.R. Individual aerosol particles from biomass burning in southern Africa: 1. Compositions and size distributions of carbonaceous particles. *J. Geophys. Res. Atmos.* **2003**, *108*, D13. [CrossRef]

76. Colbeck, I.; Atkinson, B.; Johar, Y. The morphology and optical properties of soot produced by different fuels. *J. Aerosol Sci.* **1997**, *28*, 715–723. [CrossRef]

77. Ervens, B.; Cubison, M.; Andrews, E.; Feingold, G.; Ogren, J.; Jimenez, J.; Quinn, P.; Bates, T.; Wang, J.; Zhang, Q.; et al. Ccn predictions using simplified assumptions of organic aerosol composition and mixing state: A synthesis from six different locations. *Atmos. Chem. Phys.* **2010**, *10*, 4795–4807. [CrossRef]

78. Medina, J.; Nenes, A.; Sotiropoulou, R.E.P.; Cottrell, L.D.; Ziemba, L.D.; Beckman, P.J.; Griffin, R.J. Cloud condensation nuclei closure during the international consortium for atmospheric research on transport and transformation 2004 campaign: Effects of size-resolved composition. *J. Geophys. Res. Atmos.* **2007**, *112*, D10. [CrossRef]

79. Cubison, M.; Ervens, B.; Feingold, G.; Docherty, K.; Ulbrich, I.; Shields, L.; Prather, K.; Hering, S.; Jimenez, J. The influence of chemical composition and mixing state of Los Angeles urban aerosol on CCN number and cloud properties. *Atmos. Chem. Phys.* **2008**, *8*, 5649–5667. [CrossRef]

80. Stroud, C.A.; Nenes, A.; Jimenez, J.L.; DeCarlo, P.F.; Huffman, J.A.; Bruintjes, R.; Nemitz, E.; Delia, A.E.; Toohey, D.W.; Guenther, A.B.; et al. Cloud activating properties of aerosol observed during CELTIC. *J. Atmos. Sci.* **2007**, *64*, 441–459. [CrossRef]

81. Wang, J.; Cubison, M.; Aiken, A.; Jimenez, J.; Collins, D. The importance of aerosol mixing state and size-resolved composition on CCN concentration and the variation of the importance with atmospheric aging of aerosols. *Atmos. Chem. Phys.* **2010**, *10*, 7267–7283. [CrossRef]

atmosphere

MDPI

Article

Q-Space Analysis of the Light Scattering Phase Function of Particles with Any Shape

Christopher M. Sorensen [1,*], Yuli W. Heinson [2], William R. Heinson [2], Justin B. Maughan [1] and Amit Chakrabarti [1]

[1] Department of Physics, Kansas State University, Manhattan, KS 66506, USA; maughan@phys.ksu.edu (J.B.M.); amitc@phys.ksu.edu (A.C.)
[2] Department of Energy, Environmental & Chemical Engineering, Washington University in St. Louis, St. Louis, MO 63130, USA; yuli.heinson@gmail.com (Y.W.H.); willbot1983@gmail.com (W.R.H.)
* Correspondence: sor@phys.ksu.edu; Tel.: +1-785-532-1626

Academic Editors: Swarup China and Claudio Mazzoleni
Received: 20 January 2017; Accepted: 20 March 2017; Published: 29 March 2017

Abstract: Q-space analysis is applied to the light scattering phase function of a wide variety of non-spherical and irregularly shaped particles including a great many types of dusts, fractal aggregates, spheroids, irregular spheres, Gaussian random spheres, thickened clusters and nine types of ice crystals. The phase functions were either experimental data or calculations. This analysis method uncovers many specific and quantitative similarities and differences between the scattering by various shapes and also when compared to spheres. From this analysis a general description for scattering by a particle of any shape emerges with specific details assigned to various shapes.

Keywords: light scattering; phase function; irregularly shaped particles; Q-space analysis

1. Introduction

The particles that appear in the atmosphere have a variety of shapes that can be simply divided into spheres and non-spheres. All these particles scatter and absorb light and this light-particle interaction is significant for the energy budget of the atmosphere and the Earth itself. The problem of how spherical particles interact with light was solved long ago; on the other hand a solution to describe and understand light scattering and absorption by non-spherical particles can be very challenging. Nevertheless, remarkable analytical and numerical methods have been developed and computational hardware has allowed for ever increasing speed for large scale calculations. Moreover, comprehensive experimental studies of scattering both in the lab and the field have occurred. However, given a solution or a set of data for scattering, the problem remains what to do with it. That is, if the pattern cannot be described, how can one quantitatively describe the scattering pattern and distinguish one pattern from another? Furthermore, if the pattern cannot be described, how can one know the physics responsible for the pattern?

Some time ago, we demonstrated [1–3] that the angular scattering patterns for spherical particles are best viewed as a function of the magnitude of the scattering wave vector

$$q = 2k \sin (\theta/2) \tag{1}$$

where $k = 2\pi/\lambda$, λ is the wavelength of light, θ is the scattering angle and the scale for plotting should be logarithmic. This plotting yields a distinctly different perspective than plotting scattered intensity versus linear θ. Properties that make q a viable independent variable are that its inverse is a length scale that probes the lengths inherent to the particle. It is also the Fourier variable in the mathematical description of diffraction which is physically demonstrated as the limit where the refractive index of a particle approaches one. We must also stress that the logarithmic axis for q is essential as well as q itself because so much of our world progresses geometrically rather than arithmetically. We call this procedure of plotting scattered intensity versus q double logarithmically "Q-space analysis". We shall see that Q-space analysis has descriptive abilities that can compare the similarities and differences of the scattering by different types of particles. It also leads to physical interpretation of the scattering mechanism.

This paper is concerned with the problem of how to describe and compare the angular scattering patterns, the phase functions, either observed or calculated for arbitrarily shaped particles. The foundation of this project is a comprehensive description of spherical particle scattering as viewed from Q-space, and that is where we start.

2. Q-Space Analysis Applied to Spheres

Properties universal to scattering by all particles becomes more apparent during Q-space analysis by rescaling both the q-axis and the intensity axis. When the effective radius R of the particle is known, the plot can be improved by plotting versus the dimensionless variable qR. Improvement can also be made by normalizing the differential scattering cross section, which is proportional to the scattered intensity and the phase function, by the Rayleigh differential cross section of the particle [4–6].

$$dC_{sca,\,Ray,\,sph}/d\Omega = k^4 R^6 F(m) \qquad (2)$$

where $k = 2\pi/\lambda$ and

$$F(m) = \left| \frac{m^2 - 1}{m^2 + 2} \right|^2 \qquad (3)$$

The function $F(m)$ is the square of the Lorentz-Lorenz function of the complex refractive index $m = n + i\kappa$. Equation (2) is proportional to the Rayleigh scattered intensity I_{Ray}.

Our work has shown that the scattering by spheres is well parameterized by the internal coupling parameter [7]

$$\rho' = 2kR\sqrt{F(m)} = 2kR \left| \frac{m^2 - 1}{m^2 + 2} \right| \qquad (4)$$

This parameter is similar to the well-known phase shift parameter $\rho = 2kR|m - 1|$ [4,5] but does a better job in describing the evolution of the scattering away from the diffraction limit, where $\rho' = \rho = 0$, and acting as a quasi-universal parameter for the scattering, as will be demonstrated below.

Figure 1 shows Q-space analysis applied to scattering by spheres (Mie scattering) for $\rho' = 3, 10, 30, 100$, and 1000 comprised from three refractive indices, $m = 1.1, 1.5$ and 2.0 and radii ranging from 0.25 to 647 microns. The envelope of the diffraction limit, also called the Rayleigh Debye-Gans (RDG) limit, which occurs when $\rho' \to 0$, is also shown as a dashed line. Because the scattered intensity, I, is proportional to the differential cross section, the Rayleigh normalized differential cross section is represented by a Rayleigh normalized scattered intensity, I/I_{Ray}. The Mie scattering calculation was averaged over a log-normal size distribution with a geometric width 1.2 to eliminate the interference ripples.

The scattering curves in Figure 1 with the same ρ' fall nearly on top of each other even though the size R and refractive index m vary widely for each ρ' [2,7,8]. This demonstrates the quasi-universal parameterization afforded by ρ'. Figure 1 also shows that the scattering evolves from the RDG, diffraction limit with increasing ρ'. This evolution is described with the following features:

1. For all ρ', a forward scattering lobe of constant intensity appears when $qR < 1$.
2. With increasing qR, the scattering begins to decrease in the Guinier regime [9] near $qR \simeq 1$.
3. After the Guinier regime, power law functionalities begin to appear. For $\rho' < 1$, the RDG limit, a $(qR)^{-4}$ functionality follows the Guinier. When ρ' gets large, $\rho' \geq 30$, a $(qR)^{-3}$ functionality (2d Fraunhofer diffraction [8], see below) appears after the Guinier regime.

4. The $(qR)^{-3}$ regime is followed by a "hump" regime centered near $qR \simeq \rho'$ which then crosses over to approximately touch the $(qR)^{-4}$ functionality of the RDG limit when $qR \geq \rho'$.

5. Connecting the Guinier regime and the hump regime with an equal tangent line gives a $(qR)^{-2}$ functionality which dominates, albeit briefly and imperfectly, when $\rho' < 10$. We call the region between the Guinier regime and the backscattering the "power law regime".

6. At largest qR near $2kR$ (which corresponds to $\theta = 180°$), enhanced backscattering occurs involving "rainbows" and the glory.

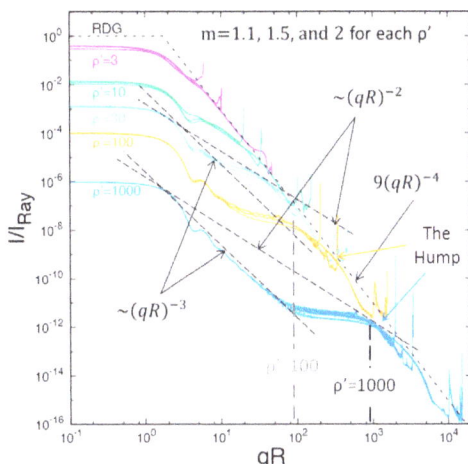

Figure 1. Q-space analysis of spheres for the RDG limit ($\rho' \rightarrow 0$, dashed line) and $\rho' = 3, 10, 30, 100,$ and 1000. Rayleigh normalized scattering intensity is plotted versus qR. The scattering curves with the same ρ' fall on top of each other (despite widely ranging R and m). Various power laws and the Hump (for $\rho' = 100$ and 1000) are indicated. A small size distribution (20% log-normal size distribution) has been applied to eliminate the interference ripples. From [10].

The Rayleigh normalized forward scattered intensity for spheres, $I(0)/I_{Ray}$, also displays quasi-universal behavior with ρ' as shown in Figure 2 and described as follows (feature number 7):

$$I(0) = I_{Ray} \text{ when } \rho' \leq 1$$
$$I(0) = I_{Ray}/\rho'^2 \text{ when } \rho' \geq 10$$

$$(5)$$

Between these limits, a ripple structure ensues. Note that from Equations (2)–(4)

$$\left(dC_{sca, Ray, sph}/d\Omega \right)/\rho'^2 = k^2 R^4 /4$$

$$(6)$$

The right hand side of Equation (6) is the forward scattering (diffraction) from a 2D circular obstacle of radius R. Now recall Babinet's principle that states the intensity diffraction pattern through an aperture is identical to diffraction by an obstacle of the same size and shape [11]. Thus, Equation (6) is the circular aperture result.

A non-zero imaginary part of the refractive index, κ, will cause absorption. However, significant absorption, such that the scattering phase function changes, is governed by another universal parameter $\kappa k R$ [12]. When $\kappa k R < 0.1$, absorption is relatively insignificant; when $\kappa k R > 3$, absorption is significant and does not alter the scattering further with increasing $\kappa k R$. It is straightforward to show that $(\kappa k R)^{-1}$ is the ratio of the absorption skin depth to the particle radius.

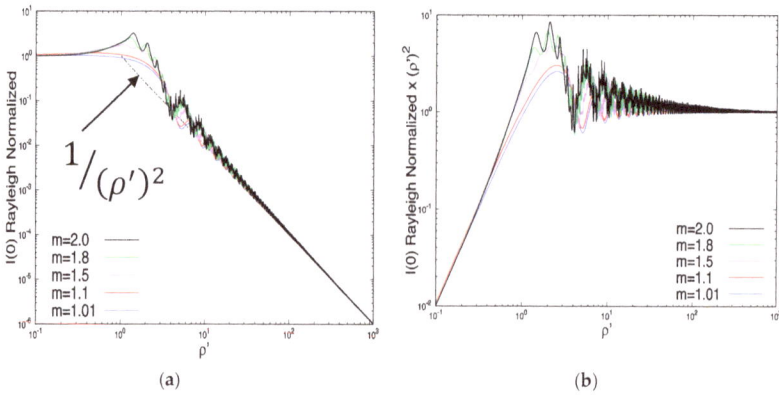

Figure 2. (a) Rayleigh normalized forward scattering, $I(0)/I_{Ray}$, versus ρ'; and (b) Rayleigh normalized forward scattering multiplied by ρ'^2 versus ρ'. From [10].

Figure 3 shows the scattering by a narrow distribution of large, refractive spheres with a mean $\rho' = 235$ and $\kappa kR = 0$ (without absorption) and $\kappa kR = 10$ (with significant absorption). This figure demonstrates that scattering by spheres with significant absorption has lost most of the hump and all of the rainbow and glory structure leaving mostly the $(qR)^{-3}$ power law. The figure also allows a comparison of the Q-space and traditional points of view (Figure 3a,b, respectively).

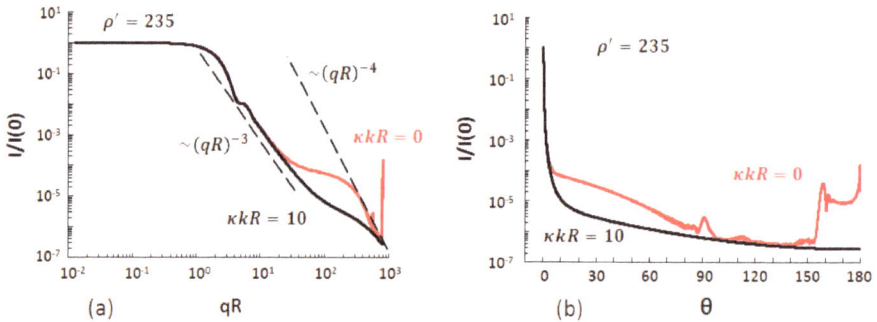

Figure 3. Comparison between the scattered light from spheres with $\rho' = 235$ (e.g., real refractive index $n = 1.5$ and $kR = 400$) and $\kappa kR = 0$, thus without absorption, and $\kappa kR = 10$, thus with significant absorption (If $kR = 400$, $\kappa = 0.025$). (a) Normalized intensity plotted logarithmically versus qR (From [10]). (b) Normalized intensity plotted linearly versus the scattering angle θ.

Thus, we conclude our survey of features uncovered by Q-space analysis for spheres with:

8. When the refractive index, $m = n + i\kappa$, is complex, a second universal parameter comes into play, κkR. When $\kappa kR < 0.1$, non-zero κ has very little effect on the scattering; otherwise κ does have an effect with the same effect for different kR and real part of the refractive index n if κkR is the same. The forward scattering when $q < R^{-1}$ is not affected, but the hump and backscattering, including possible rainbows and glories incur substantial changes. For large size parameters, kR, disappearance of the hump correlates with the scattering cross section decreasing from approximately $2\pi R^2$ to πR^2 and the absorption cross section increasing from 0 to πR^2 when $\kappa kR \geq 3$ [13].

These features of scattering by spheres, as calculated from the Mie equations, form a foundation for comparisons to scattering by irregularly shaped particles.

3. Experimental Data

3.1. Dusts

The term "dust" refers to powders of solid particles with sizes ranging roughly from one to a few hundred microns. Examples include mineral particles from deserts and agricultural regions, road dusts, plant fragments, volcanic ash and dusts that occur in the workplace. Described here are some characteristic data from experiments that have measured scattered light from a variety of dusts.

3.1.1. Amsterdam-Granada Data Set

The Amsterdam-Granada group has provided an extensive data set for light scattering from aerosolized dusts [14]. These data were obtained in the lab with light of wavelengths 441.6 nm and 632.8 nm scattered from the aerosols at angles from either 3° or 5° to 177°. Figure 4 shows an example of Saharan dust.

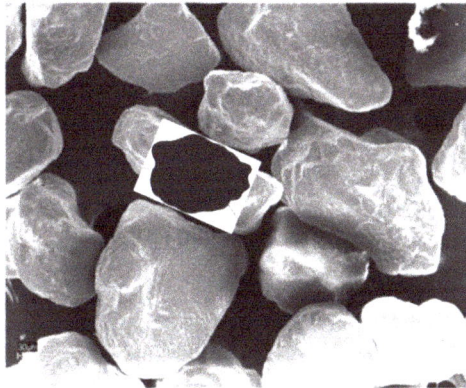

Figure 4. Saharan dust particles from [15] with a model Gaussian random sphere for comparison. Scale bar at lower left is 30 microns.

Figure 5 demonstrates the differences between plotting the scattered light intensity versus scattering angle and versus the log of the scattering wave vector magnitude q, i.e., Q-space analysis [16]. The same data are plotted on both the left and right hand sides of the figure. The data are for scattering of unpolarized light from aerosolized Libyan sand [15]. Plotting versus linear angle yields a non-descript curve; plotting versus log q (Q-space analysis) yields a straight line followed by enhanced backscattering. The power of Q-space analysis is apparent.

The bulk of the data in Figure 5, ignoring the enhance backscattering for $q > 10$ micron^{-1}, follow a straight line to imply a power law with q with an exponent magnitude of 1.68. The Q-space plot shows no change from linearity at small q down to $q \simeq 0.9$ μm^{-1} where the data end. By analogy with spherical particle scattering, we expect at small q a Guinier regime. Since no Guinier regime is seen, the implication is that the size of the dust particles is greater than the inverse of the smallest available $q \simeq 0.9$ μm, i.e., greater than $q^{-1} \simeq 1.1$ μm. This is consistent with Figure 4.

Our group [17] applied Q-space analysis to the entire Amsterdam-Granada data set; a total of 43 aerosol data sets available on the website. Examples are given in Figure 6. Remarkably, in all cases, Q-space analysis revealed plots with extensive linear regions hence very linear power laws with q.

Very often enhanced back scattering was observed at large q. Some of the data showed at small q the onset of a Guinier regime. Power law exponents were in the range 1.49 to 2.47.

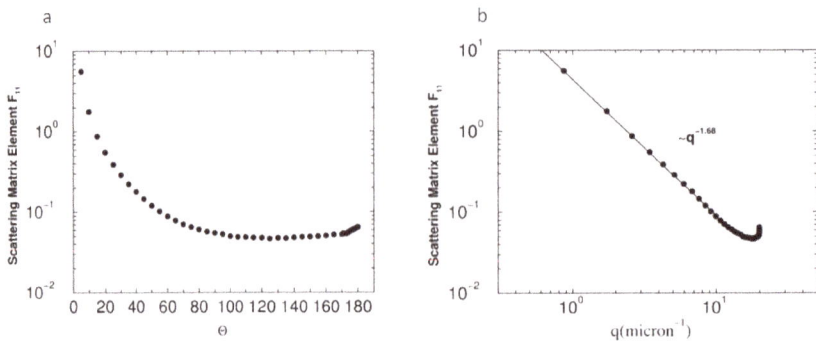

Figure 5. The F11 Mueller matrix element (scattered intensity for unpolarized incident light) for light scattering from Saharan dust: (**a**) scattering plotted vs. scattering angle θ; and (**b**) scattering plotted vs. the scattering wave vector magnitude, q. From [3].

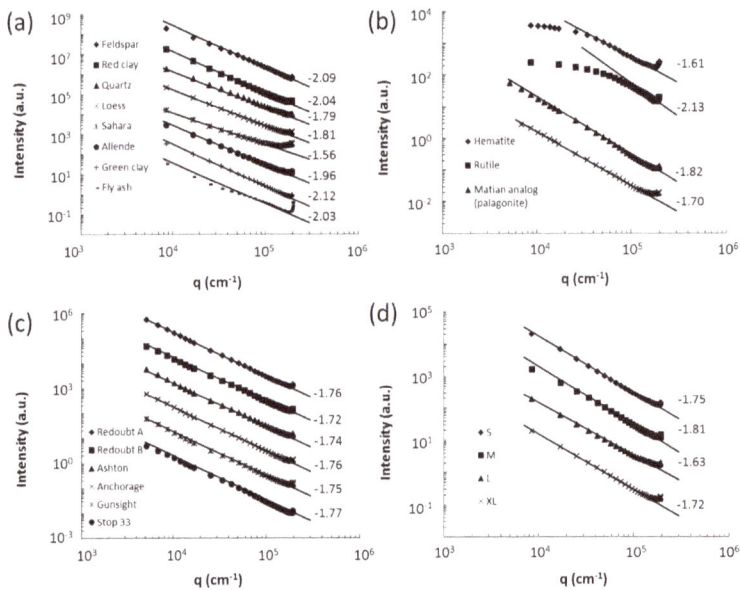

Figure 6. Q-space plots of the S_{11} matrix element (scattered intensity for unpolarized incident light) for light scattering from some of the Amsterdam-Granada data set [14]. (**a**) Feldspar, Redy clay, Quartz, Loess, Sahara [18], Allende [19], Green clay, and Fly ash [20] measured at $\lambda = 441.6$ nm; (**b**) Hematite [21,22], Rutile [22], Martian analog (palagonite) [23], and Sahara sand (Libya) [15] measured at $\lambda = 632.8$ nm; (**c**) Volcanic ash (Redoubt A, Redoubt B, Spurr Ashton, Spurr Anchorage, Spurr Gunsight, Spurr Stop 33) [24] measured at $\lambda = 632.8$ nm; and (**d**) Olivine S, Olivine M, Olivine L, and Olivine XL [19] measured at $\lambda = 441.6$ nm. In all graphs the lines are power law fits to the data and the number to the right of the plot is the power law exponent. From [17].

The most significant feature common to dusts and spheres is a power law regime. However, the power laws with spheres are complex, whereas the dusts display a single power law. Enhanced backscattering occurs for both, but again it is simple for dusts and complex for spheres. A forward scattering lobe and Guinier regimes are expected for dusts. A caveat to this comparison is that the dusts are very polydisperse, the consequences of which have not been explored.

3.1.2. Arizona Road Dust

Our group conducted light scattering measurements on Arizona Road Dust (AZRD) over an angular range of $0.32° \leq \theta \leq 157°$ at a wave length of 532 nm [25]. The scattering apparatus was specifically designed to have access to very small angles to obtain the Guinier regime for large particles. Figure 7a shows an optical microscope picture of the AZRD. Figure 7b shows an optical microscope picture of Al_2O_3 abrasive powders to be described below.

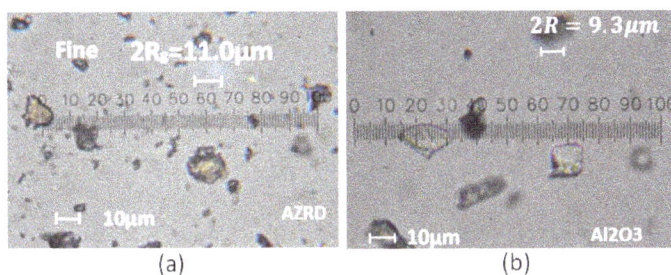

(a)   (b)

Figure 7. Optical microscope pictures of: (**a**) Arizona Road Dust (AZRD) from [25]; and (**b**) Al_2O_3 abrasive powders from [26]. Sizes, $2R_g$ and $2R$, infered from Guinier analysis of the scattered light are given.

Figure 8 shows the Q-space analysis of the scattering. One sees a hint of a constant forward scattering lobe followed by a Guinier regime, a power law, and at largest q, enhanced backscattering. Note the importance of small angles to gain the Guinier regime for these micron size particles.

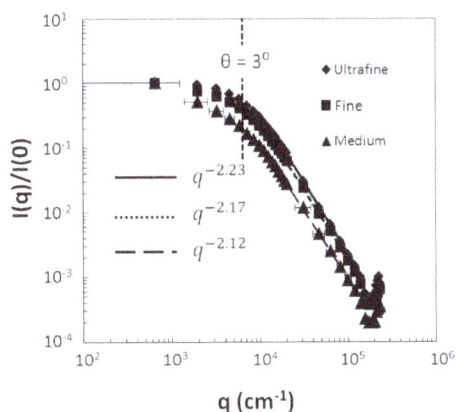

Figure 8. Scattered intensity from Arizona Road Dust for unpolarized incident light versus q normalized at the smallest q. Dashed vertical line indicates where the scattering angle is 3°. From [25].

Analysis of the Guinier regime involved an iterative procedure, see [26]. That analysis yielded an estimate of mean radii of gyrations of 2.7 µm, 5.5 µm, and 9.7 µm for the Ultrafine, Fine, and Medium dust samples, respectively. These measurements were consistent with optical measurements of the mean particle size when weighted by the tendency of bigger particles to scatter more light proportional to the square of their mean effective radius. The power slopes yield exponent magnitudes of 2.23, 2.17, and 2.12 for the Ultrafine, Fine, and Medium dust samples, respectively, each with an error of ±0.05.

We now find that the Arizona Road Dusts and the dusts in the Amsterdam-Granada data set are similar and all show significant power law regimes. The AZRD also shows a Guinier regime and it is reasonable to conclude that the Amsterdam-Granada dust would too if light was collected at angles smaller than 3°. All the dusts show enhanced back scattering.

3.1.3. Al_2O_3 Abrasive Grits

Our group also studied light scattering by irregularly shaped Al_2O_3 abrasive powders of various grit sizes [26] with the same apparatus as for the AZRD. These grits were chosen because the size could be systematically varied with the grit number while the material and average shape remained the same. An optical microscope picture of the 600 grit size is given in Figure 7b.

Figure 9 shows the Q-space analysis of the scattering for all six abrasive dusts studied. The scattering shows forward scattering, Guinier, power law, and enhanced backscattering regimes. The exponents of the power laws for Al_2O_3 abrasives decrease with increasing size.

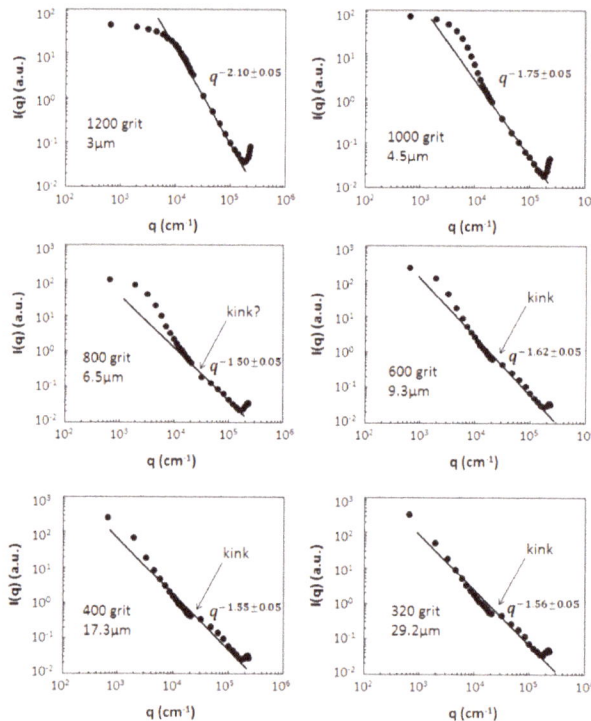

Figure 9. Scattered intensity (arbitrary units) versus the magnitude of the scattering wave vector q for six different sizes of Al_2O_3 abrasive grits. The manufacturer's size is labeled in each figure. Power law regimes are indicated by straight lines and the power law is labeled. From [26].

The Guinier regime slips away as the size increases because the size crosses from smaller than the smallest measured q inversed to larger than the smallest measured q inversed. The smallest q is $q = 6.6 \times 10^2$ cm^{-1} (at $\theta = 0.32°$), thus $q^{-1} = 15$ µm. A rather convoluted analysis of the Guinier regime for the 1200, 1000, and 800 grits led to mean perimeter diameters of 5.2 µm, 8.4 µm, and 14.4 µm, respectively. These are about a factor of two bigger than the sizes claimed by the manufacturer and labeled in Figure 2. However, these measurements, although about a factor of two larger than the sizes claimed by the manufacturer, are consistent with these values when effects for light scattering weighting are included.

Similar to the Amsterdam-Granada and Arizona road dust particles, the abrasive grits show a power law regime. However, unlike the Amsterdam-Granada and Arizona road dust particles, the largest three abrasives, for which $\rho' \geq 100$, show a kink in the power law. The reason for this kink is uncertain.

3.2. Fractal Aggregates

Fractal aggregates have scaling dimensions less than the Euclidean dimensions of space. Thus the mass scaling dimension is $D_m < d$ and the surface scaling dimension is $D_s < d - 1$; where d is the spatial dimension, typically $d = 3$. Light scattering by fractal aggregates has been reviewed in 2001 [27]. Often the fractal aggregate structure is well described by the diffusion limited cluster-cluster aggregation (DLCA) model with a fractal dimension of $D = D_m = D_s = 1.78$.

Light scattering by fractal aggregates can be well described by the Rayleigh-Debye-Gans (RDG) limit although deviations on the order of tens of percent can occur [27]. This is also the diffraction limit where $\rho' \to 0$. This limit is obtained more so for lager aggregates because the overall density of the aggregate decreases with increasing size. This occurs because $D = D_m < d$, e.g., $D = 1.78 < 3$.

Figure 10 contains some early data from our lab showing light scattering from soot in a methane/oxygen premixed flame. The optical wavelength was $\lambda = 514.5$ nm. Soot is typically composed of roughly spherical monomers with diameters of approximately 30 nm. These monomers aggregate together via diffusion limited cluster-cluster aggregation to yield aggregates with the DLCA morphology and fractal dimension of $D \simeq 1.78$. The refractive index of soot is not precisely defined but is something like $m = 1.6 + 0.6i$. In our experiment, a flat flame burner with a circular porous frit and an overhead stagnation plate was used. Such a flame is quasi-one-dimensional so that the soot is uniform in the horizontal direction across the flame but grows in size due to aggregation with distance above the flame, the height above burner, h. One can see in Figure 10 that the scattering increases and the Guinier regime moves to smaller q with increasing h; both indicating growing aggregates.

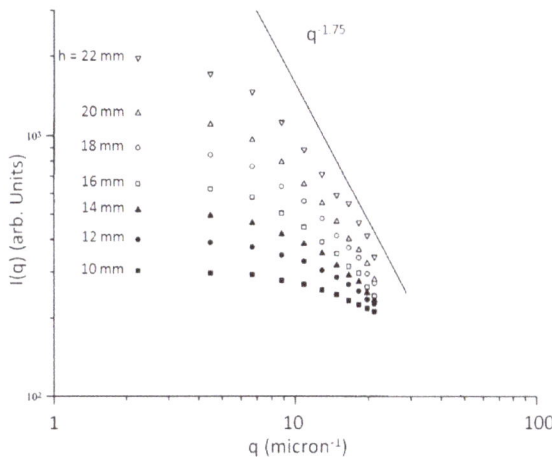

Figure 10. Scattered light intensity versus q for in-flame soot as a function of height above the burner surface, h. The flame was a methane/oxygen premixed flame with a carbon/oxygen ratio of 0.75 [28].

Overall, one sees a constant forward scattering lobe at small q, followed by a Guinier regime and finally a developing power law regime at largest q. The power law is approaching $q^{-1.75}$ to imply a fractal dimension of $D = 1.75$ for the soot aggregates in the flame. No enhanced backscattering is seen, but note that the maximum scattering angle used was 120°; hence such scattering could have been missed.

4. Theoretical Calculations

4.1. Spheroids

Mixtures of randomly oriented spheroids have been proposed as models for desert dust light scattering [29]. The spheroidal shape can be obtained by rotating an ellipse about either its major axis to yield a prolate spheroid or its minor axis to yield an oblate spheroid. Prolate spheroids are elongated balls like rugby balls, whereas oblates are squished balls like disks. If the axis of rotation has a semi-diameter of a, the spheroid will have two other semi-diameters that are equal to which we can assign a length of b. Then, the shape can be described by the aspect ratio $f = a/b$. If $f > 1$, the spheroid is prolate. If $f < 1$, the spheroid is oblate. If $f = 1$, the spheroid is a sphere.

Here, we present our initial studies of spheroidal particle light scattering analyzed via Q-space analysis. The orientationally averaged light scattered by the spheroids was calculated using a well-known and widely tested T-matrix (TM) code found freely available on NASA's Goddard Institute website [30,31]. Indices of refraction were $m = 1.3$, 1.4, 1.5 and volume equivalent radii, R_{veq}, were in the range 1 µm to 6 µm. The wavelength was $\lambda = 0.532$ µm. The structure factor was calculated with a numerical Fourier transform.

For many of these non-spherical shapes it was found that the normalization by spherical particle Rayleigh scattering, Equation (2), and use of the spherical particle internal coupling parameter, Equation (4), was insufficient to achieve a good description of the scattering in Q-space. This occurs because these equations strictly hold only for spheres. Recently we have generalized these equations for non-spherical shapes [32]. The results apply to orientationally averaged scattering which is the most common situation in practice. Briefly, in general, the Rayleigh differential cross section for any shape is [4]

$$dC_{sca,\, Ray}/d\Omega = k^4 V^2 |\alpha(m)|^2 \tag{7}$$

In Equation (7), V is the volume of the particle and $\alpha(m)$ the average volume polarizability, which is a function of the complex index of refraction m with functionality dependent upon shape. The same reformulation led to a general definition of the internal coupling for any arbitrary shape as

$$\rho' = 2\pi k \frac{V}{A}\, \alpha(m) \tag{8}$$

where A is the projected area of the scattering object in the direction of the incident light.

To calculate $|\alpha(m)|$ for an arbitrary shape, the discrete dipole approximation (DDA) was used [33]. In DDA the index m, volume V, wavelength λ and thus k and arbitrary shape are set; DDA then calculates the differential scattering cross-section. Equation (7) implies that a plot of this scattering cross-section divided by k^4 versus V^2 has a slope of $|\alpha(m)|^2$.

The Rayleigh normalization and ρ' for spheroids used this newly developed method. Figure 11 shows scattering for prolate and oblate spheroids with two-to-one aspect ratios. Comparison to Figure 1 for spheres shows very strong similarities. Although we do not present here an explicit analysis of the forward scattering, one can quantitatively verify from Figure 11 and the values of ρ' that feature 7 holds for these spheroids. All eight of the features for spheres listed above are present for spheroids.

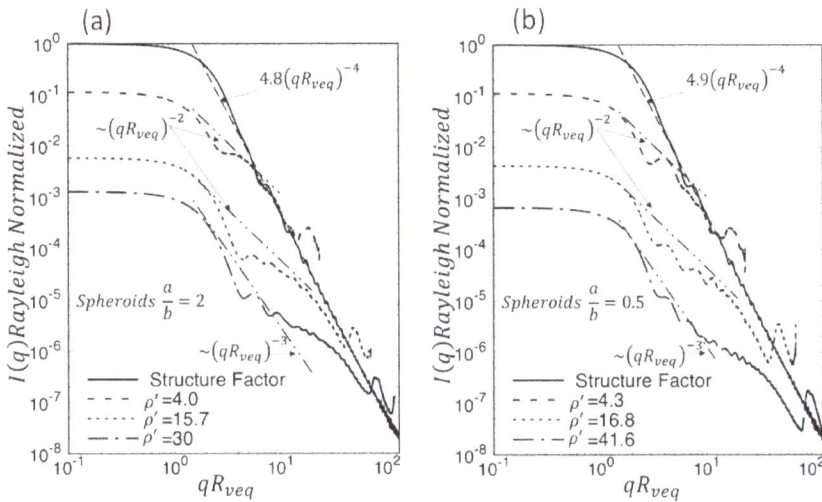

Figure 11. Rayleigh normalized scattered intensity versus qR_{veq} (volume equivalent radius) for spheroids: (**a**) prolate spheroids; and (**b**) oblate spheroids with aspect ratios $f = 2$ and $1/2$, respectively. Plots with different generalized internal coupling parameters ρ' are given. The structure factor is when $\rho' = 0$.

4.2. Irregular Spheres

Our group has applied Q-space analysis to irregular spheres [34]. These were created with an algorithm that started with a sphere and then perturbed it in various ways [35–37]. Four different types of irregular spheres were produced: strongly damaged spheres, rough surface spheres, pocked spheres, and agglomerated debris particles, and their family portrait is in Figure 12. For each type, three refractive indices were used: $m = 1.313$, $1.5 + 0.1i$ and $1.6 + 0.0005i$. The size parameter kR ranged from 2 to 14 with steps of 2. For a wavelength of 0.532 µm, these values correspond to radii of $R = 0.17$ to 1.2 µm.

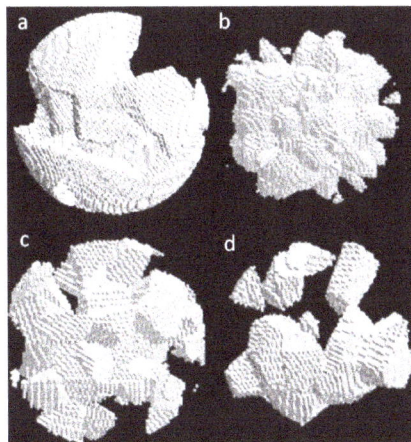

Figure 12. Irregular spheres: (**a**) strongly damaged; (**b**) rough surface; (**c**) pocked spheres; and (**d**) agglomerated debris particles.

Figure 13 shows a Q-space analysis of the scattering by agglomerated debris particles [34] calculated via DDA [38–40]. Before we deal with the light scattering, note that the plot, Figure 13a, displays the structure factor, $S(q)$, of the particle. The structure factor can be described in a few equivalent ways: it is the square of the Fourier transform of the particle's structure; it is the diffraction pattern for waves emanating from the particle; and it is the $m \to 1.00$ limit, hence $\rho' = 0$ limit, for the light scattering from the particle. The structure factor shows a forward scattering lobe, a Guinier regime, and a power law regime. The apparent enhanced backscattering is consequence of the real space cubic lattice spacing of approximately $R_g/100$ that the real space particle was represented on and hence of no consequence. The power law regime has an exponent of -4. When dealing with structure factors, the power law regime is usually referred to as the Porod regime [9] and the exponent is—$(d + 1)$ where d is the spatial dimension of the particle, typically $d = 3$. More generally, the Porod regime exponent is—$(2D_m - D_s)$ where D_m and D_s are the mass and surface scaling dimensions of the particle, respectively [41]. We can conclude that the particle has scaling dimensions of $D_m = 3$ and $D_s = 2$.

Figure 13b shows the orientationally averaged scattered light intensity for a refractive index of $m = 1.6 + 0.0005i$ for seven different size parameters hence seven different internal coupling parameters as marked. In this work the spherical forms for the Rayleigh scattering and internal coupling parameter were used. These finite refractive indices change the $\rho' = 0$ structure factor in the Porod regime. The power law remains for about one order of magnitude in qR, but the slope decrease with increasing size parameter kR. This occurs as the internal coupling parameter ρ' increasing from less than one to nearly six, and we infer that the increased internal coupling is the cause of the Porod regime slope change. Similar results were obtained for pocked and strongly damaged spheres, but the power law regime for the rough spheres had strong ripples that masked any possible power law. Another feature is the occurrence of some enhanced backscattering with increasing ρ'. Finally, note the dip near $qR \simeq 3.5$.

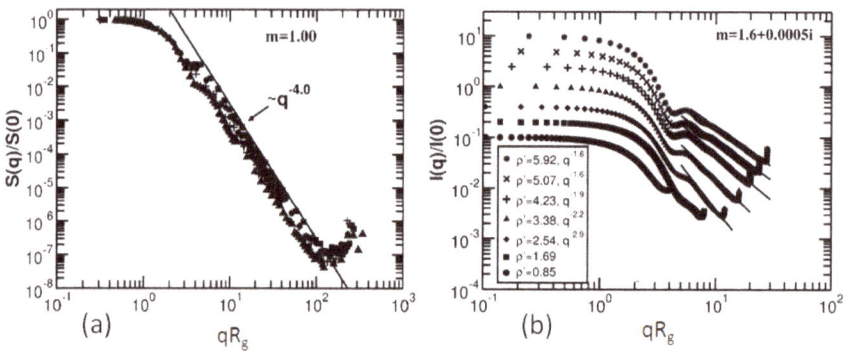

Figure 13. Q-space analysis of the scattering by agglomerated debris particles of different internal coupling parameters, ρ'. (**a**) The structure $S(q)$ factor ($m \to 1.00$). Seven size parameters from $kR = 2$ to 14 are superimposed. (**b**) Light scattering with refractive index as marked at seven different internal coupling parameters ρ'. When a power law could be fit, the fit line is shown and the power law q^{-x} is labeled in the key. These plots were shifted up and down for visibility. Adapted from [34].

Figure 14 shows the behavior of the Rayleigh normalized forward scattered intensity as a function of the internal coupling parameter. This plot is analogous to Figure 2a for spheres. All four irregular spheres with all three refractive indices, $m = 1.313, 1.5 + 0.1i$ and $1.6 + 0.0005i$ are plotted in the figure. Within the scatter of the data, Figure 14 shows a universal functionality of the forward scattering with the internal coupling parameter, and the functionality is very similar to that found for spheres (Figure 2).

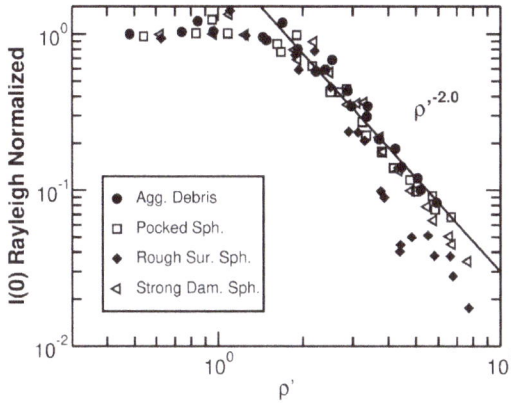

Figure 14. Rayleigh normalized forward scattering versus the internal coupling parameter ρ' for four different types of irregular spheres and three different refractive indices, $m = 1.313$, $1.5 + 0.1i$ and $1.6 + 0.0005i$ (hence a total of 12 sets of data).

4.3. Gaussian Random Spheres

A Gaussian Random Sphere (GRS) can be made to have shapes similar to dust particles. GRS's are based on smooth, random fluctuations relative to an underlying spherical shape [42]. The insert of Figure 4 shows an example of a GRS. GRS's are characterized by three parameters: (1) σ, the relative standard deviation of the distribution of deviations from a perfect sphere in the radial direction; (2) ν, the power law index of the covariance function which controls the number of bumps and dips in the tangential direction; and (3) R, the mean radius from which deviations occur and that sets the overall size of the particle.

Figure 15 shows the light scattering properties of GRSs [43] calculated using a discrete dipole approximation algorithm [38–40]. The GRSs had $\sigma = 2$ and $\nu = 3$ and the scattering was orientationally averaged. Scattering features include a forward scattering lobe, a Guinier regime near $qR_{eq} \simeq 1$, a small dip near a $qR_{eq} \simeq 3$, a power law regime, and hints at enhanced backscattering. The exponent of the power law is -4 when the internal coupling parameter ρ' is small. This is consistent with this limit being the structure factor and the particles have mass and surface scaling dimensions of $D_m = 3$ and $D_s = 2$, respectively. With increasing ρ', the exponent magnitude decreases. Note that in this work the spherical form for the internal coupling parameter was used with no significant error.

Included in Figure 15 are plots of scattering by perfect spheres as calculated with the Mie equations. The spheres have a modest size distribution with geometric width of 1.2 to eliminate interference ripples. The spheres and the GRS display similar scattering behavior with ρ' except when $\rho' \geq 2$, in the backscattering regime. There the GRS scattering spans a dip that appears in the sphere scattering near $qR \simeq kR$ to $2kR$, but does not have the sharp increase at $2kR$ (which corresponds to $\theta = 180°$) that the spheres have. When $\rho' \geq 5$, a simple enhanced backscattering appears.

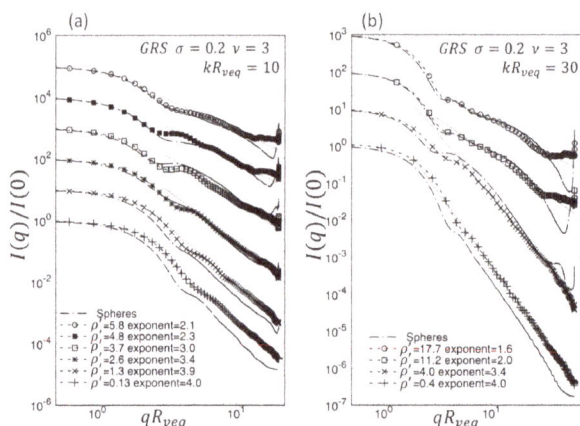

Figure 15. Forward normalized scattered intensity of GRSs compared to spheres for (**a**) $kR_{veq} = 10$, $m = 1.01-1.5$; and (**b**) $kR_{veq} = 30$, $m = 1.01-1.5$. The dashed lines with data points are the computational data from DDA calculations. Dash-dot lines are Mie calculations for a size distribution of spheres with ρ' values of the most probable radius equivalent to the ρ' values of the GRSs, and a geometric deviation in the sphere size distribution of $\sigma = 1.2$. Solid lines are power laws $(qR_{eq})^{-x}$ with exponent x given in the legend. Note that the curves have been shifted up by factors of 10 to separate them. From [43].

4.4. Thickened Clusters

Some clusters found in the atmosphere are rather large and appear denser than DLCA aggregates. An example is given in Figure 16a. In an attempt to study the light scattering properties of such clusters, we have constructed clusters with a computer algorithm. The construction process starts with a 3D, cubic point lattice. Added to these points at random are spheres with diameter equal to the lattice spacing. Ultimately, as more spheres are added, the lattice percolates; a percolation cluster with fractal dimension $D \simeq 2.5$ is formed. Clusters so conceived were then used as the backbone for the dust particle. To make the fractal dimension match the spatial dimension of three, the backbone cluster was thickened by filling the neighboring sites. Figure 16b shows an example of a thickened percolation cluster.

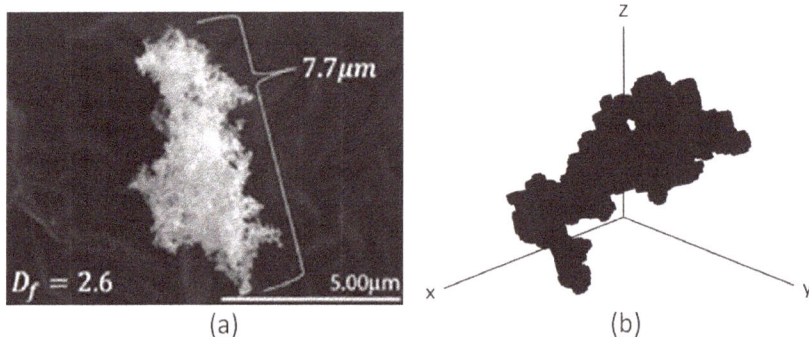

Figure 16. (**a**) SEM images of a typical soot superaggregate obtained from sampling of smoke plumes from the Nagarhole, India forest fire [44]. (**b**) A thickened percolation cluster, from [17].

The rotationally averaged scattered intensity was calculated using a DDA code developed by our group [17,43,45]. All thickened cluster DDA calculations were done with 100 random orientations on particles comprised of $(3–6) \times 10^5$ dipoles. All DDA runs for these particles were at $k|m|d \leq 0.6$ where d is the dipole spacing, well below the commonly cited standard for DDA accuracy of $k|m|d \leq 1$ [46]. To further insure the accuracy of our DDA runs, we compared the equivalent sized spheres set at the same dipole resolution, λ, and m to the results from Mie solutions. We found the error between the numerical methods reported C_{sca} values were always less than 10% and at the smallest ρ' the error was ca. 1%. It should also be noted that previous work has shown that spherical particles produce the largest errors in DDA and the error for randomly shaped 3D objects is expected to be much smaller [46,47]. Then with the application of Q-space analysis as shown in Figure 17, one finds a constant forward scattering lobe, a Guinier regime, a minor dip near $qR \simeq 3$, followed by power law regimes. There is no enhanced backscattering, but note that $\rho' < 12$. As before, the magnitude of the exponents decrease with increasing ρ'. Note that in this work the spherical form for the internal coupling parameter was used with no significant error.

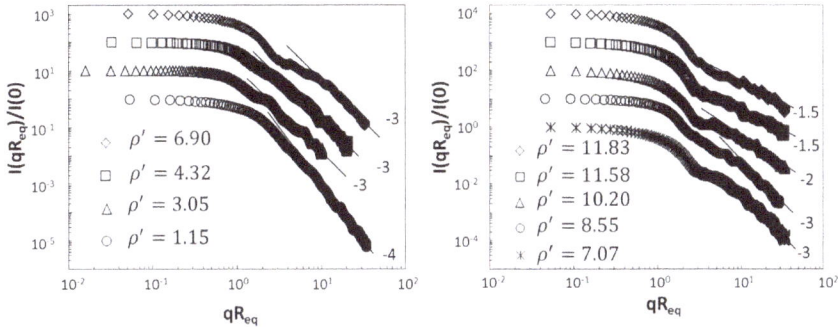

Figure 17. Scattered intensity versus qR_{eq} for thickened percolation clusters. Lines are the power law fits and the numbers to the right of the plots are the exponents of the power law. A different multiplication factor is applied to the intensity for each plot for clarity.

4.5. Ice Crystals

Our group has recently applied Q-space analysis to a variety of ice crystal shapes [10]. Such crystals occur in the atmosphere, for example, in cirrus clouds. Nine crystal shapes were studied: droxtal, solid column, 8-column aggregate, plate, 5-plate aggregate, 10-plate aggregate, hollow column, hollow column rosette, and solid column rosette all with three degrees of surface roughness, namely, $\sigma = 0.0$ (smooth), $\sigma = 0.03$ (moderately rough), and $\sigma = 0.5$ (severely rough). The scattering calculations are described in [48].

Figure 18 presents the Q-space analysis of the angular scattering functionality, proportional to the phase function, for the ice crystals. The shape generalized Rayleigh scattering and internal coupling parameters were used [32]. The optical wavelength was $\lambda = 0.53$ μm, the maximum dimensions of the various shapes were $D = 2$, 6 and 20 μm and the refractive index was $m = 1.31$. R_{eq} is volume equivalent radius. Corresponding ρ's for maximum dimensions $D = 2$, 6, 20 μm are labeled next to the scattering curves.

Figure 19 presents the Rayleigh normalized forward scattering for the ice crystals. In most cases the behavior of the forward scattering for the ice crystal is very similar to that for spheres, Figure 2b. However, Droxal shows unexplained computational problems in the range $5 \leq \rho' \leq 50$. 5-plate and 10-plate show variation with the smoothness with the roughest closest to the sphere behavior. Despite these differences, the general similarity to spheres is striking and had not been previously recognized.

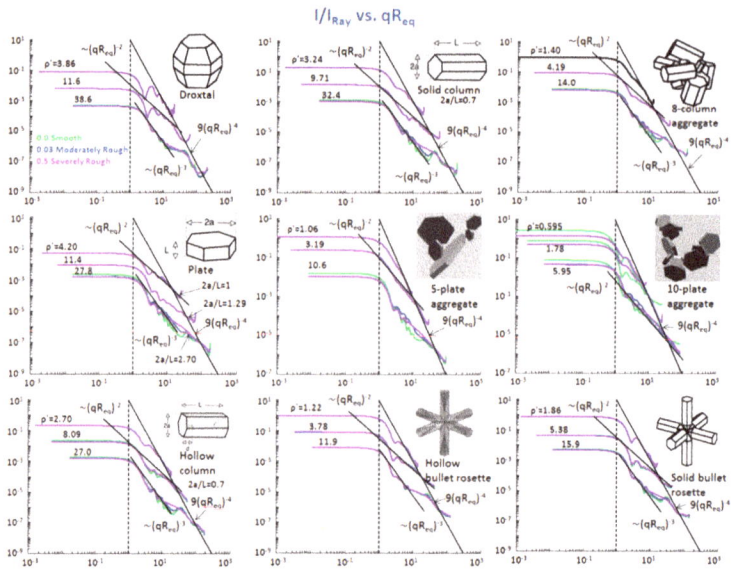

Figure 18. Q-space analysis of scattering by nine different ice crystal shapes. Plotted is the Rayleigh normalized scattering intensity, I/I_{Ray}, versus qR_{eq}. Three different values of the internal coupling parameter ρ' and three values of roughness, 0.0 (smooth), 0.03 (moderately rough) and 0.5 (severely rough) are shown for each shape. Lines for various power laws are marked. Pictures of the shapes are also included. Adapted from [10].

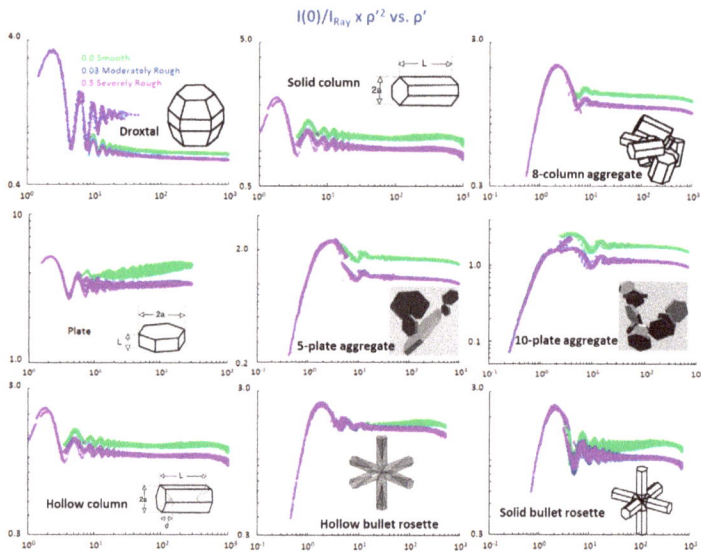

Figure 19. Rayleigh normalized forward scattering intensity, $I(0)/I_{Ray}$, multiplied by the square of the internal coupling parameter, ρ'^2, versus ρ' for nine different ice crystal shapes with three different values of roughness, 0.0 (smooth), 0.03 (moderately rough) and 0.5 (severely rough). Included are pictures of the shapes. These plots were adapted from [10] with values for $\rho' > 10^3$ deleted due to suspected computational problems. The interested reader is referred to [10].

Figure 20 illustrates the effects of the imaginary part of the refractive index κ and extreme aspect ratio.

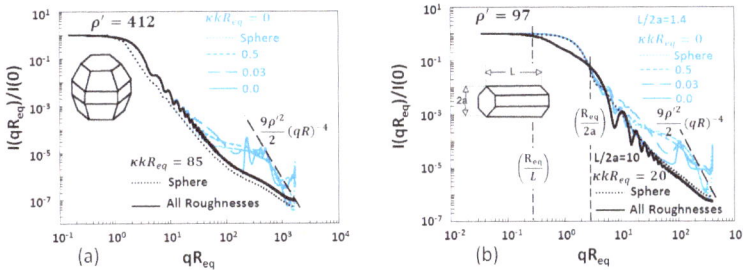

Figure 20. Comparisons between the light scattered from: (**a**) a droxtal ice crystal; and (**b**) two solid column structures with different aspect ratios, with and without absorption and, in each case, volume equivalent spheres. All the ice crystals and the spheres with significant absorption, as indicated by $\kappa k R > 1$, lose the hump and the glory. Note also the two Guinier regimes for the solid columns with $\kappa k R_{eq} = 20$ match $q R_{eq} = (L/R_{eq})^{-1}$ and $q R_{eq} = (2a/R_{eq})^{-1}$. Adapted from [10].

The Q-space analysis of Figures 18–20 shows many features in common for scattering by spheres and the ice crystals which evolve with the internal coupling parameter ρ' in a similar manner. These features are:

1. The forward scattering lobe when $q R_{eq} < 1$ behavior is very similar to that of spheres when the Rayleigh scattered intensity and internal coupling parameter are generalized for these shapes.
2. Both spheres and ice crystals have a Guinier regime near $q R_{eq} \simeq 1$. Unlike spheres, however, crystals with large aspect ratios can show two Guinier regimes.
3. Both spheres and ice crystals have a complex power law regime beyond the Guinier regime when $1 \leq q R_{eq} < 2kR$. This regime includes a $(q R_{eq})^{-3}$ functionality, for non-aggregate crystals, that starts to occur with large ρ' very likely due to 2d diffraction from the projected crystal shape. Similar to spheres, hump structure also appears centered near ρ'. At larger $q R_{eq}$, there is a tendency to approach the spherical particle diffraction limit (RDG) of $9(q R_{eq})^{-4}$. Aggregate ice crystals have a more uniform power law regime similar to fractal aggregates.
4. The parameter $\kappa k R$, plays the same role for both shapes by removing the hump near $q R_{eq} \simeq \rho'$.
5. In many cases, the ice crystals have enhanced backscattering near $q R_{eq} \simeq 2kR_{eq}$, $(\theta = 180°)$ similar but not the same as for spheres.
6. The evolution of the scattering evolves away from the 3D diffraction with increasing ρ' for all shapes including spheres.

Surface roughness plays a minor role in these features except in the hump region where smooth surfaces give a wavy structure to the hump.

5. Discussion

We have reviewed the light scattering properties of spheres, a great many types of dusts, fractal aggregates, spheroids, irregular spheres, Gaussian random spheres, thickened clusters and nine types of ice crystals. Our perspective has been the novel Q-space perspective in which the scattered intensity is plotted versus the magnitude of the scattering wave vector q on a logarithmic scale rather than the conventional linear plot versus the scattering angle θ. We find or infer that the scattering for *all these shapes* have the following same features:

1. A forward scattering lobe of constant intensity, i.e., q and θ independent, appears when $qR < 1$. This condition is equivalent to $\theta < \lambda/2\pi R$. The magnitude of the forward scattering is that of its

generalized Rayleigh scattering, I_{Ray}, when $\rho' \leq 1$, and I_{Ray}/ρ'^2 when $\rho' \geq 10$. We remark that for spheres approximately half of the total scattered light occurs when $qR < \pi$ ($\theta < \lambda/2R$) [6,13,49] and when $\kappa kR < 0.1$, and nearly all the scattered light appears in this forward lobe when $\kappa kR > 10$.

2. A Guinier regime near $qR \simeq 1$ ($\theta \approx \lambda/2\pi R$).
3. A power law regime when $1 \leq qR \leq 1.5kR$ ($\lambda/2\pi R \leq \theta \leq 90$ to $100°$). This power law regime can be very complex as for spheres, spheroids, Gaussian random spheres and ice crystals, or it can be a single power law as for many of the dusts, fractal aggregates, irregular spheres, and thickened clusters. In all cases the power law regime evolves with the internal coupling parameter ρ'.

Other features that often occur are:

4. A dip near $qR \simeq 3$ to 4 ($\theta \simeq \lambda/2R$) immediately after the Guinier regime appears for all shapes except the dusts and the DLCA aggregates. Recall that the dusts samples were polydisperse and this could smooth away any dip present in a single size scattering.
5. A $(qR)^{-3}$ regime at large ρ' appears for spheres, spheroids and the non-aggregate ice crystals. This is due to the onset of 2d Fraunhofer diffraction, thus it is expected that all non-aggregate shapes would have this regime at ρ'.
6. A "hump" regime centered near $qR \simeq \rho'$ when $\rho' \geq 30$ for spheres, spheroids and, remarkably, ice crystals. This hump disappears when $\kappa kR \geq 3$. We expect this hump to appear as the 2D Fraunhofer diffraction, $(qR)^{-3}$ regime appears at large ρ' for all shapes except aggregates.
7. An enhanced backscattering regime appears when $qR \geq 1.5$ ($\theta \geq 110°$) for all shapes except the ice crystal aggregates, DLCA and thickened aggregates. A caveat is that the data for the DLCA aggregate was limited to $\theta \leq 120°$. The backscattering appears as ρ' increases there being no enhanced backscattering in the diffraction limit when $\rho' = 0$. Typically it appears when $\rho' > 10$.

The power law of feature 3 is quite uniform (linear) for all shapes in the diffraction, RDG, $\rho' = 0$ limit, and for the dusts and aggregates at finite ρ'. It is approximately uniform for Gaussian random and irregular spheres and small spheres at finite ρ' but complex for spheres and non-aggregate ice crystals. When reasonably uniform, we have observed the power law exponent magnitude decreases with increasing ρ'. This behavior is shown in Figure 21.

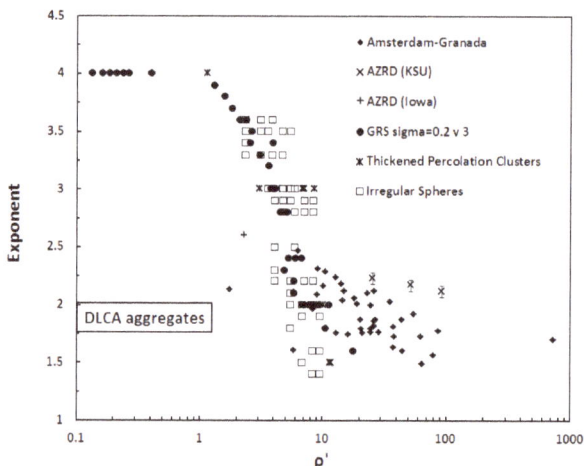

Figure 21. The exponents of the power laws versus the internal coupling parameter ρ'. DLCA fractal aggregates typically have $\rho' < 1$ and an exponent equal to the fractal dimension $D \simeq 1.8$. Amsterdam-Granada data are from [14]; AZRD (Iowa) data are from [50]. Graph is from [17].

Figure 21 shows that all the particles with a uniform power law except the DLCA fractal aggregates follow on the same trend with the internal coupling parameter ρ' regardless of the detail of their structure. The implication is that ρ' is a universal parameter for any shape much like it is for spheres as displayed in Figure 1. The magnitude of the exponents start from 4 when ρ' is small and decrease until the trend levels off to 1.75 ± 0.25 when $\rho' \geq 10$.

As alluded to above, in the $\rho' \to 0$, diffraction, RDG limit the power law regime in general obeys [41]

$$I(q) \sim q^{-(2D_m - D_s)} \tag{9}$$

where D_m and D_s are the mass and surface scaling dimensions of the scattering particle, respectively. Thus, Figure 21 not only shows an interesting trend in the exponents when the exponent description is viable but also indicates that there are two classes (at least) of non-spherical particles: fractal, with scaling exponents not directly related to the Euclidean dimensions, and non-fractals with canonical Euclidean scaling dimensions.

At this time, we cannot offer a complete explanation for the power laws found empirically above. For the Amsterdam-Granada dust we are suspicious that the broad polydispersity of the samples might have some effect with regard to smoothing the plots in Q-space. Furthermore, although we understand and can calculate the power law exponents in the $\rho' \to 0$ limit, we have no explanation for their values when $\rho' > 1$ nor their behavior as a function of ρ'.

6. Conclusions

When viewed from Q-space, the scattering phase function of any particle has distinguishing characteristics that can quantitatively describe the scattering. The major characteristics are common to all shapes, thus providing a universal description of particulate light scattering. For all shapes, the scattering evolves with increasing ρ' from the diffraction limit where the internal coupling parameter $\rho' = 0$. This evolution is universal for spheres, and we present some evidence that it is universal for other shapes as well. However, this assertion needs further testing.

Acknowledgments: This work was supported by the National Science Foundation under grant no. AGM 1261651 and the Army Research Laboratory under grant no. W911NF-14-1-0352.

Author Contributions: C.M.S. and A.C. conceived and designed the research; Y.W.H. performed the experiments and analyzed the data with C.M.S.; W.R.H. and J.B.M. performed theoretical calculations and analyzed the results with C.M.S. and A.C.; and C.M.S. wrote the paper.

Conflicts of Interest: The authors declare no conflict of interest.

References

1. Sorensen, C.M.; Fischbach, D.J. Patterns in Mie scattering. *Opt. Commun.* **2000**, *173*, 145–153. [CrossRef]
2. Berg, M.J.; Sorensen, C.M.; Chakrabarti, A. Patterns in Mie scattering: Evolution when normalized by the rayleigh cross section. *Appl. Opt.* **2005**, *44*, 7487–7493. [CrossRef] [PubMed]
3. Sorensen, C.M. Q-space analysis of scattering by particles: A review. *J. Quant. Spectrosc. Radiat.* **2013**, *131*, 3–12. [CrossRef]
4. Van de Hulst, H.C. *Light Scattering by Small Particles*; Wiley: New York, NY, USA, 1957; p. 470.
5. Kerker, M. *The Scattering of Light, and Other Electromagnetic Radiation*; Academic Press: New York, NY, USA, 1969; p. 666.
6. Bohren, C.F.; Huffman, D.R. *Absorption and Scattering of Light by Small Particles*; Wiley: New York, NY, USA, 1983; p. 350.
7. Heinson, W.R.; Chakrabarti, A.; Sorensen, C.M. A new parameter to describe light scattering by an arbitrary sphere. *Opt. Commun.* **2015**, *356*, 612–615. [CrossRef]
8. Heinson, W.R.; Chakrabarti, A.; Sorensen, C.M. Crossover from spherical particle Mie scattering to circular aperture diffraction. *J. Opt. Soc. Am. A* **2014**, *31*, 2362–2364. [CrossRef] [PubMed]
9. Guinier, A.; Fournet, G. *Small-Angle Scattering of X-rays*; Wiley: New York, NY, USA, 1955; p. 268.

10. Heinson, Y.W.; Maughan, J.B.; Ding, J.C.; Chakrabarti, A.; Yang, P.; Sorensen, C.M. Q-space analysis of light scattering by ice crystals. *J. Quant. Spectrosc. Radiat.* **2016**, *185*, 86–94. [CrossRef]

11. Hecht, E. *Optics*, 4th ed.; Addison-Wesley: Reading, MA, USA, 2002; p. 698.

12. Wang, G.; Chakrabarti, A.; Sorensen, C.M. Effect of the imaginary part of the refractive index on light scattering by spheres. *J. Opt. Soc. Am. A* **2015**, *32*, 1231–1235. [CrossRef] [PubMed]

13. Sorensen, C.M.; Maughan, J.B.; Chakrabarti, A. The partial light scattering cross section of spherical particles. *J. Opt. Soc. Am. A.* accepted for publication.

14. Munoz, O.; Moreno, F.; Guirado, D.; Dabrowska, D.D.; Volten, H.; Hovenier, J.W. The amsterdam-granada light scattering database. *J. Quant. Spectrosc. Radiat.* **2012**, *113*, 565–574. [CrossRef]

15. Munoz, O.; Volten, H.; Hovenier, J.W.; Nousiainen, T.; Muinonen, K.; Guirado, D.; Moreno, F.; Waters, L.B.F.M. Scattering matrix of large saharan dust particles: Experiments and computations. *J. Geophys. Res. Atmos.* **2007**, *112*. [CrossRef]

16. Sorensen, C.M. Q-space analysis of scattering by dusts. *J. Quant. Spectrosc. Radiat.* **2013**, *115*, 93–95. [CrossRef]

17. Heinson, Y.W.; Maughan, J.B.; Heinson, W.R.; Chakrabarti, A.; Sorensen, C.M. Light scattering q-space analysis of irregularly shaped particles. *J. Geophys. Res. Atmos.* **2016**, *121*, 682–691.

18. Volten, H.; Munoz, O.; Rol, E.; de Haan, J.F.; Vassen, W.; Hovenier, J.W.; Muinonen, K.; Nousiainen, T. Scattering matrices of mineral aerosol particles at 441.6 nm and 632.8 nm. *J. Geophys. Res. Atmos.* **2001**, *106*, 17375–17401. [CrossRef]

19. Munoz, O.; Volten, H.; de Haan, J.F.; Vassen, W.; Hovenier, J.W. Experimental determination of scattering matrices of olivine and allende meteorite particles. *Astron. Astrophys.* **2000**, *360*, 777–788.

20. Munoz, O.; Volten, H.; de Haan, J.F.; Vassen, W.; Hovenier, J.W. Experimental determination of scattering matrices of randomly oriented fly ash and clay particles at 442 and 633 nm. *J. Geophys. Res. Atmos.* **2001**, *106*, 22833–22844. [CrossRef]

21. Shkuratov, Y.; Ovcharenko, A.; Zubko, E.; Volten, H.; Munoz, O.; Videen, G. The negative polarization of light scattered from particulate surfaces and of independently scattering particles. *J. Quant. Spectrosc. Radiat.* **2004**, *88*, 267–284. [CrossRef]

22. Munoz, O.; Volten, H.; Hovenier, J.W.; Min, M.; Shkuratov, Y.G.; Jalava, J.P.; van der Zande, W.J.; Waters, L.B.F.M. Experimental and computational study of light scattering by irregular particles with extreme refractive indices: Hematite and rutile. *Astron. Astrophys.* **2006**, *446*, 525–535. [CrossRef]

23. Laan, E.C.; Volten, H.; Stam, D.M.; Munoz, O.; Hovenier, J.W.; Roush, T.L. Scattering matrices and expansion coefficients of martian analogue palagonite particles. *Icarus* **2009**, *199*, 219–230. [CrossRef]

24. Munoz, O.; Volten, H.; Hovenier, J.W.; Veihelmann, B.; van der Zande, W.J.; Waters, L.B.F.M.; Rose, W.I. Scattering matrices of volcanic ash particles of Mount ST. Helens, Redoubt, and Mount Spurr volcanoes. *J. Geophys. Res. Atmos.* **2004**, *109*. [CrossRef]

25. Wang, Y.; Chakrabarti, A.; Sorensen, C.M. A light-scattering study of the scattering matrix elements of arizona road dust. *J. Quant. Spectrosc. Radiat.* **2015**, *163*, 72–79. [CrossRef]

26. Heinson, Y.W.; Chakrabarti, A.; Sorensen, C.M. A light-scattering study of Al_2O_3 abrasives of various grit sizes. *J. Quant. Spectrosc. Radiat.* **2016**, *180*, 84–91. [CrossRef]

27. Sorensen, C.M. Light scattering by fractal aggregates: A review. *Aerosol Sci. Technol.* **2001**, *35*, 648–687. [CrossRef]

28. Sorensen, C.M.; Oh, C.; Schmidt, P.W.; Rieker, T.P. Scaling description of the structure factor of fractal soot composites. *Phys. Rev. E* **1998**, *58*, 4666–4672. [CrossRef]

29. Dubovik, O.; Sinyuk, A.; Lapyonok, T.; Holben, B.N.; Mishchenko, M.; Yang, P.; Eck, T.F.; Volten, H.; Munoz, O.; Veihelmann, B.; et al. Application of spheroid models to account for aerosol particle nonsphericity in remote sensing of desert dust. *J. Geophys. Res. Atmos.* **2006**, *111*. [CrossRef]

30. Mishchenko, M.I.; Travis, L.D. Capabilities and limitations of a current FORTRAN implementation of the *T*-matrix method for randomly oriented, rotationally symmetric scatterers. *J. Quant. Spectrosc. Radiat.* **1998**, *60*, 309–324. [CrossRef]

31. Mishchenko, M.I.; Travis, L.D.; Mackowski, D.W. *T*-matrix computations of light scatteringby nonspherical particles: A review. *J. Quant. Spectrosc. Radiat.* **2010**, *111*, 1704–1744.

32. Maughan, J.B.; Chakrabarti, A.; Sorensen, C.M. Rayleigh scattering and the internal coupling parameter for arbitrary shapes. *J. Quant. Spectrosc. Radiat.* **2017**, *189*, 339–343.

33. Yurkin, M.A.; Hoekstra, A.G. The discrete dipole approximation: An overview and recent developments. *J. Quant. Spectrosc. Radiat.* **2007**, *106*, 558–589. [CrossRef]

34. Sorensen, C.M.; Zubko, E.; Heinson, W.R.; Chakrabarti, A. Q-space analysis of scattering by small irregular particles. *J. Quant. Spectrosc. Radiat.* **2014**, *133*, 99–105. [CrossRef]

35. Zubko, E.; Shkuratov, Y.; Kiselev, N.N.; Videen, G. Dda simulations of light scattering by small irregular particles with various structure. *J. Quant. Spectrosc. Radiat.* **2006**, *101*, 416–434. [CrossRef]

36. Zubko, E.; Shkuratov, Y.; Mishchenko, M.; Videen, G. Light scattering in a finite multi-particle system. *J. Quant. Spectrosc. Radiat.* **2008**, *109*, 2195–2206. [CrossRef]

37. Zubko, E.; Kimura, H.; Shkuratov, Y.; Muinonen, K.; Yamamoto, T.; Okamoto, H.; Videen, G. Effect of absorption on light scattering by agglomerated debris particles. *J. Quant. Spectrosc. Radiat.* **2009**, *110*, 1741–1749.

38. Purcell, E.M.; Pennypacker, C.R. Scattering and absorption of light by nonspherical dielectric grains. *Astrophys. J.* **1973**, *186*, 705–714. [CrossRef]

39. Yurkin, M.A.; Maltsev, V.P.; Hoekstra, A.G. The discrete dipole approximation for simulation of light scattering by particles much larger than the wavelength. *J. Quant. Spectrosc. Radiat.* **2007**, *106*, 546–557.

40. Yurkin, M.A.; Hoekstra, A.G. The discrete-dipole-approximation code ADDA: Capabilities and known limitations. *J. Quant. Spectrosc. Radiat.* **2011**, *112*, 2234–2247. [CrossRef]

41. Oh, C.; Sorensen, C.M. Scaling approach for the structure factor of a generalized system of scatterers. *J. Nanopart Res.* **1999**, *1*, 369–377. [CrossRef]

42. Muinonen, K.; Zubko, E.; Tyynela, J.; Shkuratov, Y.G.; Videen, G. Light scattering by gaussian random particles with discrete-dipole approximation. *J. Quant. Spectrosc. Radiat.* **2007**, *106*, 360–377. [CrossRef]

43. Maughan, J.B.; Sorensen, C.M.; Chakrabarti, A. Q-space analysis of light scattering by gaussian random spheres. *J. Quant. Spectrosc. Radiat.* **2016**, *174*, 14–21. [CrossRef]

44. Chakrabarty, R.K.; Beres, N.D.; Moosmuller, H.; China, S.; Mazzoleni, C.; Dubey, M.K.; Liu, L.; Mishchenko, M.I. Soot superaggregates from flaming wildfires and their direct radiative forcing. *Sci. Rep. UK* **2014**, *4*, 5508. [CrossRef] [PubMed]

45. Berg, M.J.; Sorensen, C.M. Internal fields of soot fractal aggregates. *J. Opt. Soc. Am. A* **2013**, *30*, 1947–1955.

46. Zubko, E.; Petrov, D.; Grynko, Y.; Shkuratov, Y.; Okamoto, H.; Muinonen, K.; Nousiainen, T.; Kimura, H.; Yamamoto, T.; Videen, G. Validity criteria of the discrete dipole approximation. *Appl. Opt.* **2010**, *49*, 1267–1279. [CrossRef] [PubMed]

47. Yurkin, M.A.; de Kanter, D.; Hoekstra, A.G. Accuracy of the discrete dipole approximation for simulation of optical properties of gold nanoparticles. *J. Nanophoton.* **2010**, *4*, 041585.

48. Yang, P.; Bi, L.; Baum, B.A.; Liou, K.N.; Kattawar, G.W.; Mishchenko, M.I.; Cole, B. Spectrally consistent scattering, absorption, and polarization properties of atmospheric ice crystals at wavelengths from 0.2 to 100 µm. *J. Atmos. Sci.* **2013**, *70*, 330–347.

49. Brillouin, L. The scattering cross section of spheres for electromagnetic waves. *J. Appl. Phys.* **1949**, *20*, 1110–1125. [CrossRef]

50. Curtis, D.B.; Meland, B.; Aycibin, M.; Arnold, N.P.; Grassian, V.H.; Young, M.A.; Kleiber, P.D. A laboratory investigation of light scattering from representative components of mineral dust aerosol at a wavelength of 550 nm. *J. Geophys. Res. Atmos.* **2008**, *113*, 693–702. [CrossRef]

atmosphere

MDPI

Article

Quantifying Impacts of Aerosol Mixing State on Nucleation-Scavenging of Black Carbon Aerosol Particles

Joseph Ching [1,*], Matthew West [2] and Nicole Riemer [3]

[1] Atmospheric Environment and Applied Meteorology Research Department, Meteorological Research Institute, Japan Meteorological Agency, 1-1 Nagamine, Tsukuba, Ibaraki 305-0052, Japan
[2] Department of Mechanical Science and Engineering, University of Illinois at Urbana-Champaign, 1206 W. Green St., Urbana, IL 61801, USA; mwest@illinois.edu
[3] Department of Atmospheric Sciences, University of Illinois at Urbana-Champaign, 1301 W. Green St., Urbana, IL 61801, USA; nriemer@illinois.edu
* Correspondence: jching@mri-jma.go.jp; Tel.: +81-29-853-8934

Received: 11 October 2017; Accepted: 4 January 2018; Published: 11 January 2018

Abstract: Recent observational studies suggest that nucleation-scavenging is the principal path to removing black carbon-containing aerosol from the atmosphere, thus affecting black carbon's lifetime and radiative forcing. Modeling the process of nucleation-scavenging is challenging, since black carbon (BC) forms complex internal mixtures with other aerosol species. Here, we examined the impacts of black carbon mixing state on nucleation scavenging using the particle-resolved aerosol model PartMC-MOSAIC. This modeling approach has the unique advantage that complex aerosol mixing states can be represented on a per-particle level. For a scenario library that comprised hundreds of diverse aerosol populations, we quantified nucleation-scavenged BC mass fractions. Consistent with measurements, these vary widely, depending on the amount of BC, the amount of coating and coating material, as well as the environmental supersaturation. We quantified the error in the nucleation-scavenged black carbon mass fraction introduced when assuming an internally mixed distribution, and determined its bounds depending on environmental supersaturation and on the aerosol mixing state index χ. For a given χ value, the error decreased at higher supersaturations. For more externally mixed populations ($\chi < 20\%$), the nucleation-scavenged BC mass fraction could be overestimated by more than 1000% at supersaturations of 0.1%, while for more internally mixed populations ($\chi > 75\%$), the error was below 100% for the range of supersaturations (from 0.02% to 1%) investigated here. Accounting for black carbon mixing state and knowledge of the supersaturation of the environment are crucial when determining the amount of black carbon that can be incorporated into clouds.

Keywords: black carbon; nucleation-scavenging; aerosol mixing state; cloud microphysics; particle-resolved model

1. Introduction

Black-carbon-containing aerosol particles are emitted from the combustion of fossil fuel, biomass, and biofuel [1–5]. As one of the most important types of absorbing aerosol, black carbon (BC) exerts a multitude of impacts on the climate system [5], ranging from the local to the regional and global scales. BC-containing aerosols modify the radiative budget directly by scattering and absorbing solar radiation, and indirectly by modifying cloud microphysical properties. Owing to their ability to absorb solar radiation, BC-containing aerosols can heat the surrounding atmosphere and desiccate clouds when present as interstitial aerosol [6–9] or when residing inside cloud droplets after serving

Atmosphere **2018**, *9*, 17; doi:10.3390/atmos9010017   170   www.mdpi.com/journal/atmosphere

as cloud condensation nuclei [10,11]. Furthermore, they modify the snow albedo [12] after being deposited on Arctic snow after long-range transport [13], and warm the atmosphere where they reside, hence altering the local stability of the atmosphere [14]. This subsequently promotes accumulation of pollutants and impacts local air quality [15].

BC is one of the short-lived climate forcers, and it has been suggested that reducing BC emissions will mitigate global warming, at least temporarily [16–18]. According to [5], the best estimate of BC forcing including all forcing mechanisms is +0.71 W m^{-2} with 90% uncertainty bounds from +0.17 W m^{-2} to +2.1 W m^{-2} [5]. Ref. [5] attributes this substantial uncertainty range to the lack of knowledge of cloud interactions with both black carbon and co-emitted organic carbon, which motivates this study. The climate impacts of BC are determined by its atmospheric burden as well as its spatial and temporal distributions. These, in turn, are governed by the amount of emissions, the transport, and the removal from the atmosphere. Recent measurements by [19,20] indicated that nucleation-scavenging of BC-containing aerosols is the major removal pathway of BC from the atmosphere. Nucleation-scavenging is the process of aerosol particles acting as cloud condensation nuclei (CCN), being thus incorporated into cloud droplets. This leads to their removal if the cloud precipitates. It is important to note that, in general, clouds disperse several times before they indeed precipitate, hence nucleation-scavenging does not necessarily lead directly to the removal of aerosol.

Freshly emitted BC-containing particles, which usually do not contain hygroscopic material, require a high environmental supersaturation to activate [21]. This makes them poor CCN. However, their CCN activity can increase during transport in the atmosphere [20]. The processes contributing to the increase in CCN activity includes condensation of secondary aerosol outside cloud droplets, the addition of hygroscopic mass as a result of aqueous-phase chemistry within cloud droplets, coagulation with more hygroscopic particles, and heterogeneous oxidation reactions. The processes included in this study are the formation of secondary aerosol (gas-to-particle conversion) and coagulation.

Indeed, previous measurements of nucleation-scavenged BC mass fractions cover the entire range from practically zero to 100%, and vary with the environment where the measurements were taken. Some of these studies are compiled in Table 1, which is based on the studies by [22,23]. Note that most of the studies listed reported scavenging fractions based on bulk mass measurements, and the particle size ranges may differ between the studies.

Table 1. Nucleation-scavenged black carbon (BC) mass fraction measurements from previous campaigns. This table is adapted from [22,23].

Sampling Site	Citations	Environment	Average Scavenged BC Mass Fraction
Po Valley, Italy	[24]	Urban	0.06
Kleiner Feldberg, Germany	[25]	Rural	0.15
Puy de Dome, France	[26]	Mid altitude (1465 m)	0.33
Mt. Sonnblick, Austria	[27]	Mid altitude (3106 m)	0.45
Rax, Austria	[22]	Mid altitude (1644 m)	0.54
Great Dun Fell, UK	[28]	Rural-Coastal	0.57
Jungfraujoch, Switzerland	[23]	High altitude (3850 m)	0.61
Mt. Sonnblick, Austria	[29]	High altitude (3106 m)	0.74
Spitzbergen, Norway	[30]	Arctic	0.80
Mt. Soledad, La Jolla, USA	[20]	Marine-Coastal	0.01–0.1
Tokyo, Japan	[19]	Urban	0.1–1.0

Here, we investigate the question of what role aerosol mixing state plays in determining how much black carbon is incorporated into cloud droplets. We are using the term "aerosol mixing state" here in the sense of Winkler [31], who defined it as the distribution of chemical compounds over the particle population.

The evolution of mixing state is challenging to represent in aerosol models that use sectional or modal approaches. This is owing to their inherent assumption that particles within modes or size bins have the same composition. Here, we use a particle-resolved model to simulate the evolution of an aging aerosol population. The unique advantage of this approach is that the composition of each individual particle is tracked over time, and hence there are no approximating assumptions about mixing state.

Earlier particle-resolved modeling studies by [32] showed that the CCN activity of BC-containing aerosol can increase substantially after emission, either because hygroscopic species condense on the BC-containing particles, or because the BC-containing particles coagulate with other, more hygroscopic particles. Ref. [32] further concluded that the BC mixing state is an important factor for determining the CCN activation properties of BC, and neglecting realistic mixing state information leads to significant errors in the BC mass fraction that can undergo nucleation-scavenging [32–34]. Building on our studies [32–34], in this work, we quantified the BC mass fraction that can be incorporated into cloud droplets by nucleation-scavenging for a wide range of scenarios, and determined the error that is introduced by assuming internal mixture. This approximates the assumptions made in state-of-the-art regional and global models. Our aim was to understand how this error is related to aerosol mixing state, expressed in terms of the mixing state index χ, and how this relationship depends on the environmental supersaturation at which the nucleation-scavenged mass fraction was evaluated.

2. Methodology

2.1. PartMC-MOSAIC: A Particle-Resolved Approach to Simulated Aerosol Dynamics and Chemistry

In this study, we applied the particle-resolved model PartMC-MOSAIC (Particle Monte Carlo [35]-Model for Simulating Aerosol Interactions and Chemistry, [36]), which simulates the composition evolution of individual particles in an aerosol population within a well-mixed volume in the atmosphere. A detailed model description of PartMC-MOSAIC is provided in [35]. Here, we give a brief summary.

The PartMC module uses a stochastic Monte Carlo sampling model to handle Brownian coagulation among particles, particle emission into the volume, and the mixing of particles with background atmosphere. PartMC is coupled to the aerosol chemistry module MOSAIC [36] to simulate gas phase chemistry (CBM-Z [37]), gas-particle partitioning, and aerosol thermodynamics. The CBM-Z gas phase mechanism includes a total of 77 gas phase species. The aerosol species treated in MOSAIC are sulfate, nitrate, ammonium, chloride, carbonate, methanesulfonic acid, sodium, calcium, "other inorganic mass" (representing species such as SiO_2, metal oxides, and other unmeasured or unknown inorganic species present in aerosols), black carbon, primary organic carbon (POA), and secondary organic carbon (SOA). The formation of SOA is simulated with the SORGAM (Secondary Organic Aerosol Model) scheme [38], and includes 15 reaction products of aromatic precursors, higher alkenes, α-pinene, and limonene. Aqueous-phase chemistry or the oxidation of organic aerosol and associated changes in particle composition and CCN activity are not yet included in PartMC, but will be the focus of future model development.

The PartMC-MOSAIC model output consists of the composition vector of each computational particle, i.e., we store the mass of each aerosol species that is contained in each particle. Based on this information, per-particle quantities such as the particle sizes, their critical supersaturation, and optical properties are readily computed for assumed morphologies. Information about the particle population such as the bulk mass concentration of any aerosol species, the aerosol number concentration, and the aerosol size distribution can also easily be constructed.

Using the particle-resolved approach, simulating particle growth is free of numerical diffusion in mass composition space. Importantly, we do not prescribe in any way how the chemical species are distributed amongst the particles in the population. This means that, at any given time during the simulation, the aerosol mixing state is the result of the evolution of particle compositions, rather than a part of the modeling assumptions. The simulation results from PartMC-MOSAIC can therefore serve as a benchmark for representing aerosol mixing state, and the model was used for this purpose in previous studies [32–34, 39–42]. Assumptions are being made regarding the particle shape as we assume spherical particles to calculate coagulation rates and mass transfer rates. It would be interesting to expand the framework to account for non-spherical particles as this might impact the aerosol dynamics processes, as for example shown in [43]. However, considerable further model development is required to quantify these effects.

2.2. Setup of Idealized Urban Plume Scenarios

The basis for the analyses in this paper is a scenario library of eight urban plume scenarios described in [33, 34]. These scenarios were constructed to investigate the aging process of carbonaceous aerosol from combustion sources in urban environments. The term "aging" refers to the transition from CCN-inactive to CCN-active owing to coagulation with other aerosol particles or condensation of secondary organic or inorganic aerosol material [44]. We used approximately 10,000 computational particles for each simulation, with the exact number of particles varying over the course of the simulated period. In [32], we quantified the 95% confidence intervals for the results by performing an ensemble of runs, and concluded that this particle number is sufficient to obtain accurate statistics.

While the input parameters varied between scenarios to represent a range of different aging conditions, the general setup for each scenario was the same, and followed [39]. The total simulation time was 48 h. During the first 12 h, gas and aerosol emissions entered the air parcel. Background aerosol particles were introduced over the entire simulation time owing to dilution with background air. Tables 2–4 specify the details of the initial and background conditions as well as the emissions for aerosol particles and gas phase species for the base case. We assume that all particles from a given emission source have initially the same composition, with the mass fractions listed in Table 2. For example, particles from gasoline emissions are assumed to consist of an internal mixture of 80% POA and 20% BC. The different sub-populations are initially externally mixed. After entering the simulation, the particle composition evolves as described in Section 2.1.

Table 2. Size distribution and composition of aerosol emissions prescribed in simulations. BC stands for black carbon and POA stands for primary organic carbon.

Emission	Emission Strength $(m^{-2} s^{-1})$	Mean Diameter (μm)	Geometric Standard Deviation	Composition by Mass
Meat cooking	9×10^6	0.0864	1.9	100% POA
Diesel vehicles	1.6×10^8	0.05	1.7	30% POA, 70% BC
Gasoline vehicles	5×10^7	0.05	1.7	80% POA, 20% BC

Table 3. Size distribution and composition of initial conditions and background aerosols prescribed in simulations.

Initial / Background	Number Concentration (m^{-3})	Mean Diameter (μm)	Geometric Standard Deviation	Composition by Mass
Aitken mode	1.8×10^9	0.02	1.45	49.6% $(NH_4)_2SO_4$ 49.6% SOA 0.8% BC
Accumulation mode	1.5×10^9	0.116	1.65	49.6% $(NH_4)_2SO_4$ 49.6% SOA 0.8% BC

Table 4. Initial conditions and background mixing ratios (second column) and twelve-hour average emission (third column) prescribed in simulations of gaseous species for the ten environmental scenarios.

Chemical Species	Mixing Ratio (ppbv)	Emission Flux (nmol m^2 s^{-1})
Nitrogen oxide	0.1	15.9
Nitrogen dioxide	1.0	0.84
Nitric acid	1.0	-
Ozone	50.0	-
Hydrogen peroxide	1.1	-
Carbon monoxide	80	291.3
Sulfur dioxide	0.8	2.51
Ammonia	0.5	6.11
Hydrogen chloride	0.7	-
Methane	2200	-
Ethane	1.0	-
Formaldehyde	1.2	1.68
Methanol	0.12	0.28
Methyl hydrogen peroxide	0.5	-
Acetaldehyde	1.0	0.68
Paraffin carbon	2.0	96.0
Acetone	1.0	1.23
Ethene	0.2	7.28
Terminal olefin carbons	2.3×10^{-2}	2.43
Internal olefin carbons	3.1×10^{-4}	2.43
Toluene	0.1	4.04
Xylene	0.1	2.41
Lumped organic nitrate	0.1	-
Peroxyacetyl nitrate	0.8	-
Higher organic acid	0.2	-
Higher organic peroxide	2.5×10^{-2}	-
Isoprene	0.5	0.23
Alcohols	-	3.45

The following conceptual model guided this setup, summarized by Figure 1 in [39]. After the simulation starts at 6:00 a.m. LST (local solar time), the Lagrangian air parcel represents a volume of air in the polluted, well-mixed boundary layer during the daytime. After sunset, the air parcel represents the polluted air that remains in the nocturnal residual layer. This layer is decoupled from the stable surface layer, and hence we discontinue emissions after 12 h. We continue to simulate the aerosol aging for a second day of simulation without adding fresh emissions to capture the effects of a longer processing time. This corresponds to an air parcel being advected over the ocean for another day, again decoupled from the stable marine surface layer.

As shown in previous studies [44,45], the aging rate, i.e., the conversion of BC-containing particles from hydrophopic to hygroscopic, is determined by both condensation of secondary aerosol and coagulation with more hygroscopic particles. The relative importance of these two processes depends on the particular environmental conditions. For our base case scenario, condensation dominated during the daytime, while coagulation dominated during the nighttime.

To change the rate of the individual aging processes and their relative magnitudes, we obtained the other scenarios by changing the emission rate of the particles containing BC (100%, 25% and 2.5% of the base case), the background particle number concentrations (100% and 10% of the base case) and the gaseous emission rate (50% and 25% of the base case). This also created a range of BC mass concentrations, spanning a range from 0.05 to 3.6 μg m^{-3}, consistent with observations listed in the Environmental Protection Agency (EPA) Report to Congress on BC [46].

As in [34], we focus here on the *aerosol state* at different stages of aging, rather than the *temporal evolution*. As the basis for our analysis of BC scavenging, we therefore used the hourly aerosol states from all scenarios, which comprised $48 \times 8 = 384$ populations covering a wide range of mixing states. These can be seen as possible populations present at cloud base, entering the cloud.

To quantify mixing state, we used the mixing state index χ. This concept originated from information theoric entropy measures and is detailed in [1]. In short, the mixing state index χ is calculated based on the mass fraction of each of the aerosol species contained in each individual particle, and the total bulk mass fractions of each of the aerosol species contained in that aerosol population. The mixing state index χ ranges from 0% to 100%, where 0% indicates that the aerosol population is completely externally mixed, and 100% indicates that the population is completely internally mixed. The definition of "species" for calculating the mass fractions depends on the application. Since we focus on aerosol activation, we grouped the aerosol model species according to their hygroscopicity into two surrogate species; BC and POA form one surrogate species as their hygroscopicity is very low. All other (more hygroscopic) model species form the second surrogate species. The 384 aerosol populations that form our dataset cover a range of mixing states from $\chi = 7.5\%$ to $\chi = 88\%$ (compare to Figure 2 in [34]).

Figure 1 shows four example populations where we chose the two-dimensional number density in terms of dry diameter and BC mass fraction, $n(D, w_{BC})$, for illustration purposes. The black carbon mass fraction is defined for each individual particle i as

$$w_{BC,i} = \frac{m_i^{BC}}{m_i},\tag{1}$$

where m_i^{BC} and m_i are the mass of BC contained in particle i and the total dry particle mass, respectively. The two-dimensional number concentration distribution $n(D, w_{BC})$ is then defined by

$$n(D, w_{BC}) = \frac{\partial^2 N(D, w_{BC})}{\partial \log_{10} D \, \partial w_{BC}},\tag{2}$$

where $N(D, w_{BC})$ is the two-dimensional cumulative number distribution, which equals the number concentration of particles with total dry diameter less than D and black carbon dry mass fraction less than w_{BC}.

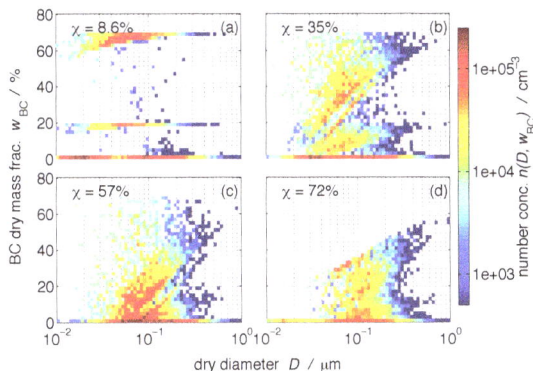

Figure 1. Number concentration in two-dimensional space of dry diameter and black carbon mass fraction, $n(D, w_{BC})$, for four example populations that differ in mixing state index χ.

Figure 1a shows a population that contains very fresh carbonaceous emissions. We observe three distinct subpopulations that differed in their BC content, namely particles that originated from diesel vehicles ($w_{BC} = 70\%$), particles from gasoline vehicles ($w_{BC} = 30\%$), and particles with zero or very low BC content from meat cooking emissions and background aerosol (compare to Tables 2 and 3).

Not surprisingly, this population is characterized by a low mixing state index, $\chi = 8.6\%$, indicating that it is very externally mixed.

Figure 1b,c show populations that experienced some aging, but fresh emissions were still introduced, which maintained the range of BC mass fractions up to 70%. The mixing state indexes for these populations were $\chi = 35\%$ and $\chi = 57\%$, respectively. Figure 1d is an example of a population where fresh emissions had ceased to enter. The maximum BC mass fractions in this population ranged between only 20% and 50% depending on particle size because secondary aerosol formation took place while fresh BC-containing particles were not replenished. The mixing state index of this population is higher than any of the other three examples ($\chi = 72\%$), meaning that the population has moved more towards an internal mixture. The diagonal features in these graphs arise when condensation of secondary aerosol material occurs on the particles. The BC mass fraction of small particles decreases faster than that of larger particles. The scatter that forms in between the main sub-populations is due to particle coagulation.

In [34], we examined the relationship between χ and the error incurred in CCN concentration predictions when assuming internal mixture. Here, we applied the same framework to quantify the error in predicting the BC mass fraction that can undergo nucleation-scavenging.

2.3. Framework of Error Quantification

For this study, we used a similar error quantification framework to [34], but we tailored it to quantify the relationship between aerosol mixing state and the error in nucleation-scavenged mass fraction of black carbon aerosols when assuming internal mixture. Figure 2 outlines the approach. For each population, we calculated the critical supersaturations, s_c, for each particle according to [44] using the dimensionless hygroscopicity parameter κ [47]. The overall κ for a particle is the volume-weighted average of the κ values of the constituent species. We assumed $\kappa = 0.65$ for all salts formed from the $NH_4^+ - SO_4^{2-} - NO_3^-$ system. For all SOA model species, we assumed $\kappa = 0.1$, and, for POA and BC, we assumed $\kappa = 0.001$ and $\kappa = 0$, respectively [47].

Figure 2a shows the distribution density function $\partial^2 N(D, s_c)/\partial \log_{10} D\, \partial s_c$ in terms of particle dry diameter D and per-particle critical supersaturation s_c for the same population as shown in Figure 1b. Figure 2a illustrates that both particle size and composition determine critical supersaturation; as the particle size increases, s_c decreases, and at a given size, a range of s_c values exist owing to differences in particle composition.

Given a certain environmental supersaturation threshold s_{env}, we can separate the population into CCN active and CCN inactive particles. In Figure 2a, we used $s_{env} = 0.3\%$ as an example. The CCN number concentration can be readily determined by summing over all particle dry diameters D and over the supersaturation range from 0 to s_{env}.

Recall that the model tracks the BC content of the individual active/non-active particles. Therefore, to determine the nucleation-scavenged BC mass fraction at a certain environmental supersaturation, we summed over the black carbon mass concentrations associated with the CCN at that particular supersaturation. This quantity is the total nucleation-scavenged BC mass concentration, $m_{ns,BC}(s_{env})$. We then calculated the ratio of nucleation-scavenged BC mass concentration to the total BC mass concentration of the whole aerosol population, m_{BC}. This ratio is defined as the nucleation-scavenged BC mass fraction, and denoted as $f_{BC}(s_{env})$:

$$f_{BC}(s_{env}) = \frac{m_{ns,BC}(s_{env})}{m_{BC}}. \tag{3}$$

We evaluated this quantity for each aerosol population at 50 supersaturation values from 0.02% to 1% in steps of 0.02%. Figure 2c shows the spectrum of $f_{BC}(s_{env})$ (red line) that corresponds to the population shown in Figure 2a.

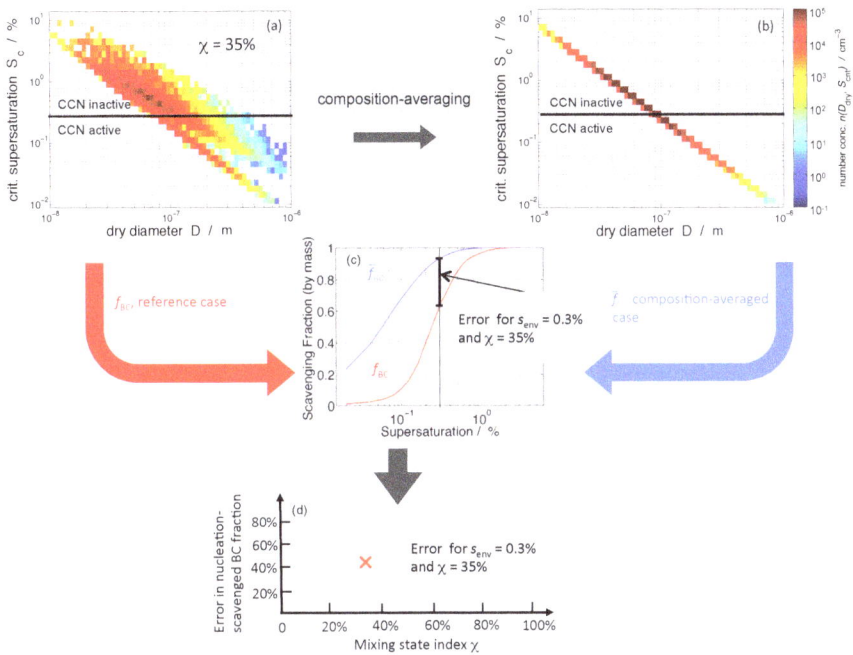

Figure 2. Conceptual framework for error quantification; see Section 2.3 for details. (**a**) $\partial^2 N(D, s_c)/\partial \log_{10} D\, \partial s_c$ for the reference case; (**b**) $\partial^2 N(D, s_c)/\partial \log_{10} D\, \partial s_c$ after composition averaging; (**c**) resulting nucleation-scavenged black carbon mass fractions; (**d**) error in nucleation-scavenged BC fraction as a function of mixing state.

Figure 2b shows the aerosol population after composition averaging. The properties of the composition averaging operation are detailed in [32], Appendix B1. In brief, it assigns each particle a composition equal to the average composition of the population, and preserves the bulk aerosol mass concentrations, the number concentration, and the particle diameters, but modifies the per-particle compositions. The mixing state index of such a population is 100%. Because all particles were assigned the same composition (equal to the average composition), the spread in critical supersaturations at a given size vanishes. This alters the CCN concentration and also the amount of BC mass that is associated with those CCN. By this operation, some particles are assigned a higher critical supersaturation, while others are assigned a lower critical supersaturation compared to the original value.

The blue line in Figure 2c shows the spectrum $\bar{f}_{BC}(s_{env})$ corresponding to the composition-averaged population shown in Figure 2b. The difference in the two spectra represents the error that is incurred in f_{BC} by assuming internal mixture, which will depend on the supersaturation threshold at which CCN activity is evaluated.

We define the relative error in nucleation-scavenged BC mass fraction, $\Delta f_{BC}(\Pi, \bar{\Pi}, s_{env})$, evaluated for the particle populations Π and $\bar{\Pi}$ before and after composition-averaging, respectively, and for a particular environmental supersaturation s_{env}, as

$$\Delta f_{BC}(\Pi, \bar{\Pi}, s_{env}) = \frac{\bar{f}_{BC}(\bar{\Pi}, s_{env}) - f_{BC}(\Pi, s_{env})}{f_{BC}(\Pi, s_{env})}, \tag{4}$$

where $\tilde{f}_{BC}$ and f_{BC} refer to the composition-averaged and the particle-resolved populations, respectively. We calculated the relative error in nucleation-scavenged BC mass fraction for each of the 384 populations and will discuss the relationship between Δf_{BC} and χ in Section 3.2.

In Section 3.1, we will use a size-resolved version of f_{BC} to relate our results to ambient observations by [19,20]. This is calculated by sorting all particles into 150 logarithmically spaced size bins from 0.1 nm to 10 μm based on their BC core diameter assuming spherical shape. The size-resolved nucleation-scavenged BC mass fraction, f_{BC}^j, is then defined as the ratio of nucleation-scavenged BC mass concentration in bin j, $m_{ns,BC}^j(s_{env})$, to the total BC mass concentration in bin j, m_{BC}^j. This gives

$$f_{BC}^j(s_{env}) = \frac{m_{ns,BC}^j(s_{env})}{m_{BC}^j}. \tag{5}$$

Note that, as the number of size bins becomes large and their width small, the size-resolved mass-based fraction f_{BC}^j approaches the size-resolved number fraction of nucleation-scavenged BC-containing particles, which is the quantity reported by [19,20]. We have included the derivation of this fact in the Supplemental Material, as well as Figures S1 and S2 that confirm this fact.

3. Results

3.1. Size-Resolved Nucleation-Scavenged BC Mass Fraction

Figure 3 shows the size-resolved nucleation-scavenged BC mass fraction, f_{BC}^j, as a function of BC core diameter D_{BC} at four selected supersaturations, 0.1%, 0.3%, 0.4%, and 0.6%. These supersaturation values are representative of conditions ranging from stratus clouds to convective clouds. The solid red lines and the shading indicate the average of f_{BC}^j over all 384 cases and the corresponding standard deviation. We used BC core size here as an independent variable to connect our results to observational findings by [19,20]. The range of measurements by [19,20] are summarized by the blue and orange rectangles, respectively. Note that these authors report different individual datasets in their studies, which we did not replicate here, since only a qualitative comparison is possible.

For all supersaturations, the quantity f_{BC}^j tended to increase with BC core diameter, especially within the size ranges that were accessed by [19,20], as indicated by the shaded rectangles in Figure 3. For any given supersaturation, the spread of f_{BC}^j originated from mixing state differences within the 384 aerosol populations. For example, for particles of BC core diameter between 100 nm and 108 nm at a supersaturation of 0.1%, the average nucleation-scavenged BC mass fraction for all our scenarios is 58%, with a standard deviation of $\pm$33%. Given a certain BC core diameter, as the environmental supersaturation threshold increased, f_{BC}^j also increased. Taking again the example of BC core diameters between 100 nm and 108 nm, the average value for f_{BC}^j increased from 57% for $s_{env} = 0.1\%$ to 96% for $s_{env} = 0.6\%$. The shape of these curves is determined by the underlying mass size distributions of the BC-containing particles, shown in the Supplemental Material, Figure S3.

The nucleation-scavenged BC mass fraction has been determined observationally by [19,20]. Ref. [20] used a single particle soot photometer (SP2) to measure the size distribution of the BC-containing particles activated to form stratocumulus cloud droplets at a marine boundary layer site in California, while [19] carried out measurements of BC in ambient particles before precipitation and in rainwater after precipitation using an SP2 in the Tokyo metropolitan area.

The PartMC-MOSAIC simulation results agree qualitatively with the two observational studies [19,20] in that f_{BC}^j increases with BC core diameter. The measurements of f_{BC}^j by [20] are significantly lower than both the measurements by [19] and our modeling results. The activated fractions of BC-containing particles were less than 0.2 for BC core diameter ranging between 70 and 220 nm for the two cloud case studies reported in [20] (Figure 6a,f therein). In contrast, Ref. [19] found that the nucleation-scavenged number fraction increased from 0.6 to 0.9 when BC core diameter increased from 200 to 350 nm (Figure 3 therein). The discrepancy between [20] and our model results can possibly be attributed to the fact that the measurements were performed in a marine environment, while the PartMC-MOSAIC simulations represent urban environments of different pollution levels. In addition, Ref. [20] cautioned that the activation fractions reported should be considered as lower limits due to imperfect instrumental detection efficiency. In contrast, due to similar urban environments in which the simulations and the measurements by [19] took place, PartMC-MOSAIC results agree better with the observations by [19].

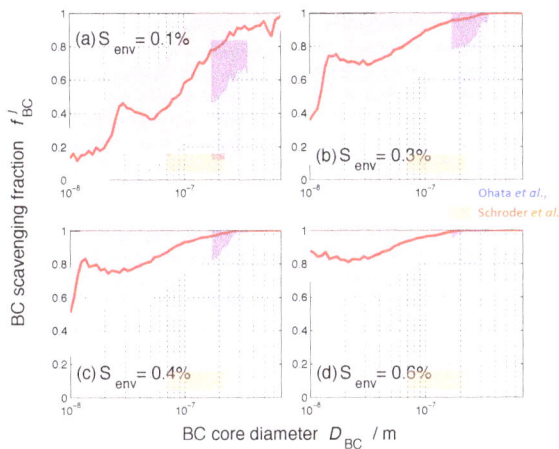

Figure 3. The size-resolved nucleation-scavenged black carbon (BC) mass fraction, f_{BC}^j, as a function of black carbon core diameter, D_{BC}, at four selected supersaturation values: 0.1%, 0.3%, 0.4%, and 0.6%. The solid line represents the average over the 384 populations, and the color shading indicates one standard deviation. The orange and blue shaded boxes indicate the values that were obtained from field studies by [19,20], respectively.

Figure 4 shows the same data as Figure 3, but displays f_{BC} as a function of supersaturation for selected BC core diameter ranges. From the combination of Figures 3 and 4, we conclude that, for a given core size, the nucleation-scavenged BC fraction varies drastically depending on the environmental supersaturation. Moreover, for any given BC core size and environmental supersaturation, the nucleation-scavenged BC fraction spans a wide range, which reflects the fact that different amounts of coating can exist for any given core size depending on the history of the particle. This implies that the core size by itself is not a good predictor for the amount of BC that can be incorporated into cloud droplets.

Figure 5 shows the nucleation-scavenged BC mass fraction as a function of particle dry diameter D. As before, the solid line represents the average for the 384 populations, and the color shading represents one standard deviation. For any given environmental supersaturation, the range of nucleation-scavenged BC fractions is largest when the slope of the spectrum is large. For example, for a 100-nm particle at a supersaturation of 0.3%, f_{BC}^j is on average 0.71 with a standard deviation of 0.31. This variability is introduced by the different particle compositions for any given population. The size range with the largest sensitivity to composition depends on the environmental supersaturation. Just as the core size was not a good predictor by itself for the amount of BC that can be incorporated in cloud droplets, the dry diameter alone is also not sufficient to predict f_{BC}^j.

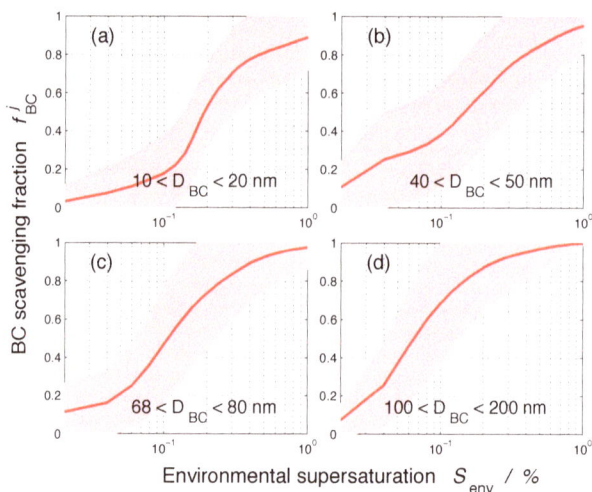

Figure 4. The nucleation-scavenged BC mass fraction, f_{BC}^j, for selected size ranges as a function of supersaturation. The solid line represents the average over the 384 populations, and the color shading indicates one standard deviation.

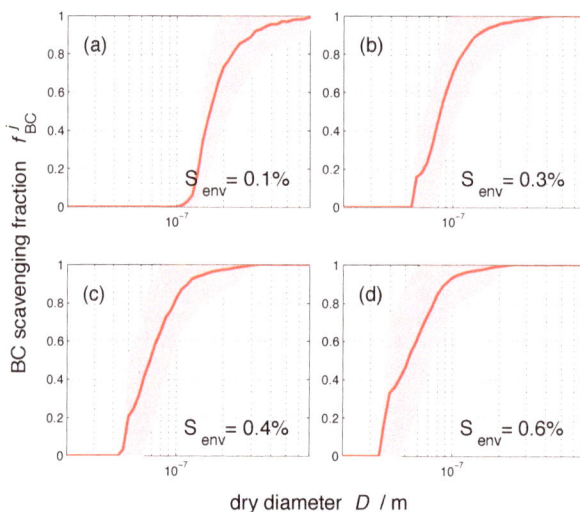

Figure 5. The size-resolved nucleation-scavenged BC mass fraction, f_{BC}^j, as a function of dry diameter, D, at four selected supersaturation values: 0.1%, 0.3%, 0.4%, and 0.6%. The solid line represents the average over the 384 populations, and the color shading indicates one standard deviation.

3.2. Nucleation-Scavenged BC Mass Fraction and Aerosol Mixing State

After presenting the range of f_{BC}^j values and their dependence on BC core diameter, dry diameter, and environmental supersaturation threshold, we will now turn to quantifying the importance of mixing state in determining the nucleation-scavenged BC mass fraction f_{BC} (not size-resolved).

Figure 6 shows the nucleation-scavenged BC mass fraction, f_{BC}, for all 384 populations as a function of mixing state parameter χ. As expected, one χ value can correspond to a range of f_{BC} values.

As the environmental supersaturation increases, low values for f_{BC} only occur for more externally mixed populations. For example, for s_{env} of 0.1%, we find f_{BC} of less than 20% for χ values up to 70%. In contrast, for s_{env} of 0.6%, f_{BC} reaches as low as 20% only for the population with the lowest χ of 8%.

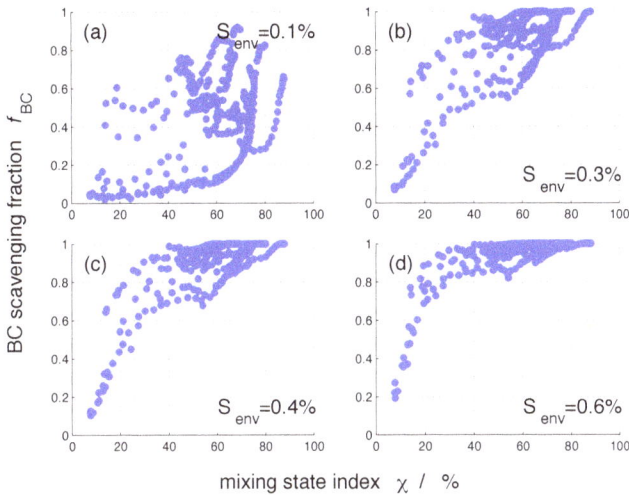

Figure 6. Nucleation-scavenged BC mass fraction, f_{BC}, for the 384 aerosol populations for selected supersaturations as a function of mixing state index χ. The definition of mixing state index is described in Section 2.2.

The four panels of Figure 7 show the absolute value of the error in nucleation-scavenged BC mass fraction, Δf_{BC} (Equation (4)), at four selected supersaturations (0.1%, 0.3%, 0.4% and 0.6%) for all 384 aerosol populations. For the sake of clarity, we used a logarithmic scale for the ordinate, and we took the absolute value of Δf_{BC} before taking its logarithmic value to avoid handling negative Δf_{BC} values, which represent an underestimation of f_{BC} after composition averaging. These cases are indicated by red dots in Figure 7.

The maximum values for Δf_{BC} were on the order of several thousand percent and occurred for the most externally mixed populations. For such populations, the corresponding value for f_{BC} was small, 10% or less. As expected, the error decreased when the aerosol population became more internally mixed. For more internally-mixed aerosol populations, particles have more similar compositions than for more externally-mixed populations. Consequently, the modifications of the composition averaging procedure are less severe. This is consistent with the previous findings regarding errors in CCN concentrations presented in [34]. For any given χ value, a range of Δf_{BC} existed due to the different extent of modifications to the per-particle composition by composition averaging [34].

As χ increased, the error decreased more quickly for larger environmental supersaturations. For example, for a low supersaturation value such as $s_{env} = 0.1\%$, Δf_{BC} dropped below 10% only for populations with mixing state index χ larger than 84%. In constrast, for $s_{env} = 0.6\%$, this was the case already for populations with χ around 60%. It follows that the highest errors Δf_{BC} can be expected for aerosol populations having small χ values at low environmental supersaturations.

We can also display these results by showing the error $|\Delta f_{BC}|$ as a function of supersaturation as in Figure 8. The four panels separate the populations according to their mixing state index. This figure shows that for quite internally mixed populations ($\chi > 75\%$), the error was larger than 100% only for supersaturations of 0.1% or lower, while for quite externally mixed populations ($\chi < 20\%$), the error could reach above 100% even at supersaturations of 1%.

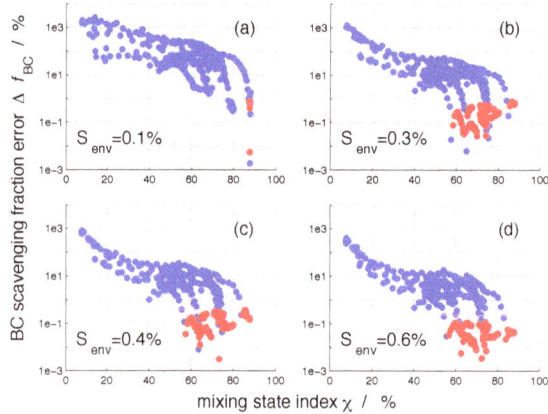

Figure 7. The absolute value of the error in nucleation-scavenged black carbon mass fraction, $|\Delta f_{BC}|$, for the 384 aerosol populations for selected supersaturations as a function of mixing state index χ. The definition of the error in the BC nucleation-scavenging fraction is given by Equation (4). The definition of mixing state index is described in Section 2.2. The red dots indicate cases where Δf_{BC} is negative, and the blue dots represent cases where Δf_{BC} is positive.

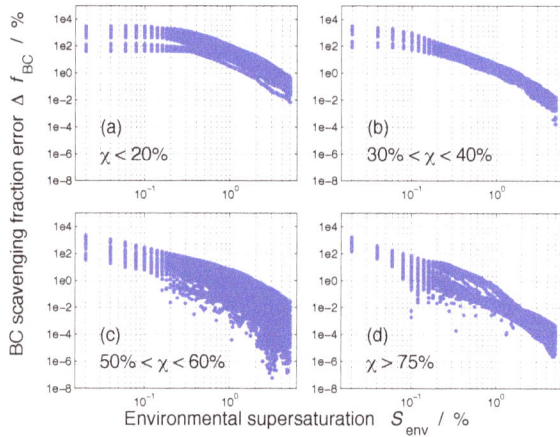

Figure 8. The absolute value of the error in nucleation-scavenged black carbon mass fraction, $|\Delta f_{BC}|$, for the 384 aerosol populations for selected ranges of mixing state index, χ, as a function of environmental supersaturation, s_{env}.

Since we averaged the chemical composition of all the particles in the population, irrespective of the particle-size, the Δf_{BC} values presented in Figure 7 represent an upper estimate of Δf_{BC}. Figure S4 in the Supplemental Material shows the error Δf_{BC} when the composition averaging is performed within size bins, rather than for the entire population. Here, we used five size bins per decade.

For completeness, we show number-based nucleation scavenging fractions, f_{BC}^N and associated errors in the Supplemental Material (Figures S5 and S6). We conclude that the error in the nucleation-scavenged BC mass fraction, f_{BC}, exceeded the error of the nucleation-scavenged BC number fraction, f_{BC}^N, for more externally-mixed populations ($\chi < 20\%$), in some cases by over an order of magnitude. For more internally-mixed populations ($\chi > 60\%$), Δf_{BC} was comparable to Δf_{BC}^N, if not smaller. These results are consistent to the ones shown in [34], where the error in CCN/CN ratio was quantified, which is a similar metric to f_{BC}^N.

4. Conclusions

This study presents an analysis of the nucleation-scavenged BC mass fraction, f_{BC}, simulated by the particle-resolved model PartMC-MOSAIC for a wide range of environmental conditions, and quantifies the error that is introduced when calculating f_{BC} and assuming internal mixture. PartMC-MOSAIC simulates changes in particle composition due to coagulation and the formation of inorganic and organic secondary aerosol species. Other aging processes, such as aerosol in-cloud processing or heterogeneous reactions, were not included in this study.

The BC mass fraction that can be incorporated in cloud droplets depends on the size of the BC cores, the amount of coating of secondary aerosol, and the environmental supersaturation. In qualitative consistency with observational findings, our results show that f_{BC} increases for larger black carbon core sizes. However, the existence of coatings and the environmental supersaturation introduces a large variability.

To determine the error in f_{BC} when assuming internal mixture, we applied the framework developed in [34], and related the mixing state index χ [1] to the error Δf_{BC}. The error Δf_{BC} depended on both environmental supersaturation and χ. In general, for a given χ value, Δf_{BC} was smaller at higher supersaturation. As χ increased, Δf_{BC} decreased more quickly for higher supersaturations.

As aerosol populations become increasingly internally mixed, Δf_{BC} decreases. For example, for more externally mixed populations ($\chi < 20\%$), f_{BC} could be overestimated by more than 1000% at supersaturations of 0.1%, while for $\chi > 75\%$, Δf_{BC} is below 100% for the range of supersaturations (from 0.02% to 1%) investigated here.

Since nucleation-scavenging is the principal process for removing BC-containing aerosols [19,20], capturing the evolution of BC aerosol mixing state in models is essential for determining the atmospheric lifetime, burden, vertical profile, and long-range transport of BC-containing aerosols in the atmosphere. In this regard, recent advancements in modeling aerosol mixing state could improve the understanding of the vertical variation of BC mixing state and the associated wet deposition process [48]. Meanwhile, measurements of vertical profiles of BC-containing aerosols would be valuable in evaluating regional and global models' simulated BC spatial distributions [49]. Accounting for aerosol mixing state in evaluating the spatial and temporal distribution of BC-containing aerosols would be a benefit, if not a strict prerequisite, to further constrain the radiative forcing uncertainty of BC-containing aerosols in future studies.

Supplementary Materials: The following are available online at www.mdpi.com/2073-4433/9/1/17/s1, Figure S1: Size-resolved nucleation-scavenged BC number fraction, as a function of black carbon core diameter, at four selected supersaturation values; Figure S2: Nucleation-scavenged BC number fraction for selected size ranges as a function of supersaturation; Figure S3: BC mass size distributions as a function of BC core diameter for the nucleation-scavenged BC-containing particles, and for all BC-containing particles for selected supersaturations; Figure S4: Absolute value of the error in black carbon nucleation-scavenged mass fraction, $|\Delta f_{BC}|$ with size-resolved composition-averaged aerosol populations and for composition-averaging of the entire population, at four selected supersaturation values as a function of mixing state index χ; Figure S5: Total nucleation-scavenged BC number fraction, f_{BC}^N, as a function of aerosol mixing state index χ, at four selected supersaturation values; Figure S6: Error in total nucleation-scavenged BC number fraction, Δf_{BC}^N, as a function of aerosol mixing state index χ, at four selected supersaturation values.

Acknowledgments: Joseph Ching's research was supported by the Environmental Research and Technology Development Fund (5-1605) of the Environmental Restoration and Conservation Agency (ERCA), Japan. Matthew West acknowledges an National Science Foundation Civil, Mechanical and Manufacturing Innovation (NSF CMMI) CAREER grant 1150490 and Department of Energy, Atmospheric System Research (DOE ASR) grant DE-SC0011771. Nicole Riemer was supported by an National Science Foundation, Atmospheric and Geospace Sciences NSF AGS CAREER grant 1254428.

Author Contributions: J.C., M.W., and N.R. designed the experiments, analyzed the model results, and wrote the paper. J.C. performed the model simulations.

Conflicts of Interest: The authors declare no conflict of interest. The funding sponsors had no role in the design of the study; in the collection, analyses, or interpretation of data; in the writing of the manuscript, and in the decision to publish the results.

References

1. Riemer, N.; West, M. Quantifying aerosol mixing state with entropy and diversity measures. *Atmos. Chem. Phys.* **2013**, *13*, 11423–11439.
2. Pöschl, U. Atmospheric Aerosols: Ccomposition, Transformation, Climate and Health Effects. *Angew. Chem. Int. Ed.* **2005**, *44*, 7520–7540.
3. Hansen, A.D.; Bodhaine, B.A.; Dutton, E.G.; Schnell, R.C. Aerosol black carbon measurement at the South Pole: Initial results 1986–1987. *Geophys. Res. Lett.* **1988**, *15*, 1193–1196.
4. Hansen, J.; Nazarenko, L. Soot climate forcing via snow and ice albedos. *PNAS* **2004**, *101*, 423–428.
5. Bond, T.C.; Doherty, S.J.; Fahey, D.W.; Forester, P.M.; Berntsen, T.; DeAngelo, B.J.; Flanner, M.G.; Ghan, S.; Kärcher, B.; Koch, D.; et al. Bounding the role of black carbon in the climate system: A scientific assessment. *J. Geophys. Res.* **2013**, *118*, 5380–5552.
6. Hansen, J.; Sato, M.; Ruedy, R. Radiative forcing and climate response. *J. Geophys. Res. Atmos.* **1997**, *102*, 6831–6864.
7. Ackerman, A.S.; Toon, O.B.; Stevens, D.E.; Heymsfield, A.J.; Ramanathan, V.; Welton, E.J. Reduction of tropical cloudiness by soot. *Science* **2000**, *288*, 1042–1047.
8. Koch, D.; Schulz, M.; Kinne, S.; McNaughton, C.; Spackman, J.R.; Balkanski, Y.; Bauer, S.; Berntsen, T.; Bond, T.C.; Boucher, O.; et al. Evaluation of black carbon estimations in global aerosol models. *Atmos. Chem. Phys.* **2009**, *9*, 9001–9026.
9. Koch, D.; Genio, A.D.D. Black carbon semi-direct effects on cloud cover: Review and synthesis. *Atmos. Chem. Phys.* **2010**, *10*, 7685–7696.
10. Chylek, P.; Hallett, J. Enhanced absorption of solar radiation by cloud droplets containing soot particles in their surface. *Q. J. R. Meteorol. Soc.* **1992**, *118*, 167–172.
11. Wendisch, M.; Mertens, S.; Ruggaber, A.; Nakajima, T. Vertical profiles of aerosol and radiation and the influence of a temperature inversion: Measurements and radiative transfer calculations. *J. Appl. Meteorol.* **1996**, *35*, 1703–1715.
12. Flanner, M.G.; Zender, C.S.; Randerson, J.T.; Rasch, P.J. Present day climate forcing and response from black carbon in snow. *J. Geophys. Res.* **2007**, *112*, doi:10.1029/2006JD008003.
13. Samset, B.H.; Myhre, G.; Herber, A.; Kondo, Y.; Li, S.M.; Moteki, N.; Koike, M.; Oshima, N.; Schwarz, J.P.; Balkanski, Y.; et al. Modelled black carbon radiative forcing and atmospheric lifetime in AeroCom Phase II constrained by aircraft observations. *Atmos. Chem. Phys.* **2014**, *14*, 12465–12477.
14. Wilcox, E.M.; Thomas, R.M.; Praveen, P.S.; Pistone, K.; Bender, F.A.M.; Ramanathan, V. Black carbon solar absorption suppresses turbulence in the atmospheric boundary layer. *Proc. Natl. Acad. Sci. USA* **2015**, *113*, 11794–11799.
15. Ding, A.; Huang, X.; Nie, W.; Sun, J.; Kerminen, V.M.; Petäjä, T.; Su, H.; Cheng, Y.; Yang, X.Q.; Wang, M.; et al. Enhanced haze pollution by black carbon in megacities in China. *Geophys. Res. Lett.* **2016**, *43*, 2873–2879.
16. Chen, W.; Lee, Y.H.; Adams, P.J.; Nenes, A.; Seinfeld, J.H. Will black carbon mitigation dampen aerosol indirect forcing? *Geophys. Res. Lett.* **2010**, *37*, L09801, doi:10.1029/2010GL042886.
17. Bond, T.C.; Sun, H. Can reducing black carbon emissions counteract global warming? *Environ. Sci. Technol.* **2005**, *39*, 5921–5926.
18. Jacobson, M.Z. Short-term effects of controlling fossil-fuel soot, biofuel soot and gases, and methane on climate, Arctic ice, and air pollution health. *J. Geophys. Res.* **2010**, *115*, doi:10.1029/2009JD013795.
19. Ohata, S.; Moteki, N.; Mori, T.; Koike, M.; Kondo, Y. A key process controlling the wet removal of aerosols: New observational evidence. *Sci. Rep.* **2016**, *6*, 34113, doi:10.1038/srep34113.
20. Schroder, J.C.; Hanna, S.J.; Modini, R.L.; Corrigan, A.L.; Kreidenweis, S.M.; Macdonald, A.M.; Noone, K.J.; Russell, L.M.; Leaitch, W.R.; Bertram, A.K. Size-resolved observations of refractory black carbon particles in cloud droplets at a marine boundary layer site. *Atmos. Chem. Phys.* **2015**, *15*, 1367–1383.

21. Weingartner, E.; Burtscher, H.; Baltensperger, H. Hygroscopic properties of carbon and diesel soot particles. *Atmos. Environ.* **1997**, *31*, 2311–2327.

22. Hitzenberger, R.; Berner, A.; Giebl, H.; Drobesch, K.; Kasper-Giebl, A.; Loeflund, M.; Urban, H.; Puxbaum, H. Black carbon (BC) in alpine aerosols and cloud water—Concentrations and scavenging efficiencies. *Atmos. Environ.* **2001**, *35*, 5135–5141.

23. Cozic, J.; Verheggen, B.; Mertes, S.; Connolly, P.; Bower, K.; Petzold, A.; Baltensperger, U.; Weingartner, E. Scavenging of black carbon in mixed phase clouds at the high alpine site Jungfraujoch. *Atmos. Chem. Phys.* **2007**, *7*, 1797–1807.

24. Hallberg, A.; Ogren, J.; Noone, K.; Heintzenberg, J.; Berner, A.; Solly, I.; Kruisz, C.; Reischl, G.; Fuzzi, S.; Facchini, M.; et al. Phase partitioning for different aerosol species in fog. *Tellus B* **1992**, *44*, 545–555.

25. Hallberg, A.; Noone, K.; Ogren, J.; Svenningsson, I.; Flossmann, A.; Wiedensohler, A.; Hansson, H.C.; Heintzenberg, J.; Anderson, T.; Arends, B.; et al. Phase partitioning of aerosol particles in clouds at Kleiner Feldberg. *J. Atmos. Chem.* **1994**, *19*, 107–127.

26. Sellegri, K.; Laj, P.; Dupuy, R.; Legrand, M.; Preunkert, S.; Putaud, J.P. Size-dependent scavenging efficiencies of multicomponent atmospheric aerosols in clouds. *J. Geophys. Res. Atmos.* **2003**, *108*, doi:10.1029/2002JD002749.

27. Kasper-Giebl, A.; Koch, A.; Hitzenberger, R.; Puxbaum, H. Scavenging efficiency of 'aerosol carbon' and sulfate in supercooled clouds at Mt. Sonnblick (3106 m asl, Austria). *J. Atmos. Chem.* **2000**, *35*, 33–46.

28. Gieray, R.; Wieser, P.; Engelhardt, T.; Swietlicki, E.; Hansson, H.C.; Mentes, B.; Orsini, D.; Martinsson, B.; Svenningsson, B.; Noone, K.; et al. Phase partitioning of aerosol constituents in cloud based on single-particle and bulk analysis. *Atmos. Environ.* **1997**, *31*, 2491–2502.

29. Hitzenberger, R.; Berner, A.; Kromp, R.; Kasper-Giebl, A.; Limbeck, A.; Tscherwenka, W.; Puxbaum, H. Black carbon and other species at a high-elevation European site (Mount Sonnblick, 3106 m, Austria): Concentrations and scavenging efficiencies. *J. Geophys. Res. Atmos.* **2000**, *105*, 24637–24645.

30. Heintzenberg, J.; Leck, C. Seasonal variation of the atmospheric aerosol near the top of the marine boundary layer over Spitsbergen related to the Arctic sulphur cycle. *Tellus B Chem. Phys. Meteorol.* **1994**, *46*, 52–67.

31. Winkler, P. The growth of atmosphierc aerosol particles as a function of the relative humidity—II. an improved concept of mixed nuclei. *Aerosol Sci.* **1973**, *4*, 373–387.

32. Ching, J.; Riemer, N.; West, M. Impacts of black carbon mixing state on black carbon nucleation scavenging: Insights from a particle-resolved model. *J. Geophys. Res. Atmos.* **2012**, *117*, doi:10.1029/2012JD018269.

33. Ching, J.; Riemer, N.; West, M. Impacts of black carbon particles mixing state on cloud microphysical properties: Sensitivity to environmental conditions. *J. Geophys. Res. Atmos.* **2016**, *121*, 5990–6013.

34. Ching, J.; Fast, J.; West, M.; Riemer, N. Metrics to quantify the importance of mixing state for CCN activity. *Atmos. Chem. Phys.* **2017**, *17*, 7445–7458.

35. Riemer, N.; West, M.; Zaveri, R.; Easter, R. Simulating the evolution of soot mixing state with a particle-resolved aerosol model. *J. Geophys. Res. Atmos.* **2009**, *114*, D09202, doi:10.1029/2008JD011073.

36. Zaveri, R.A.; Easter, R.C.; Fast, J.D.; Peters, L.K. Model for Simulating Aerosol Interactions and Chemistry (MOSAIC). *J. Geophys. Res. Atmos.* **2008**, *113*, D13204, doi:10.1029/2007JD008782.

37. Zaveri, R.A.; Peters, L.K. A new lumped structure photochemical mechanism for large-scale applications. *J. Geophys. Res. Atmos.* **1999**, *104*, 30387–30415.

38. Schell, B.; Ackermann, I.J.; Binkowski, F.S.; Ebel, A. Modeling the formation of secondary organic aerosol within a comprehensive air quality model system. *J. Geophys. Res.* **2001**, *106*, 28275–28293.

39. Zaveri, R.; Barnard, J.; Easter, R.; Riemer, N.; West, M. Particle-resolved simulation of aerosol size, composition, mixing state, and the associated optical and cloud condensation nuclei activation properties in an evolving urban plume. *J. Geophys. Res. Atmos.* **2010**, *115*, D17210, doi:10.1029/2009JD013616.

40. Kaiser, J.; Hendricks, J.; Righi, M.; Riemer, N.; Zaveri, R.A.; Metzger, S.; Aquila, V. The MESSy aerosol submodel MADE3 (v2. 0b): Description and a box model test. *Geosci. Model Dev.* **2014**, *7*, 1137–1157.

41. Fierce, L.; Bond, T.C.; Bauer, S.E.; Mena, F.; Riemer, N. Black carbon absorption at the global scale is affected by particle-scale diversity in composition. *Nat. Commun.* **2016**, *7*, 12361, doi:10.1038/ncomms12361.

42. Fierce, L.; Riemer, N.; Bond, T.C. Toward reduced representation of mixing state for simulating aerosol effects on climate. *Bull. Am. Meteorol. Soc.* **2017**, *98*, 971–980.

43. Tian, J.; Brem, B.; West, M.; Bond, T.; Rood, M.; Riemer, N. Simulating aerosol chamber experiments with the particle-resolved aerosol model PartMC. *Aerosol Sci. Technol.* **2017**, *51*, 856–867.

Atmosphere **2018**, *9*, 17

44.   Riemer, N.; West, M.; Zaveri, R.; Easter, R.   Estimating black carbon aging time-scales with a particle-resolved aerosol model. *J. Aerosol Sci.* **2010**, *41*, 143–158.

45.   Fierce, L.; Riemer, N.; Bond, T.C. Explaining variance in black carbon's aging timescale. *Atmos. Chem. Phys.* **2015**, *15*, 3173–3191.

46.   United States Environmental Protection Agency. *Report to Congress on Black Carbon*; Technical Report EPA-450/R-12-001; United States Environmental Protection Agency: Washington, DC, USA, 2012.

47.   Petters, M.D.; Kreidenweis, S.M. A single parameter representation of hygroscopic growth and cloud condensation nucleus activity. *Atmos. Chem. Phys.* **2007**, *7*, 1961–1971.

48.   Curtis, J.H.; Riemer, N.; West, M. A single-column particle-resolved model for simulating the vertical distribution of aerosol mixing state: WRF-PartMC-MOSAIC-SCM v1.0. *Geosci. Model Dev.* **2017**, *10*, 4057–4079.

49.   Oshima, N.; Kondo, Y.; Moteki, N.; Takegawa, N.; Koike, M.; Kita, K.; Matsui, H.; Kajino, M.; Nakamura, H.; Jung, J.; et al. Wet removal of black carbon in Asian outflow: Aerosol Radiative Forcing in East Asia (A-FORCE) aircraft campaign. *J. Geophys. Res.* **2012**, *117*, doi:10.1029/2011JD016552.

Article

Machine Learning to Predict the Global Distribution of Aerosol Mixing State Metrics

Michael Hughes [1], John K. Kodros [2], Jeffrey R. Pierce [2], Matthew West [3] and Nicole Riemer [1,*]

[1] Department of Atmospheric Sciences, University of Illinois at Urbana-Champaign, Urbana, IL 61801, USA; hughes18@illinois.edu

[2] Department of Atmospheric Sciences, Colorado State University, Fort Collins, CO 80523, USA; jkodros@atmos.colostate.edu (J.K.K.); jeffrey.pierce@colostate.edu (J.R.P.)

[3] Department of Mechanical Science and Engineering, University of Illinois at Urbana-Champaign, Urbana, IL 61801, USA; mwest@illinois.edu

*   Correspondence: nriemer@illinois.edu

Received: 20 November 2017; Accepted: 5 January 2018; Published: 9 January 2018

Abstract: Atmospheric aerosols are evolving mixtures of chemical species. In global climate models (GCMs), this "aerosol mixing state" is represented in a highly simplified manner. This can introduce errors in the estimates of climate-relevant aerosol properties, such as the concentration of cloud condensation nuclei. The goal for this study is to determine a global spatial distribution of aerosol mixing state with respect to hygroscopicity, as quantified by the mixing state metric χ. In this way, areas can be identified where the external or internal mixture assumption is more appropriate. We used the output of a large ensemble of particle-resolved box model simulations in conjunction with machine learning techniques to train a model of the mixing state metric χ. This lower-order model for χ uses as inputs only variables known to GCMs, enabling us to create a global map of χ based on GCM data. We found that χ varied between 20% and nearly 100%, and we quantified how this depended on particle diameter, location, and time of the year. This framework demonstrates how machine learning can be applied to bridge the gap between detailed process modeling and a large-scale climate model.

Keywords: aerosol modeling; mixing state; machine learning

1. Introduction

Field measurements show that individual aerosol particles are a complex mixture of a wide variety of species, such as soluble inorganic salts and acids, insoluble crustal materials, trace metals, and carbonaceous materials [1,2]. To characterize this mixture, the term "aerosol mixing state" is frequently used. This, in general, comprises both the distribution of chemical compounds across the aerosol population ("population mixing state") and the distribution of chemical compounds within and on the surface of each particle ("morphological mixing state").

Both the population mixing state and the morphological mixing state are of importance for aerosol impacts, including chemical reactivity, cloud condensation nuclei (CCN) activity, and aerosol optical properties [3]. However, the morphological mixing state is beyond the scope of this study. We will focus here exclusively on the population mixing state, and refer to it for brevity as "mixing state". In this context, the terms "internal" and "external" mixture are frequently used. An external mixture consists of particles that each contain only one species, which may be different for different particles. In contrast, an internal mixture describes a particle population where different species are present within one particle. If all particles consist of the same species mixture, and the relative abundances are identical, the term "fully internal mixture" is commonly used. Considering that aerosol populations contain particles of many different sizes, we can define these terms for the entire populations (comprising all particle sizes) or for individual size ranges. An aerosol population might be approximately internally mixed for a certain size range, but the internal mixture assumption might not be fulfilled if a large size range is considered. While mixing state can impact both CCN properties and optical properties, here we target CCN properties, and interpret aerosol "species" in terms of hygroscopicity.

An example of an external mixture is shown in Figure 1a, which represents a particle population consisting of six particles, with the blue and the red color symbolizing two different aerosol species with different hygroscopicities. A fully internal mixture is shown in Figure 1d. In reality, aerosol populations assume mixing states that are neither fully externally nor internally mixed, as depicted by Figure 1b,c. Note that each of the four populations in Figure 1 contains the same total amounts of the two species, but the species distribution amongst the particles differs.

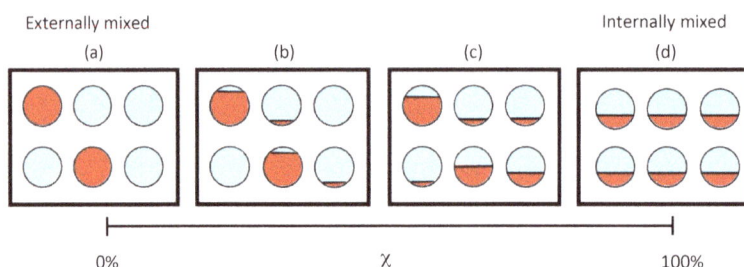

Figure 1. Schematic of aerosol mixing states for four different aerosol populations that have the same bulk composition. The blue and red color represent aerosol species with different hygroscopicity: (**a**) fully external mixture; (**b,c**) intermediate mixing states; and (**d**) internal mixture. The mixing state metric χ measures the degree of internal mixing, ranging from 0% to 100%.

Aerosol mixing state is challenging to represent in atmospheric aerosol models. The most rigorous approach is the particle-resolved approach by Riemer et al. [4], which explicitly resolves population mixing state. However, this method is too computationally demanding for routine use in spatially-resolved regional or global chemical transport models. Instead of resolving the full aerosol mixing state, regional and global models therefore use distribution-based methods, commonly known as modal and sectional models [5–7]. An inherent assumptions of these methods is that within one mode or within one size section, the aerosol particles are assumed to be internally mixed. This assumption can lead to misprediction in climate-relevant aerosol properties such as CCN concentrations and optical properties [8–12].

To illustrate this concept, Figure 2 shows the global distribution of the fraction of hygroscopic species (sulfate, ammonium, sea-salt, and aged organics) as simulated by GEOS-Chem-TOMAS for the month of January 2010 for particles of ~358 nm. For areas where this fraction is close to 100% (oceans) or close to 0% (parts of the Saharan desert), the aerosol consists essentially of only hygroscopic or only non-hygroscopic species, respectively, so mixing state is not an issue in these areas. However, there are many regions such as the continental US or Europe where the fraction is between the two extremes. For these regions, the question is, given the local conditions, what degree of internal/external mixing is most likely? Our approach seeks to answer this question for different particle sizes, different geographic locations, and different seasons.

To quantify the degree of internal/external mixture Riemer and West [13] introduced the mixing state index χ. This is a scalar quantity that varies between 0% for completely external mixtures and 100% for completely internal mixtures, as indicated by Figure 1. It can be calculated from per-particle species mass fractions (see Section 2.1), which requires either simulations with computationally expensive, high-detail aerosol models [13,14] or observations with a sophisticated suite of instruments [15,16].

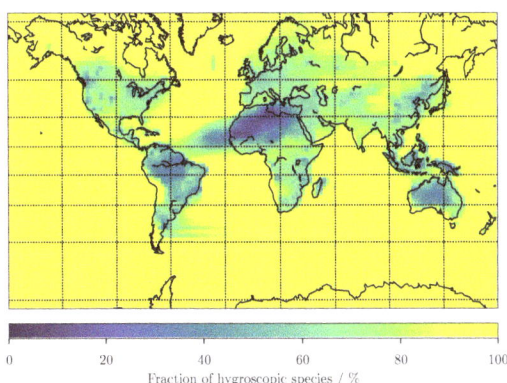

Figure 2. Global distribution of fraction of hygroscopic species as simulated by GEOS-Chem-TOMAS for the month of January for particles of ~358 nm.

Ching et al. [14] quantified the relationship of mixing state index χ and the error in CCN concentrations when neglecting mixing state information by assuming a fully internal mixture. The study shows that for more externally mixed populations (χ below 20%) neglecting mixing state leads to errors up to 150%, whereas for populations with χ larger than 75%, the error vanishes (Figure 3). To establish this relationship, Ching et al. [14] used particle-resolved simulations from a 0-D box model scenario library that represented a suite of idealized urban plume scenarios. Thus far, no studies have calculated spatial distributions of the mixing state parameter. This, however, is important for understanding where global models may need to take mixing state into account.

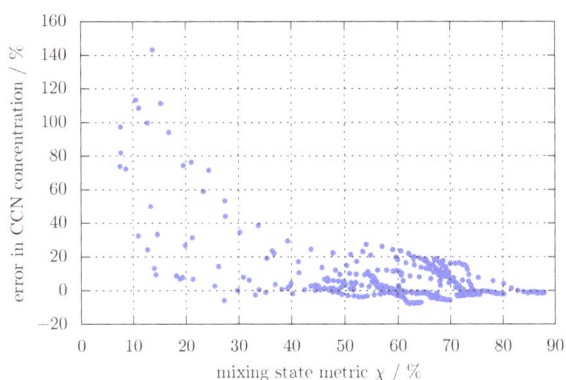

Figure 3. Relative error in CCN concentration when neglecting aerosol mixing state as a function of aerosol mixing state index χ. Each dot represents an aerosol population from Ching et al. [14]. CCN concentration was evaluated at a supersaturation of 0.6%.

The goal of this study is therefore to produce the first global distribution of mixing state parameter χ. This will allow us to map out areas on the globe where low χ values can be expected—these are the areas where we expect large errors in CCN prediction when using a simplified aerosol model that does not or not fully resolve aerosol mixing state. Conversely, it is informative to delineate areas where the mixing state approaches an internal mixture, as for these areas assuming an internal mixture would be appropriate for CCN predictions.

As mentioned before, it is currently not feasible to directly run a particle-resolved aerosol model on a global scale, which would be needed to create a global map of χ directly. We therefore propose an approach that combines particle-resolved modeling and output from a global chemical transport model with machine learning techniques, as outlined in Figure 4. This involves the construction of a scenario library of particle-resolved simulations using the PartMC-MOSAIC, which cover a wide range of conditions that are expected to be encountered in different environments around the globe. This dataset is then used to train a model of χ using machine learning techniques. Importantly, the features of this model are dictated by the list of variables that are known to the global scale model, in our case GEOS-Chem-TOMAS [17,18].

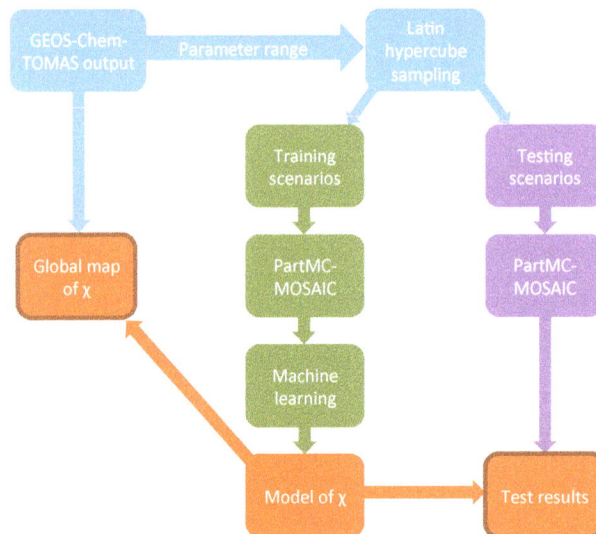

Figure 4. Schematic of the learning architecture used to train, test, and use the machine-learning model.

Many examples exist in the aerosol modeling literature where parameters for coarser models were derived on the basis of box model simulations that capture certain microphysical processes in detail [19]. However, the choice of the explanatory variables (features) and the fitting of the coarse model were typically done "by hand". This approach works well if the relevant parameter space is low-dimensional so that a few features can be identified that govern a certain process. In our case, there are many relevant variables that could potentially influence χ, hence machine learning methods represent an appropriate tool.

The remainder of the paper is structured as follows: Section 2 describes the tools and methods that are used this study, including the mixing state metric χ, the particle-resolved aerosol model PartMC-MOSAIC, the dataset from the global model GEOS-Chem-TOMAS, the simulations that yield the training and testing dataset, and the machine learning methods. Section 3 presents the global maps of mixing state parameter χ as obtained from the machine learning procedure. Section 4 concludes our results and provides a perspective for future work.

2. Methods

2.1. Mixing State Metric χ

We quantified aerosol mixing state with the framework discribed in Riemer and West [13], specifically using the mixing state metric χ. This was inspired by diversity metrics used in other disciplines such as ecology [20], economics [21], neuroscience [22], and genetics [23].

Given a population of N aerosol particles, each consisting of some amounts of A distinct aerosol species, the mixing state metrics can be determined if the masses of species a in particle i are known, denoted by μ_i^a, for $i = 1, \ldots, N$, and $a = 1, \ldots, A$. From this quantity, all other related quantities can be calculated, as described by Riemer and West [13] and here listed in Table 1. The diversity metrics can then be constructed as summarized in Table 2.

Table 1. Aerosol mass and mass fraction definitions and notation, used to construct the diversity metrics shown in Table 2. The number of particles in the population is N, and the number of species is A. This table is taken from Riemer and West [13].

Quantity	Meaning
μ_i^a	Mass of species a in particle i
$\mu_i = \sum_{a=1}^{A} \mu_i^a$	Total mass of particle i
$\mu^a = \sum_{i=1}^{N} \mu_i^a$	Total mass of species a in population
$\mu = \sum_{i=1}^{N} \mu_i$	Total mass of population
$p_i^a = \dfrac{\mu_i^a}{\mu_i}$	Mass fraction of species a in particle i
$p_i = \dfrac{\mu_i}{\mu}$	Mass fraction of particle i in population
$p^a = \dfrac{\mu^a}{\mu}$	Mass fraction of species a in population

Table 2. Definitions of aerosol mixing entropies, particle diversities, and mixing state index. In these definitions, we take $0 \ln 0 = 0$ and $0^0 = 1$. This table is taken from Riemer and West [13].

Quantity	Name	Units	Range	Meaning
$H_i = \sum_{a=1}^{A} -p_i^a \ln p_i^a$	Mixing entropy of particle i	—	0 to $\ln A$	Shannon entropy of species distribution within particle i
$H_\alpha = \sum_{i=1}^{N} p_i H_i$	Average particle mixing entropy	—	0 to $\ln A$	average Shannon entropy per particle
$H_\gamma = \sum_{a=1}^{A} -p^a \ln p^a$	Population bulk mixing entropy	—	0 to $\ln A$	Shannon entropy of species distribution within population
$D_i = e^{H_i} = \prod_{a=1}^{A} (p_i^a)^{-p_i^a}$	Particle diversity of particle i	Effective species	1 to A	Effective number of species in particle i
$D_\alpha = e^{H_\alpha} = \prod_{i=1}^{N} (D_i)^{p_i}$	Average particle (alpha) species diversity	Effective species	1 to A	Average effective number of species in each particle
$D_\gamma = e^{H_\gamma} = \prod_{a=1}^{A} (p^a)^{-p^a}$	Bulk population (gamma) species diversity	Effective species	1 to A	Effective number of species in the bulk
$\chi = \dfrac{D_\alpha - 1}{D_\gamma - 1}$	Mixing state index	—	0% to 100%	Degree to which population is externally mixed ($\chi = 0\%$) versus internally mixed ($\chi = 100\%$)

Based on the per-particle mass fractions, the particle diversity D_i can be calculated, which can be interpreted as the number of "effective species" of particle i. For a particle consisting of A species, the particle diversity D_i can be maximally A, which occurs when all A species are present in equal mass fractions. From the D_i values of all particles, we can determine the population-level quantities

D_α and D_γ, with D_α being the average effective number of species in each particle, and D_γ being the effective number of species in the bulk. The mixing state index χ is defined as

$$\chi = \frac{D_\alpha - 1}{D_\gamma - 1}. \tag{1}$$

The mixing state index χ varies from 0% (a fully externally mixed population) to 100% (a fully internally mixed population). Since χ has the intuitive interpretation of the "degree of internal mixing", it can be used as a metric for error quantification, i.e., to determine the magnitude of error that is introduced in estimating aerosol impacts when neglecting mixing state information. This was shown by Ching et al. [14] for the example of CCN concentration, as illustrated in Figure 3.

The definition of "species" for calculating the mass fractions depends on the application. It can refer to individual chemical species, as in the studies by Riemer and West [13], Healy et al. [15], O'Brien et al. [16], Giorio et al. [24], and Fraund et al. [25]. Alternatively, it can refer to species groups, as in Dickau et al. [26] who quantified mixing state with respect to volatile and non-volatile components. Since we are concerned with CCN properties in this paper, we will group the chemical model species according to hygroscopicity, defining two species groups. Black carbon (BC), primary organic aerosol (POA), and freshly emitted mineral dust are combined into one surrogate species, since their hygroscopicities are very low. All other model species (inorganic and secondary organic aerosol species) are combined into a second surrogate species. The mixing state index χ is calculated from these two surrogate species. Note that calculating χ based on the two surrogate species does not bias the value of χ in a systematic way compared to the value based on the individual chemical species. A χ value close to 0% can be interpreted as the hygroscopic and non-hygroscopic species existing in different particles, whereas a χ value close to 100% would correspond to an aerosol population where all particles contain the same amount of hygroscopic and non-hygroscopic species.

2.2. Particle-Resolved Aerosol Modeling

A detailed model description of stochastic particle-resolved aerosol model PartMC-MOSAIC is provided by Riemer et al. [4]. In summary, PartMC (Particle-resolved Monte Carlo) is a zero-dimensional aerosol model, which explicitly tracks the composition of many individual particles within a well-mixed computational volume. This computational volume is assumed to be representative for a much larger air parcel within the planetary boundary layer. The processes of emission, dilution with the background, and Brownian coagulation are simulated with a stochastic Monte Carlo approach. To improve efficiency of the method, we use weighted particles in the sense of DeVille et al. [27] and efficient stochastic sampling methods [28].

PartMC is coupled with the aerosol chemistry model MOSAIC (Model for Simulating Aerosol Interactions and Chemistry) [29]. This includes the gas phase photochemical mechanism CBM-Z [30], the Multicomponent Taylor Expansion Method (MTEM) for estimating activity coefficients of electrolytes and ions in aqueous solutions [31], the multi-component equilibrium solver for aerosols (MESA) for solid–liquid partitioning within particles [32] and the adaptive step time-split Euler method (ASTEM) for dynamic gas–particle partitioning over the size- and composition-resolved aerosol [29]. To simulate secondary organic aerosol (SOA) the SORGAM scheme is used [33]. The CBM-Z gas phase mechanism includes 77 gas species. MOSAIC treats key aerosol species including sulfate (SO_4), nitrate (NO_3), ammonium (NH_4), chloride (Cl), carbonate (CO_3), methanesulfonic acid (MSA), sodium (Na), calcium (Ca), other inorganic mass (OIN), BC, POA, and SOA. The model species OIN represents species such as SiO_2, metal oxides, and other unmeasured or unknown inorganic species. Our SOA model species include reaction products of aromatic precursors, higher alkenes, α-pinene and limonene. In this study, PartMC includes condensation/evaporation of vapors to/from particles and coagulation between particles. It does not included nucleation in this study, and the limitations on our results will be discussed throughout.

PartMC-MOSAIC has been used in the past for process studies of mixing state impacts on aerosol properties in various environments. For example, Tian et al. [34] investigated the aging of aerosol particles in a ship plume. Ching et al. [12] quantified the response of cloud droplet number concentration to changes in emissions of black-carbon-containing particles, and Mena et al. [35] carried out plume-exit modeling to determine cloud condensation nuclei activity of aerosols from residential biofuel combustion.

2.3. GEOS-Chem-TOMAS Dataset

To provide initial concentrations of gas-phase and size-resolved aerosol-phase species in a large-scale global model, we use the Goddard Earth Observing System chemical-transport model, GEOS-Chem, version 10.01 [36] (http://acmg.seas.harvard.edu/geos/) coupled with the TwO Moment Aerosol Sectional (TOMAS) microphysics scheme [17]. We simulated the year 2010 with re-analysis meteorology fields from GEOS5 (http://gmao.gsfc.nasa.gov). Simulations included a horizontal resolution of $2° \times 2.5°$ and 47 vertical layers. GEOS-Chem includes tracers for 52 gas-phase species. Standard emission setup is described in the study by Kodros et al. [18]. We used the 15-bin version of TOMAS, with size sections ranging from approximately 3 nm to 10 μm. TOMAS includes tracers for aerosol number concentration, sulfate, organic aerosol, black carbon, sea salt, and dust. Nucleation in the simulations follows a ternary nucleation scheme involving water, sulfuric acid, and ammonia following the parameterization of Napari et al. [37], scaled with a global tuning factor of 10–5 [38,39]. When ammonia mixing ratios are less than 1 pptv, the model defaults to a binary nucleation scheme (sulfuric acid and water) [40]. Detailed descriptions of aerosol microphysics included in TOMAS can be found in Adams and Seinfeld [17], Lee et al. [41], and Lee and Adams [42]. GEOS-Chem-TOMAS has been evaluated against observed aerosol size distributions [43,44].

2.4. Design of the Training and the Testing Scenarios

At the core of the machine learning framework is the design of a scenario library of particle-resolved simulations to create a large number of aerosol populations with different compositions and different mixing states. Scenario libraries that we developed in previous work [10,12,45] focused on urban environments, and in particular on the aging process of carbonaceous aerosol by coagulation and condensation of secondary aerosol. Here, we expanded the list of aerosol types by including sea salt aerosol and dust emissions.

We did not include the process of particle nucleation in this set of training simulations because there are still significant uncertainties about the treatment of particle-level post-nucleation growth mechanisms [46]. The lack of nucleation in our training library can be expected to introduce errors into our global mixing state predictions in the smaller size bins where particles may be influenced by nucleation and growth. In particular, we expect that true χ values in the Aitken and accumulation modes will generally be lower than our predicted values in areas with pre-existing non-hygroscopic particles (e.g., from combustion) where significant nucleation occurs because freshly nucleated particles will then create a more-externally mixed population.

All scenarios used a simulation time of 24 h, starting at 6:00 a.m. local time, with output being saved every 10 min. We used 10,000 computational particles for each simulation. The initial conditions for aerosol and gas phase were the same for all scenarios and are identical to Zaveri et al. [8]. Specifically, the aerosol initial condition consisted of Aitken and Accumulation mode with internally mixed ammonium sulfate, secondary organic aerosol, and trace amounts of black carbon, as listed in Table 3. Although the initial conditions were fixed in these scenarios, these particles generally evolved substantially over the course of the simulations. However, we cannot rule out that this choice influenced our results, and we will address this in future work by introducing more variability to the design of the initial condition.

Table 3. Number concentration, N_a, of the initial aerosol population. The aerosol size distributions are assumed to be lognormal and defined by the geometric mean diameter, D_g, and the geometric standard deviation, σ_g.

Initial/Background	N_a/cm^{-3}	$D_g/\mu m$	σ_g	Composition by Mass
Aitken mode	1800	0.02	1.45	49.64% $(NH_4)_2SO_4$ + 49.64% SOA + 0.72% BC
Accumulation mode	1500	0.116	1.65	49.64% $(NH_4)_2SO_4$ + 49.64% SOA + 0.72% BC

Twenty-five input parameters were varied between scenarios to represent a range of environmental conditions with different levels of gas phase emissions and emissions of primary aerosol particles to allow for large variations in the mixing state evolution. Latin hypercube sampling [47] was used to provide an efficient sampling across this high-dimensional space. The details of our setup are listed in Table 4. The input parameter space was sampled so that the resulting distributions of simulated variables, such as gas phase and bulk aerosol concentrations, were similar to that of the corresponding distribution in the output data of GEOS-Chem-TOMAS. The distributions need not be identical, but they must be similar enough that the model that is trained from the PartMC library is not required to extrapolate far outside the parameter range on which it was trained.

Table 4. List of input parameters and their sampling ranges and procedures to construct the scenario library. See the main text for details.

	Range	Sampling Method
Environmental Variable		
RH	10–100%	uniform within specified ranges [1]
Latitude	70° S–70° N	uniform
Day of Year	1–365	uniform
Temperature	based on latitude and day of year [2]	uniform
Dilution rate	$1.5 \times 10^{-5}\,s^{-1}$	constant
Mixing height	400 m	constant
Gas phase emissions		
SO_2, NO_x, NH_3, VOC	0–100% of emissions in Riemer et al. [4]	non-uniform [3]
Carbonaceous Aerosol Emissions (one mode) [4]		
D_g	25–250 nm	uniform
σ_g	1.4–2.5	uniform
BC/OC mass ratio	0–100%	non-uniform [3]
E_a	0–$1.6 \times 10^7\,m^{-2}\,s^{-1}$	non-uniform [3]
Sea Salt Emissions (two modes) [5]		
$D_{g,1}$	180–720 nm	uniform
$\sigma_{g,1}$	1.4–2.5	uniform
$E_{a,1}$	0–$1.69 \times 10^5\,m^{-2}\,s^{-1}$	non-uniform [3]
$D_{g,2}$	1–6 μm	uniform
$\sigma_{g,2}$	1.4–2.5	uniform
$E_{a,2}$	0–$2380\,m^{-2}\,s^{-1}$	non-uniform [3]
OC fraction	0–20%	uniform
Dust Emissions (two modes) [6]		
$D_{g,1}$	80–320 nm	uniform
$\sigma_{g,1}$	1.4–2.5	uniform
$E_{a,1}$	0–$586{,}000\,m^{-2}\,s^{-1}$	non-uniform [3]
$D_{g,2}$	1–6 μm	uniform
$\sigma_{g,2}$	1.4–2.5	uniform
$E_{a,2}$	0–$2380\,m^{-2}\,s^{-1}$	non-uniform [3]
hygroscopicity (κ)	0.001–0.031	uniform

Specific information about the individual variables listed in Table 4 is as follows. (1) Relative humidity was sampled from a range of 10% to 100%, using two uniform distributions: The range 10% to 60% comprised 25% of the sampled RH values, while the range 60% to 100% made up the remaining 75%. The 60%-to-100% range was sampled more heavily because the global average RH is 73% [48]. (2) We obtained monthly global temperatures from the NCEP/NCAR reanalysis data [48] for the years 1981–2010. For each latitude ϕ and month m, we determined the mean temperature $\bar{T}(\phi, m)$ and the standard deviation $\sigma(\phi, m)$, taken over all longitudes and all 30 years in the dataset. The temperature was then uniformly sampled from a range of $\bar{T}(\phi, m) \pm 3\sigma(\phi, m)$, if $3\sigma > 8$ K, or $\bar{T}(\phi, m) \pm 8$ K otherwise. For simplicity, the sampled temperature was kept constant for the duration of the 24-h simulation. (3) The emission fluxes of aerosol and gases were sampled from a non-uniform distribution by multiplying the maximum emission rate with a random number between 0 and 1 raised to the fourth power. This ensured that our sampling space was skewed towards the lower emission rates, while still retaining some scenarios that represent highly polluted conditions. (4) The aerosol distributions for the emitted carbonaceous particles were prescribed as log-normal, with geometric mean diameter D_g and geometric standard deviation σ_g. (5) Sea salt particles were emitted wet rather than dry. For composition, a simplified mixture of 53.89% Cl^-, 38.56% Na^+, and 7.55% SO_4^{2-} by mass was, based on the mass ratio of Cl^- to SO_4^{2-} of 7.15 in seawater ([49], p. 384) and adding enough Na^+ to balance the charges. Additionally, because organic species are a substantial but variable component of sea salt aerosols Vignati et al. [50], a variable amount OC is added, making up 0% to 20% of the mass of the particles. One third of all scenarios had no sea salt emissions. (6) One third of all scenarios had no dust emissions.

A total of 1000 scenarios were created in this fashion to make up the training library. Since we are saving the output every 10 min of each 24-h simulations, this yields 144,000 particle populations for our training dataset. For testing purposes, a second library of 240 scenarios (34,560 populations) was created in the same manner to gauge the accuracy of the model, using the same distributions, but with different combinations of parameters. This provides a check against overfitting, in which the model that is learned has been fit to the stochastic noise in the training set, resulting in poor predictive performance for any other data set.

2.5. Machine Learning as Applied to PartMC

Machine learning refers to a variety of algorithms that are used to identify and model patterns in large datasets, and then use these models to make predictions. It has proven to be a diverse set of tools in the atmospheric sciences. Past applications have included interpreting remote sensing data [51], estimating uncertainty in aerosol optical depth data [52], prediction of aerosol-induced health impacts [53], and forecasting solar radiation for energy generation [54].

Our model predicts χ in a single global-model grid cell, given inputs of the GEOS-Chem-TOMAS variables in that grid cell. We present two variants of this model, one that predicts χ for the bulk aerosol population, and one that predicts χ for each size bin of the global model. A total of 34 input feature variables were used, including gas concentrations, aerosol mass concentrations, aerosol number concentration, solar zenith angle, and latitude. Note that the mass concentrations of the different aerosol species are not lumped into hygroscopic and non-hygroscopic species for this purpose, but are used individually. MOSAIC species were mapped to TOMAS species when training the model. At each horizontal location, we computed the average predicted χ over grid layers up to 840 mb.

We used gradient-boosted regression trees ([55], Chapter 10) as the machine-learning algorithm for this study, because this is a well-understood algorithm that offers good predictive accuracy with moderate computational cost and is able to perform automatic feature selection during training. Gradient boosting methods [56,57] form a prediction model as a sequence of weak prediction models, each of which fits the residual of the previous predictors in the sequence and thus serves to slightly improve the overall prediction accuracy.

For gradient-boosted regression trees, the weak prediction models are regression trees ([55], Section 9.2.2), which predict an output value as a tree of decisions on input values. For example, a single depth-2 regression tree for χ might have a first decision of "(latitude > 50°)?", and if this is true it might have a second decision of "([SO$_2$] < 30 ppb)?", and if this is false then it outputs $\chi = 0.8$. A depth-n tree allows up to n-way interactions between feature variables.

We used the implementation of gradient-boosted regression trees from scikit-learn [58]. The model was trained on the training data set and then its performance was evaluated on the testing data set (see Section 2.4). We used a least-squares loss function and all of our gradient-boosted models used 400 decision trees as submodels, as this was sufficient to obtain the best performance on the testing data set. We tested different tree depths, as shown in Figure 5 (left). Similar to many applications ([55], Chapter 10) we found that tree depths between 4 and 8 worked well, and we used depth 8 for the final model used in the remainder of the paper to give good prediction accuracy with reasonable computational speed.

The performance of our final model is shown in Figure 5 (right). In this figure, a perfect model would be the red 1:1 line. Our model has $R^2 = 0.94$ and a mean error of 1.67%. The maximum error for any testing scenario is 13.02%.

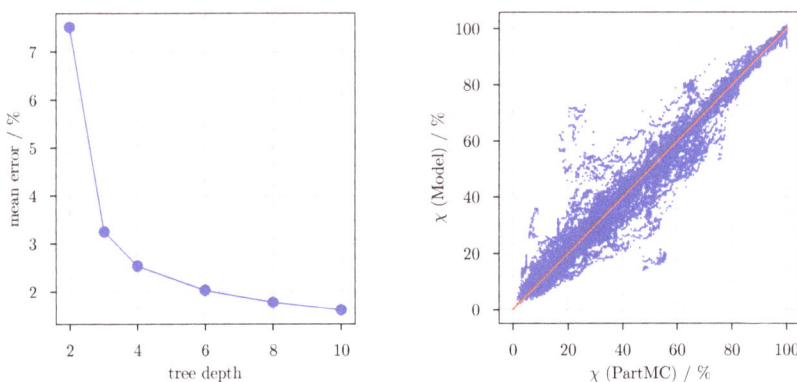

Figure 5. (**Left**) Mean error in the predicted χ values from the testing data set as a function of tree depth for the gradient boosted regression tree model; and (**Right**) true χ values versus model-predicted values for our final model (corresponding to depth 8 in the left panel).

3. Results

3.1. Predicting χ for the Bulk Aerosol Population

Using output from GEOS-Chem-TOMAS and the model for χ that was trained on particle-resolved data, we can now produce global distributions of χ. Figure 6 shows examples of such distributions using six-hourly output from GEOS-Chem-TOMAS and comparing two different dates, 06:00 UTC on 1 January 2010, and 1 June 2010. Note that χ was calculated based on the entire size range of aerosol particles and hence if coarse-mode particles and fine-mode particles have different compositions, this would result in a lower χ value (more externally mixed), even if the course and fine modes each had higher χ values (more internally mixed). Because χ is a mass-weighted quantity, the χ values for all sizes are dominated by the coarse mode mixing state.

We determined χ only for grid cells that contained between 5% and 95% hygroscopic material, hence excluding areas where essentially only one surrogate species (either hygroscopic or non-hygroscopic) was present. We see from Figure 6 that these excluded areas cover much of the oceans, and much of the Sahara and other deserts. This exclusion is because it is meaningless to discuss the mixing state between hygroscopic and non-hygroscopic material when there is essentially only a single type present.

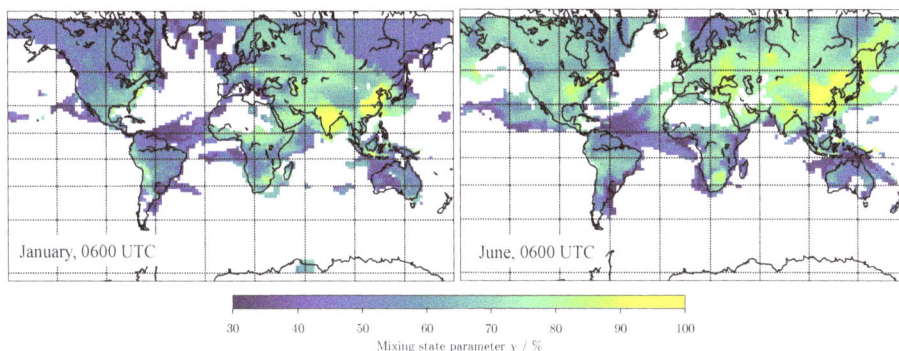

Figure 6. Global distribution of χ from the machine-learning model, at 06:00 UTC on January 1 (**left**), and June 1 (**right**), 2010. The model used to predict χ here is trained on the PartMC output that includes the entire particle population.

For both dates, the predicted χ varied from 30% to 97%. High χ values existed over industrial source regions including East China, India, and the Eastern and Midwestern United States, with χ approaching 100%. This result can be interpreted that in these regions non-hygroscopic (mainly freshly emitted carbonaceous aerosol) and hygroscopic (mainly secondary) aerosol species are mixed together within the same particle. This prediction is consistent with the fact that highly polluted areas have extremely short aging timescales for carbonaceous emissions [9,59], and so—at least on the scale of the grid resolution used here in GEOS-Chem-TOMAS—assuming an internal mixture of non-hygroscopic and hygroscopic species is appropriate. However, we note that the nucleation is frequently observed in many of these regions, and hence our training data that omitted nucleation may be overestimating χ in some of these regions.

Plumes of aerosol with relatively high χ values of around 80% can also be seen to be transported over the oceans in the outflow of continents, e.g., east of China. This was more prominent for 1 June over the Northern Hemisphere, which is consistent with a larger availability of photochemically produced secondary species that can condense on the originally non-hygroscopic carbonaceous particles, thereby moving the population towards a more internal mixture.

3.2. Predicting χ for Individual Size Bins

Rather than including the entire PartMC particle populations for the machine-learning process, we can also group the PartMC output according to particle size first, and then train a separate model for χ for each individual size category. This altered the input feature variables for the model from bulk aerosol mass concentrations and total number concentration to the mass aerosol mass concentrations and number concentration within the size range.

Choosing the TOMAS size bins, we obtained results for the testing data set, as shown in Table 5. The R^2 values are generally lower than for the case without size resolution, which is expected since for each size bin a smaller set of particles is available for learning the model. In fact, for size bins 1–6 (corresponding to dry diameters from ~3–30 nm), the R^2 value were very low, so that we only discuss the results for size bins 7 and larger (dry diameters above ~30 nm). In future work, we plan to refine these results by increasing the particle samples in the smaller size bins.

Table 5. Error statistics for prediction error in size-resolved χ.

Bin Number	7	8	9	10	11	12	13	14	15
Bin median diameter (nm)	56.3	89.4	142	225.3	357.7	567.8	901.4	2024	6424
R^2	19.68%	65.31%	79.68%	87.63%	90.87%	89.45%	81.87%	70.94%	36.42%
Mean error	12.55%	9.13%	7.21%	5.99%	5.16%	5.51%	6.91%	8.64%	11.86%

Figure 7 shows the global maps of size-resolved χ, based on GEOS-Chem-TOMAS output fields averaged for the months of January and July for size bin 8 (χ_8, bin median diameter of ~90 nm) and size bin 14 (χ_{14}, bin median diameter of ~2 μm). Other months had very similar distributions and are not shown.

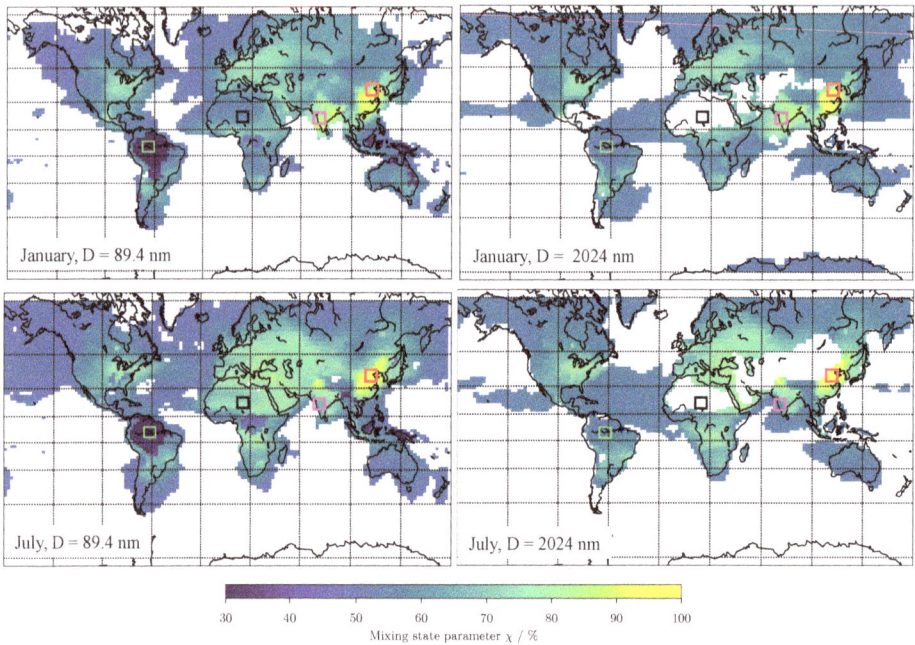

Figure 7. Global distribution of size-resolved χ values from the machine-learning model based GEOS-Chem-TOMAS inputs for the months of: January (**top**); and July (**bottom**). (**Left**) χ for size bin 8, bin median diameter is 89.4 nm. (**Right**) χ for size bin 14, bin median diameter is 2024 nm. The colored boxes show the regions over which data were averaged for display in Figure 9.

The distribution for χ_8 shows low values of approximately 20% in the Amazon basin, central Africa, and Indonesia. These are areas with large contribution of carbonaceous aerosol from biomass burning. The low χ_8 value in this size range means that the carbonaceous material is externally mixed from other (more hygroscopic) aerosol in these areas. In contrast, internally mixed aerosol is predicted for East Asia and India. For January, plumes with internally mixed aerosol extend from India into the Arabian Sea (winter Monsoon), while, for July, this is not the case (summer Monsoon).

Due to the setup of our scenario library, we need to be aware of some biases that we might introduce with our choices. By using the same initial condition for all simulations, we may underestimate χ in locations where the local emissions are relatively small but different to the initial conditions and where the conditions are not conducive for secondary aerosol formation. Conversely, the χ values for 90-nm particles in the polluted regions (Eastern US, India, Europe, and China) are likely biased high, since our scenario library does not include nucleation, as mentioned in Section 2.4. Nucleation events and growth to 90 nm are routinely observed in these areas, along with primary carbonaceous emissions at these sizes, which may be fresh or aged. Overall, nucleation events are likely to decrease χ in this size range, since a more external mixture would be created. We plan to quantify the impacts of both the initial condition choice as well as the impacts of nucleation on the machine learning procedure in future work.

Figure 8 shows the composition of the aerosol in these two size bins as a fraction of hygroscopic species. This figure confirms that for large areas over the oceans the aerosol consists of only hygroscopic material, and the 2 μm bin over the desert areas contains only non-hygroscopic material, which is the reason why χ was not determined for those areas.

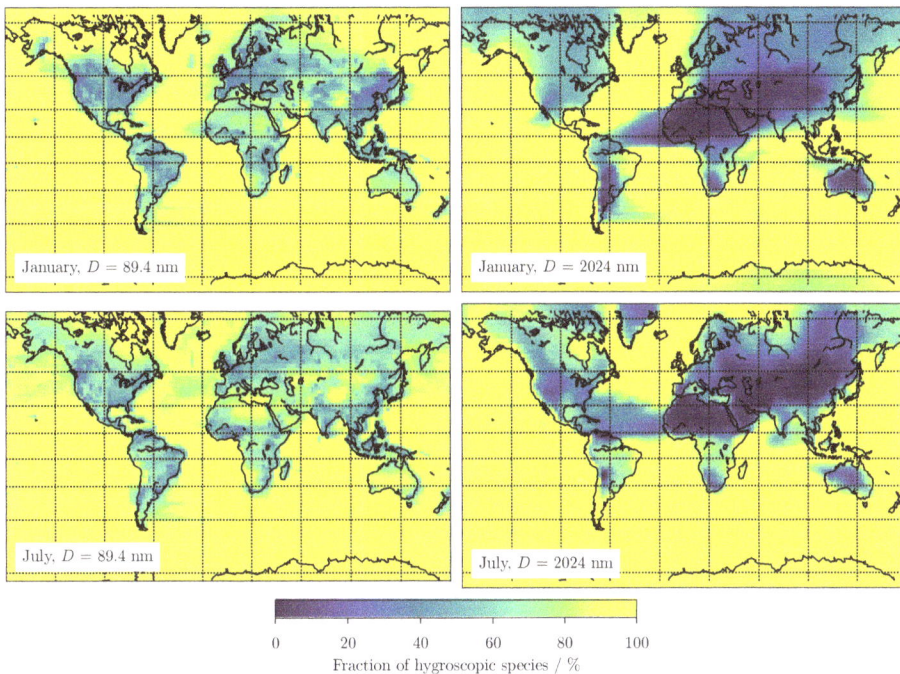

Figure 8. Global distribution of fraction of hygroscopic species as simulated by GEOS-Chem-TOMAS for the months of: January (**top**); and July (**bottom**). (**Left**) χ for size bin 8, bin median diameter is 89.4 nm. (**Right**) χ for size bin 14, bin median diameter is 2024 nm.

Figure 9 shows the size-resolved χ for selected regions, which are indicated in Figure 7 as colored boxes. This figure confirms the strong size dependence for the Amazon region while the other regions do not show a pronounced size dependence. Differences between summer and winter are noticeable for North East China (χ is higher in July), Sahara (χ is higher in July), and Central India (χ is lower in July). A possible explanation for the higher χ values in summer for North East China and the Sahara is a generally larger production of condensable gases during the Northern Hemisphere summer, which help creating a more internal mixture. The lower χ values over India during summer might be related to the Monsoon, which removes both condensable gases as well as aged aerosol.

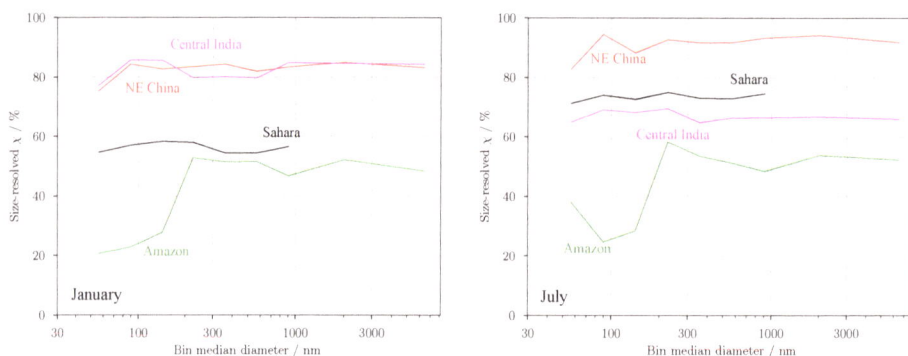

Figure 9. Size-resolved values of χ forselected regions in: January (**left**); and July (**right**). See Figure 7 for the location of these regions.

It is interesting to note that the smallest χ value is about 20%, so the classic "external mixture" with χ approaching zero is not found anywhere in these examples. We want to emphasize again that the χ values for smaller sizes in polluted regions such as North East China and Central India may be overestimated because our training data set does not include the process of nucleation. Including nucleation is generally expected to create a more external mixture, since freshly nucleated (hygroscopic) particles would co-exist with carbonacous particles in these environments.

4. Conclusions

This paper presents the first estimate of spatial distribution of aerosol mixing state over the globe as quantified by the mixing state metric χ. We defined this metric to estimate the degree to which hygroscopic and non-hygroscopic species are mixed on a per-particle basis, with $\chi = 0\%$ being completely externally mixed and $\chi = 100\%$ being completely internally mixed. We obtained this global estimate by training a machine-learning model of χ on detailed particle-resolved box model data, and then applying the model to GCM output to predict χ globally.

In some parts of the globe, the aerosol appeared to be quite externally mixed, with χ values as low as 20%, suggesting that an external-mixing assumption is likely to be valid there. This was the case for the size range below 150 nm in regions where biomass burning aerosol dominated, such as the Amazon Basin, Central Africa, and Indonesia. In contrast, the mixing state index χ reached values of 90% for polluted regions in East Asia in July, indicating that an internally-mixed assumption is appropriate for those regions, at least for the spatial resolution of the GCM that was used here. In much of the globe, however, the aerosol mixing state was not clearly internally or externally mixed, which may indicate that assuming either limiting case could lead to significant errors. Previous work by Ching et al. [14] can be used to link the global maps of χ values from this study with estimated errors for CCN concentrations. For the χ values between 30% and 100% found in this study, assuming an internal mixture would introduce an overestimation in CCN concentrations of up to 50%, with the error decreasing to a few percent for χ larger than 80%. For χ values lower than 20%, errors in CCN concentration of up to 100% can occur, but these χ values did not occur in our study. The scenarios in the study by Ching et al. [14] were focused on the aging of carbonaceous aerosol and and therefore did not encompass the full range of conditions that might be encountered around the globe. Nevertheless, they provide guidance of how the predicted distribution of χ values relates to expected errors in CCN predictions when assuming an internal mixture.

While the methodology used in this paper is effective at extrapolating high-detail simulation output to the global scale, it is important to understand the limitations of such a method. Roughly speaking, our model takes the GCM output variables in each grid cell and infers the mixing state χ value from particle-resolved box model simulations with similar corresponding state variables. This could deliver inaccurate χ estimates if there are no similar box model scenarios, if there are multiple box model scenarios that differ significantly in their χ predictions; if the comparison is inexact due to differences in the microphysics/chemistry models between the GCM and PartMC-MOSAIC; or if the matching box model scenarios had significantly different histories and therefore have misleading mixing states. For example, the lack of nucleation in the box model scenarios may well lead to somewhat overpredicted χ values in the sizes up to 90 nm in polluted regions. Additionally, we assumed a composition of our pre-existing particles in our training simulations, which may influence our results presented here.

An important issue that should be addressed in future work is the question of end-to-end verification and validation of the χ predictions. This could be accomplished by performing single-particle measurements in different locations, similar to what has been done in Healy et al. [15] for a single location in Paris during the MEGAPOLIS campaign. Another possibility would be to perform particle-resolved aerosol simulations within a 3D chemical transport model (at great computational expense) to calculate χ directly over small regions, and to compare these explicitly calculated χ values to χ predicted with machine learning.

It will be straightforward to adapt our model training to predict χ based on aerosol optical properties, rather than hygroscopicity. This would answer the question of how absorbing and non-absorbing aerosol species are mixed on a per-particle basis, which is important to capture the absorption enhancement of black-carbon-containing aerosol [60,61]. The approach presented in this paper could be generalized to other problems where particle-scale processes cannot directly be simulated within the large-scale modeling framework, but for which accurate small-scale models exist.

Acknowledgments: M.H. and N.R. were supported by NSF AGS-1254428. J.K.K. and J.R.P. acknowledge funding from NSF AGS-1559607. M.W. acknowledges funding from NSF CMMI-1150490.

Author Contributions: M.H. and N.R. designed the PartMC-MOSAIC simulations and analyzed the results; M.H. performed the PartMC model runs and the machine learning; M.W. guided the machine learning procedures; N.R. wrote the paper with contributions of all authors; and J.K.K. and J.R.P. performed the GEOS-Chem-TOMAS simulations. All authors contributed to the interpretation of the results.

Conflicts of Interest: The authors declare no conflict of interest.

References

1. Noble, C.A.; Prather, K.A. Real-time single particle mass spectrometry: A historical review of a quarter century of the chemical analysis of aerosols. *Mass Spectrom. Rev.* **2000**, *19*, 248–274.
2. Bein, K.J.; Zhao, Y.; Wexler, A.S.; Johnston, M.V. Speciation of size-resolved individual ultrafine particles in Pittsburgh, Pennsylvania. *J. Geophys. Res.* **2005**, *110*, D07S05.
3. Johnson, K.S.; Zuberi, B.; Molina, L.T.; Molina, M.J.; Iedema, M.J.; Cowin, J.P.; Gaspar, D.J.; Wang, C.; Laskin, A. Processing of soot in an urban environment: Case study from the Mexico City Metropolitan Area. *Atmos. Chem. Phys.* **2005**, *5*, 3033–3043.
4. Riemer, N.; West, M.; Zaveri, R.; Easter, R. Simulating the evolution of soot mixing state with a particle-resolved aerosol model. *J. Geophys. Res. Atmos.* **2009**, *114*, D09202.
5. Seigneur, C.; Hudischewskyj, A.B.; Seinfeld, J.H.; Whitby, K.T.; Whitby, E.R.; Brock, J.R.; Barnes, H.M. Simulation of aerosol dynamics: A comparative review of mathematical models. *Aerosol Sci. Technol.* **1986**, *5*, 205–222.
6. Wexler, A.S.; Lurmann, F.W.; Seinfeld, J.H. Modelling urban aerosols—I. Model development. *Atmos. Environ.* **1994**, *28*, 531–546.
7. Whitby, E.R.; McMurry, P.H. Modal aerosol dynamics modeling. *Aerosol Sci. Technol.* **1997**, *27*, 673–688.
8. Zaveri, R.; Barnard, J.; Easter, R.; Riemer, N.; West, M. Effect of aerosol mixing-state on optical and cloud activation properties. *J. Geophys. Res. Atmos.* **2010**, *115*, D17210.
9. Fierce, L.; Riemer, N.; Bond, T.C. Toward reduced representation of mixing state for simulating aerosol effects on climate. *Bull. Am. Meteorol. Soc.* **2017**, *98*, 971–980.

10. Fierce, L.; Bond, T.C.; Bauer, S.E.; Mena, F.; Riemer, N. Black carbon absorption at the global scale is affected by particle-scale diversity in composition. *Nat. Commun.* **2016**, *7*, 12361.

11. Ching, J.; Riemer, N.; West, M. Impacts of black carbon mixing state on black carbon nucleation scavenging: Insights from a particle-resolved model. *J. Geophys. Res. Atmos.* **2012**, *117*, doi:10.1029/2012JD018269.

12. Ching, J.; Riemer, N.; West, M. Impacts of black carbon particles mixing state on cloud microphysical properties: Sensitivity to environmental conditions. *J. Geophys. Res. Atmos.* **2016**, *121*, 5990–6013.

13. Riemer, N.; West, M. Quantifying aerosol mixing state with entropy and diversity measures. *Atmos. Chem. Phys.* **2013**, *13*, 11423–11439.

14. Ching, J.; Fast, J.; West, M.; Riemer, N. Metrics to quantify the importance of mixing state for CCN activity. *Atmos. Chem. Phys.* **2017**, *17*, 7445–7458.

15. Healy, R.; Riemer, N.; Wenger, J.; Murphy, M.; West, M.; Poulain, L.; Wiedensohler, A.; O'Connor, I.; McGillicuddy, E.; Sodeau, J.; et al. Single particle diversity and mixing state measurements. *Atmos. Chem. Phys.* **2014**, *14*, 6289–6299.

16. O'Brien, R.E.; Wang, B.; Laskin, A.; Riemer, N.; West, M.; Zhang, Q.; Sun, Y.; Yu, X.Y.; Alpert, P.; Knopf, D.A.; et al. Chemical imaging of ambient aerosol particles: Observational constraints on mixing state parameterization. *J. Geophys. Res.* **2015**, *120*, 9591–9605.

17. Adams, P.J.; Seinfeld, J.H. Predicting global aerosol size distributions in general circulation models. *J. Geophys. Res. Atmos.* **2002**, *107*, 4370.

18. Kodros, J.K.; Cucinotta, R.; Ridley, D.A.; Wiedinmyer, C.; Pierce, J.R. The aerosol radiative effects of uncontrolled combustion of domestic waste. *Atmos. Chem. Phys.* **2016**, *16*, 6771–6784.

19. Ghan, S.J.; Abdul-Razzak, H.; Nenes, A.; Ming, Y.; Liu, X.; Ovchinnikov, M.; Shipway, B.; Meskhidze, N.; Xu, J.; Shi, X. Droplet nucleation: Physically-based parameterizations and comparative evaluation. *J. Adv. Model. Earth Syst.* **2011**, *3*, doi:10.1029/2011MS000074.

20. Whittaker, R.H. Evolution and Measurement of Species Diversity. *Taxon* **1972**, *21*, 213–251.

21. Drucker, J. Industrial Structure and the Sources of Agglomeration Economies: Evidence from Manufacturing Plant Production. *Growth Chang.* **2013**, *44*, 54–91.

22. Strong, S.; Koberle, R.; de Ruyter van Steveninck, R.; Bialek, W. Entropy and Information in Neural Spike Trains. *Phys. Rev. Lett.* **1998**, *80*, 197–200.

23. Falush, D.; Stephens, M.; Pritchard, J.K. Inference of population structure using multilocus genotype data: Dominant markers and null alleles. *Mol. Ecol. Notes* **2007**, *7*, 574–578.

24. Giorio, C.; Tapparo, A.; Dall'Osto, M.; Beddows, D.C.; Esser-Gietl, J.K.; Healy, R.M.; Harrison, R.M. Local and Regional Components of Aerosol in a Heavily Trafficked Street Canyon in Central London Derived from PMF and Cluster Analysis of Single-Particle ATOFMS Spectra. *Environ. Sci. Technol.* **2015**, *49*, 3330–3340.

25. Fraund, M.; Pham, D.Q.; Bonanno, D.; Harder, T.H.; Wang, B.; Brito, J.; de Sá, S.S.; Carbone, S.; China, S.; Artaxo, P.; et al. Elemental Mixing State of Aerosol Particles Collected in Central Amazonia during GoAmazon2014/15. *Atmosphere* **2017**, *8*, 173, doi:10.3390/atmos8090173.

26. Dickau, M.; Olfert, J.; Stettler, M.E.J.; Boies, A.; Momenimovahed, A.; Thomson, K.; Smallwood, G.; Johnson, M. Methodology for quantifying the volatile mixing state of an aerosol. *Aerosol Sci. Technol.* **2016**, *50*, 759–772.

27. DeVille, R.E.L.; Riemer, N.; West, M. Weighted Flow Algorithms (WFA) for stochastic particle coagulation. *J. Comput. Phys.* **2011**, *230*, 8427–8451.

28. Michelotti, M.D.; Heath, M.T.; West, M. Binning for efficient stochastic multiscale particle simulations. *Multiscale Model. Simul.* **2013**, *11*, 1071–1096.

29. Zaveri, R.A.; Easter, R.C.; Fast, J.D.; Peters, L.K. Model for simulating aerosol interactions and chemistry (MOSAIC). *J. Geophys. Res. Atmos. (1984–2012)* **2008**, *113*, doi:10.1029/2007JD008782.

30. Zaveri, R.A.; Peters, L.K. A new lumped structure photochemical mechanism for large-scale applications. *J. Geophys. Res. Atmos. (1984–2012)* **1999**, *104*, 30387–30415.

31. Zaveri, R.A.; Easter, R.C.; Wexler, A.S. A new method for multicomponent activity coefficients of electrolytes in aqueous atmospheric aerosols. *J. Geophys. Res. Atmos. (1984–2012)* **2005**, *110*, doi:10.1029/2004JD004681.

32. Zaveri, R.A.; Easter, R.C.; Peters, L.K. A computationally efficient multicomponent equilibrium solver for aerosols (MESA). *J. Geophys. Res. Atmos. (1984–2012)* **2005**, *110*, doi:10.1029/2004JD005618.

33. Schell, B.; Ackermann, I.J.; Binkowski, F.S.; Ebel, A. Modeling the formation of secondary organic aerosol within a comprehensive air quality model system. *J. Geophys. Res.* **2001**, *106*, 28275–28293.

34. Tian, J.; Riemer, N.; West, M.; Pfaffenberger, L.; Schlager, H.; Petzold, A. Modeling the evolution of aerosol particles in a ship plume using PartMC-MOSAIC. *Atmos. Chem. Phys.* **2014**, *14*, 5327–5347.

35. Mena, F.; Bond, T.C.; Riemer, N. Plume-exit modeling to determine cloud condensation nuclei activity of aerosols from residential biofuel combustion. *Atmos. Chem. Phys.* **2017**, *17*, 9399–9415.

36. Bey, I.; Jacob, D.J.; Yantosca, R.M.; Logan, J.A.; Field, B.D.; Fiore, A.M.; Li, Q.; Liu, H.Y.; Mickley, L.J.; Schultz, M.G. Global modeling of tropospheric chemistry with assimilated meteorology: Model description and evaluation. *J. Geophys. Res. Atmos.* **2001**, *106*, 23073–23095.

37. Napari, I.; Noppel, M.; Vehkamäki, H.; Kulmala, M. Parametrization of ternary nucleation rates for H_2SO_4-NH_3-H_2O vapors. *J. Geophys. Res. Atmos.* **2002**, *107*, 4381.

38. Jung, J.; Fountoukis, C.; Adams, P.J.; Pandis, S.N. Simulation of in situ ultrafine particle formation in the eastern United States using PMCAMx-UF. *J. Geophys. Res. Atmos.* **2010**, *115*, doi:10.1029/2009JD012313.

39. Westervelt, D.; Pierce, J.; Riipinen, I.; Trivitayanurak, W.; Hamed, A.; Kulmala, M.; Laaksonen, A.; Decesari, S.; Adams, P. Formation and growth of nucleated particles into cloud condensation nuclei: model—Measurement comparison. *Atmos. Chem. Phys.* **2013**, *13*, 7645–7663.

40. Vehkamäki, H.; Kulmala, M.; Napari, I.; Lehtinen, K.E.; Timmreck, C.; Noppel, M.; Laaksonen, A. An improved parameterization for sulfuric acid–water nucleation rates for tropospheric and stratospheric conditions. *J. Geophys. Res. Atmos.* **2002**, *107*, AAC 3-1–AAC 3-10.

41. Lee, Y.; Pierce, J.; Adams, P. Representation of nucleation mode microphysics in a global aerosol model with sectional microphysics. *Geosci. Model Dev.* **2013**, *6*, 1221–1232.

42. Lee, Y.; Adams, P. A fast and efficient version of the TwO-Moment Aerosol Sectional (TOMAS) global aerosol microphysics model. *Aerosol Sci. Technol.* **2012**, *46*, 678–689.

43. Kodros, J.; Pierce, J. Important global and regional differences in aerosol cloud-albedo effect estimates between simulations with and without prognostic aerosol microphysics. *J. Geophys. Res. Atmos.* **2017**, *122*, 4003–4018.

44. Pierce, J.; Croft, B.; Kodros, J.; D'Andrea, S.; Martin, R. The importance of interstitial particle scavenging by cloud droplets in shaping the remote aerosol size distribution and global aerosol-climate effects. *Atmos. Chem. Phys.* **2015**, *15*, 6147–6158.

45. Fierce, L.; Riemer, N.; Bond, T.C. Explaining variance in black carbon's aging timescale. *Atmos. Chem. Phys.* **2015**, *15*, 3173–3191.

46. Kulmala, M.; Petäjä, T.; Ehn, M.; Thornton, J.; Sipilä, M.; Worsnop, D.; Kerminen, V.M. Chemistry of atmospheric nucleation: On the recent advances on precursor characterization and atmospheric cluster composition in connection with atmospheric new particle formation. *Annu. Rev. Phys. Chem.* **2014**, *65*, 21–37.

47. McKay, M.D.; Beckman, R.J.; Conover, W.J. A comparison of three methods for selecting values of input variables in the analysis of output from a computer code. *Technometrics* **1979**, *21*, 239–245.

48. Kalnay, E.; Kanamitsu, M.; Kistler, R.; Collins, W.; Deaven, D.; Gandin, L.; Iredell, M.; Saha, S.; White, G.; Woollen, J.; et al. The NCEP/NCAR 40-year reanalysis project. *Bull. Am. Meteorol. Soc.* **1996**, *77*, 437–471.

49. Seinfeld, J.; Pandis, S. *Atmospheric Chemistry and Physics: From Air Pollution to Climate Change*; John Wiley & Sons: Hoboken, NJ, USA, 2006.

50. Vignati, E.; Facchini, M.; Rinaldi, M.; Scannell, C.; Ceburnis, D.; Sciare, J.; Kanakidou, M.; Myriokefalitakis, S.; Dentener, F.; O'Dowd, C. Global scale emission and distribution of sea-spray aerosol: Sea-salt and organic enrichment. *Atmos. Environ.* **2010**, *44*, 670–677.

51. Camps-Valls, G.; Bruzzone, L. *Kernel Methods for Remote Sensing Data Analysis*; John Wiley & Sons: Chichester, UK, 2009.

52. Ristovski, K.; Vucetic, S.; Obradovic, Z. Uncertainty analysis of neural-network-based aerosol retrieval. *IEEE Trans. Geosci. Remote Sens.* **2012**, *50*, 409–414.

53. Lary, D.J.; Lary, T.; Sattler, B. Using machine learning to estimate global PM2.5 for environmental health studies. *Environ. Health Insights* **2015**, *9*, 41–52.

54. Lauret, P.; Voyant, C.; Soubdhan, T.; David, M.; Poggi, P. A benchmarking of machine learning techniques for solar radiation forecasting in an insular context. *Sol. Energy* **2015**, *112*, 446–457.

55. Hastie, T.; Tibshirani, R.; Friedman, J. *The Elements of Statistical Learning*, 2nd ed.; Springer: New York, NY, USA, 2009.

56. Friedman, J.H. Stochastic gradient boosting. *Comput. Stat. Data Anal.* **2002**, *38*, 367–378.

57. Mason, L.; Baxter, J.; Bartlett, P.L.; Frean, M.R. Boosting algorithms as gradient descent. In *Advances in Neural Information Processing Systems*; Solla, S.A., Leen, T.K., Müller, K., Eds.; MIT Press: Cambridge, MA, USA, 2000; Volume 12, pp. 512–518.

58. Pedregosa, F.; Varoquaux, G.; Gramfort, A.; Michel, V.; Thirion, B.; Grisel, O.; Blondel, M.; Prettenhofer, P.; Weiss, R.; Dubourg, V.; et al. Scikit-learn: Machine Learning in Python. *J. Mach. Learn. Res.* **2011**, *12*, 2825–2830.

59. Riemer, N.; West, M.; Zaveri, R.; Easter, R. Estimating black carbon aging time-scales with a particle-resolved aerosol model. *J. Aerosol Sci.* **2010**, *41*, 143–158.

60. Schnaiter, M.; Horvath, H.; Möhler, O.; Naumann, K.H.; Saathoff, H.; Schöck, O. UV-VIS-NIR spectral optical properties of soot and soot-containing aerosols. *J. Aerosol Sci.* **2003**, *34*, 1421–1444.

61. Peng, J.; Hu, M.; Guo, S.; Du, Z.; Zheng, J.; Shang, D.; Zamora, M.L.; Zeng, L.; Shao, M.; Wu, Y.S.; et al. Markedly enhanced absorption and direct radiative forcing of black carbon under polluted urban environments. *Proc. Natl. Acad. Sci. USA* **2016**, *113*, 4266–4271.

MDPI
St. Alban-Anlage 66
4052 Basel
Switzerland
Tel. +41 61 683 77 34
Fax +41 61 302 89 18
www.mdpi.com

Atmosphere Editorial Office
E-mail: atmosphere@mdpi.com
www.mdpi.com/journal/atmosphere

www.ingramcontent.com/pod-product-compliance
Lightning Source LLC
Chambersburg PA
CBHW051846210326
41597CB00033B/5795